Ceramic Materials for Electronics

ELECTRICAL ENGINEERING AND ELECTRONICS

A Series of Reference Books and Textbooks

1. Rational Fault Analysis, *edited by Richard Saeks and S. R. Liberty*
2. Nonparametric Methods in Communications, *edited by P. Papantoni-Kazakos and Dimitri Kazakos*
3. Interactive Pattern Recognition, *Yi-tzuu Chien*
4. Solid-State Electronics, *Lawrence E. Murr*
5. Electronic, Magnetic, and Thermal Properties of Solid Materials, *Klaus Schröder*
6. Magnetic-Bubble Memory Technology, *Hsu Chang*
7. Transformer and Inductor Design Handbook, *Colonel Wm. T. McLyman*
8. Electromagnetics: Classical and Modern Theory and Applications, *Samuel Seely and Alexander D. Poularikas*
9. One-Dimensional Digital Signal Processing, *Chi-Tsong Chen*
10. Interconnected Dynamical Systems, *Raymond A. DeCarlo and Richard Saeks*
11. Modern Digital Control Systems, *Raymond G. Jacquot*
12. Hybrid Circuit Design and Manufacture, *Roydn D. Jones*
13. Magnetic Core Selection for Transformers and Inductors: A User's Guide to Practice and Specification, *Colonel Wm. T. McLyman*
14. Static and Rotating Electromagnetic Devices, *Richard H. Engelmann*
15. Energy-Efficient Electric Motors: Selection and Application, *John C. Andreas*
16. Electromagnetic Compossibility, *Heinz M. Schlicke*
17. Electronics: Models, Analysis, and Systems, *James G. Gottling*

18. Digital Filter Design Handbook, *Fred J. Taylor*

19. Multivariable Control: An Introduction, *P. K. Sinha*

20. Flexible Circuits: Design and Applications, *Steve Gurley, with contributions by Carl A. Edstrom, Jr., Ray D. Greenway, and William P. Kelly*

21. Circuit Interruption: Theory and Techniques, *Thomas E. Browne, Jr.*

22. Switch Mode Power Conversion: Basic Theory and Design, *K. Kit Sum*

23. Pattern Recognition: Applications to Large Data-Set Problems, *Sing-Tze Bow*

24. Custom-Specific Integrated Circuits: Design and Fabrication, *Stanley L. Hurst*

25. Digital Circuits: Logic and Design, *Ronald C. Emery*

26. Large-Scale Control Systems: Theories and Techniques, *Magdi S. Mahmoud, Mohamed F. Hassan, and Mohamed G. Darwish*

27. Microprocessor Software Project Management, *Eli T. Fathi and Cedric V. W. Armstrong (Sponsored by Ontario Centre for Microelectronics)*

28. Low Frequency Electromagnetic Design, *Michael P. Perry*

29. Multidimensional Systems: Techniques and Applications, *edited by Spyros G. Tzafestas*

30. AC Motors for High-Performance Applications: Analysis and Control, *Sakae Yamamura*

31. Ceramic Materials for Electronics: Processing, Properties, and Applications, *edited by Relva C. Buchanan*

Additional Volumes in Preparation

Electrical Engineering-Electronics Software

1. Transformer and Inductor Design Software for the IBM PC, *Colonel Wm. T. McLyman*

2. Transformer and Inductor Design Software for the Macintosh, *Colonel Wm. T. McLyman*

Ceramic Materials for Electronics

Processing, Properties, and Applications

edited by

RELVA C. BUCHANAN

Department of Ceramic Engineering
University of Illinois at Urbana-Champaign
Urbana, Illinois

MARCEL DEKKER, INC. New York and Basel

Library of Congress Cataloging-in-Publication Data

Ceramic materials for electronics.
(Electrical engineering and electronics; 32) Includes bibliographies and index. 1. Electronic ceramics. I. Buchanan, Relva C., II. Series. TK7871.15.C4C47 1986 621.381'028 86-4585
ISBN 0-8247-7501-5

MARCEL DEKKER, INC.
270 Madison Avenue, New York, New York 10016

Current printing (last digit):
10 9 8 7 6 5 4 3 2

PRINTED IN THE UNITED STATES OF AMERICA

Preface

Ceramic materials contribute essential passive functions as components for a wide range of electronic applications related to device packaging, sensors, and magnetic uses such as recording, memory storage, and communications. The major categories of ceramics for electronics use include piezoelectrics, capacitor dielectrics, thick film hybrid components, ferrites, ceramic substrates, and glasses as well as temperature, voltage, humidity, and gas sensors. Use of these materials as finished products, because of their integral association with the dynamic growth in electronics, constitutes a major growth area within the ceramics field. However, much of this expanding use for ceramics tends to be masked by their applications as components rather than as end use products.

The material presented in this book is addressed to those who design and fabricate the ceramic parts, as well as those who utilize them as devices and components in the various systems. Accordingly, this book is intended for users of ceramics in the electronics, communications, and related fields who might wish to refresh or broaden their knowledge within allied specialties. The book should also be a valuable reference text for practicing ceramists and students in ceramic engineering and materials science. A background at the senior undergraduate level in engineering or related fields is presumed.

The many types, properties, and applications of ceramic materials used in electronics are explored in depth in the following chapters by area specialists. Detailed attention is given under each category to the influence of composition modification, raw material synthesis, processing, and developed microstructures on the unique and prescribed properties of the material, as well as on component and device possibilities and applications.

Chapter 1 considers formulations and properties of ceramic insulators including uses of glass and ceramic substrates. Basic phenomena relating to dielectric properties and conduction are also presented. Chapter 2 covers dielectric capacitor materials, their formulation, processing, and properties, with heavy emphasis on microstructure development, multilayer ceramic capacitor fabrication, and capacitor performance parameters. Chapter 3 deals with piezoelectric and electrooptic ceramic materials. Consideration is given to basic symmetry, phase and composition relationships, processing and fabrication, microstructure, properties, applications, and aging characteristics of both PZT and PLZT materials. Chapter 4 considers ferrite ceramic materials with presentation of basic magnetic phenomena, ferrite compositions, and properties in relationship to crystal structure. Heavy emphasis is placed on processing, fabrication, and testing of useful ferrite parts. The subject of ceramic sensors is covered in Chapter 5. Basic defect structure and composition effects on NTC, PTC, and gas sensors are presented, including grain boundary effects, processing, properties, and use characteristics. Chapter 6 deals with the subject of varistors as a separate category of sensor materials. Again, emphasis is on basic phenomena, composition, impurity, and processing effects on final properties. Chapter 7 covers the subject of multilayer ceramic technology, which integrates thick film use with a dense ceramic substrate system to achieve high packaging density and reliability. The very different interactive processes involved are detailed. Finally, Chapter 8 considers conducting ceramics and presents an in-depth discussion of defect structure and conduction mechanisms in ceramic oxide and other materials as related to composition, processing, and end use parameters. All chapters are well referenced in the different topical areas. It is hoped that the material presented will contribute to a heightened awareness and understanding of electronic ceramic materials and their properties.

Relva C. Buchanan

Contents

Contributors

Relva C. Buchanan, B.S., Sc.D. Professor, Department of Ceramic Engineering, University of Illinois at Urbana–Champaign, Urbana, Illinois

Gilbert Goodman, B.S., Sc.D. Manager, Materials Research and Technical Facilities, Corporate Applied Research Group, Johnson Controls, Inc., Milwaukee, Wisconsin

Gene H. Haertling, Ph.D. Vice President of Technical Staff/Manager of Research and Development, Ceramic Products Operation, Motorola, Inc., Albuquerque, New Mexico

David C. Hill, M.S. President, MLC Technologies, Inc., Attleboro, Massachusetts

Lionel M. Levinson, Ph.D. Manager, Electronic Materials Unit, Corporate Research and Development, General Electric Company, Schenectady, New York

Herbert R. Philipp, Ph.D. Physicist, Corporate Research and Development, General Electric Company, Schenectady, New York

*Thomas G. Reynolds III, Ph.D.** Manager, Research and Engineering, Ferroxcube Division, Amperex Electronic Corp., Saugerties, New York

Harry L. Tuller, B.S., M.S., Eng. Sc.D. Associate Professor of Ceramics and Electronic Materials, Departments of Materials Science and Engineering, Massachusetts Institute of Technology, Cambridge, Massachusetts

Wayne S. Young, Ph.D. Manager, Department of Materials Applications, General Products Division, IBM Corporation, San Jose, California

**Present affiliation*: Innovation Manager, Ceramics, Elcoma Passive Components, Nederlandse Philips, Eindhoven, The Netherlands

1

Properties of Ceramic Insulators

RELVA C. BUCHANAN
University of Illinois at Urbana-Champaign, Urbana, Illinois

I. INTRODUCTION

The primary function of insulation in electrical circuits is the physical separation of conductors and the regulation or prevention of current flow between them. Ancillary but important other functions are to provide mechanical support, heat dissipation, and environmental protection for the conductors. Ceramic materials which in use provide primarily these functions are classified as ceramic insulators. They include most glasses, porcelains, and oxide and nitride materials, as well as mica. The main advantage of ceramics as insulators is their capability for high-temperature operation without hazardous degradation in chemical, mechanical, or dielectric properties.

Insulators of the type described belong to the class of materials known as linear dielectrics. These are materials in which the electric displacement (D) increases in direct proportion to the electric field in the dielectric (E), the constant of proportionality being defined as the relative dielectric constant (k'). These relationships can be given as [1]

$$D = \varepsilon_0 E_a = \varepsilon_0 k'E \ (C/m^2) \tag{1}$$

where E_a is the applied field and ε_0 the permittivity (dielectric constant) of vacuum = 8.85×10^{-12} F/m. The electric displacement describes the extent to which the electric field has been altered by the presence of the dielectric. The dielectric constant, as defined, is a measure of the ability of the material to store charge relative to vacuum and is a characteristic material property.

In addition to the dielectric constant (k'), three other material properties are important in determining the insulating characteristics of the material. These are the electrical (volume) resistivity (ρ), the dissipation factor ($\tan \delta$), and the dielectric strength (DS). The electrical resistivity is simply a measure of the resistance that a unit cube of the material offers to current flow in a given (dc) field. In an ac field, the electrical resistivity and the dielectric constant are related by the dissipation factor, which measures the energy loss per cycle (usually in the form of heat) from the material. This relationship is given by

$$\sigma = \frac{1}{\rho} = \omega \varepsilon_0 k' \tan \delta \tag{2}$$

where σ is the conductivity $(\Omega\text{-m})^{-1}$, $\omega = 2\pi f$ (f = frequency), and $k' \tan \delta$ is the loss factor (k"). Equation (2) also shows the frequency dependence of the resistivity (ρ) and defines the loss factor.

The dielectric strength is a measure of the maximum voltage gradient that can be impressed across the dielectric without physical degradation of its insulating properties, leading to breakdown. Although critical to the insulating use of the dielectric, it is not a true material property since it is significantly influenced by measurement conditions, sample size, and flaws in the ceramic.

As a practical matter, ceramic materials that satisfy the following property criteria at 25°C would be classified as good insulators [2]:

Dielectric constant (k')	$\leqslant 30$
Electrical resistivity (ρ)	$\geqslant 10^{12}$ (Ω-cm)
Dissipation factor (tan δ)	$\leqslant 0.001$
Dielectric strength (DS)	$\geqslant 5.0$ kV/mm
Dielectric loss factor (k")	$\leqslant 0.03$

Table 1 lists typical dielectric values for some commonly used ceramic insulators with the frequency and temperature of measurement specified as appropriate. These dielectric properties are important in specifying the useful operating conditions for the insulators. Excluded from Table 1 are such classes of ceramic materials as ferroelectrics, ferrites, and sensor devices, which typically exhibit a nonlinear response to changes in applied field or temperature and for which insulating characteristics, although important, are often of secondary consideration.

Other properties that can significantly influence the choice of dielectric for insulation purposes are the mechanical and thermal properties of the material. Typical values for these properties are listed in Table 2 for the ceramic insulators shown in Table 1. Of these, the thermal conductivity, thermal expansion coefficient, and rupture strength exercise the most direct influence on the use of the ceramic as an insulator.

II. COMPOSITIONS

Most of the ceramic insulators listed in Table 1 can be conveniently grouped as glasses, glass ceramics, porcelains, and dense single-phase or mixed-oxide materials. This distinction largely reflects the amount of glassy phase in the ceramics, which tends to control the dielectric properties. Where high dielectric performance is required, therefore, glassy-phase content must be minimized in the insulator and mobile ion content should, as far as practicable, be

Table 1 Dielectric Properties of Ceramic Insulators

Material	Properties at 1 MHz (room temperature)			Dielectric strength (kV/mm)	Resistivity at 25°C (Ω-cm)
	Tan δ	Dielectric constant	Loss factor		
Porcelain	0.008–0.020	5.0–6.5	0.10–0.05	6.1–13.0	10^{14}
Zircon	0.001–0.0014	8.0–10.5	0.013–0.009	6.3–11.5	$>10^{14}$
Steatite	0.0008–0.0035	5.9–6.1	0.004–0.03	7.9–13.8	10^{17}
Forsterite	0.0004–0.001	5.8–6.7	0.002–0.008	7.9–11.9	10^{17}
Cordierite	0.003–0.007	4.1–5.4	0.008–0.010	5.5–9.1	10^{16}
Alumina	0.0003–0.002	8.2–10.2	0.002–0.03	9.9–15.8	10^{16}
Spinel	0.0004	7.5	0.003	11.9	10^{14}
Mullite	0.004–0.005	6.2–6.8	0.025–0.034	7.8	10^{14}
Magnesia	0.001	8.2	0.002–0.01	8.5–11.0	$>10^{14}$
BeO	0.001	5.8	0.006	9.5–13.8	$>10^{16}$
Zirconia	0.01	12.0	0.12	∼5.0	10^{9}
Thoria	0.0003	13.5	0.004	∼5.3	10^{10}

Hafnia	0.01	12	0.12		10^8
Ceria	0.0007	15	0.011		10^9
Spodumene	0.004	6.4	0.03		10^{10}
$Li_2O-Al_2O_3-SiO_2$	0.005	5.3	0.03–0.10		10^{12}
BN	0.001	4.2	0.004	35.6–55.4	10^{14}
Si_3N_4	0.0001	6.1	0.0006	15.8–19.8	10^{14}
Pyroceram	0.0017–0.013	5.5–6.3	0.01–0.07	9.9–11.9	10^{12}
Glass-bonded mica	0.0015–0.003	6.4–9.2	0.023–0.011	10.6–23.7	10^{14}
Mica	0.0002	5.4–8.7	0.004	39.5–79.1	10^{16}
Glass (general)	0.0005–0.01	4.3–8.5	0.002–0.14	7.8–13.2	10^{12}
Glass (quartz)	0.0003	3.8–5.4	0.0015	15–25.0	$10^{14}-10^{18}$
PbA1 silicate	0.001	8.2–15	0.006	8.9–16.0	10^{13}
Air or vacuum		1.0			

Table 2 Thermomechanical Properties of Ceramic Insulators

Material	Specific gravity	Thermal conductivity $\frac{\text{cal-cm}}{\text{sec-°C-cm}^2}$	Thermal coefficient of expansion (10^{-6} °C^{-1})	Tensile strength (MPa)	Transv. strength (MOR) (MPa)	Compress. strength (MPa)	Thermal shock resistance
Porcelain	2.4	0.006	6.0	48	83	352	Fair
Zircon	3.7–4.3	0.012–0.02	3.5–5.5	96	172	524	Good
Steatite	2.8	0.008	7.8–10.4	76	145	628	Moderate
Forsterite	2.8	0.004–0.01	10.6	76	145	593	Poor
Cordierite	2.0–2.9	0.008	2.3	51	117	348	Excellent
Alumina	3.6	0.06–0.07	8.0	179	421	3455	Good
Spinel	2.8	0.018	6.6	∿95	103	1710	Fair
Mullite	2.5	0.006	4.3–5.0	84	131	1400	Fair
Magnesia	2.5–3.3	0.095	12.8	90	138	1041	Fair
BeO	1.8–3.0	0.30–0.6	4.2–9.4	103	248	1559	Good
Zirconia	5.6	0.059	3.0–8.3	148	186	1376	Poor

Thoria	9.7	0.033	5.3–9.0	115	131	1524	Poor
Hafnia	9.0	0.004	6.5	90	110	1386	Poor
Ceria	7.0	0.029	10.0	88	110	1386	Poor
Spodumene	2.4	0.012	2.0	30	55	900	Good
$Li_2O-Al_2O_3-SiO_2$		0.004	-0.4–0.5	34	62	393	Excellent
BN	2.1	0.069	4.3	25	52	283	Good
$Si_3N_4(\alpha)$; H.P.	3.2	0.03–0.07	3.1	214	697	2007	Excellent
Pyroceram	2.4–2.6	0.004–0.009	0.2–4	64	248		Good
Glass-bonded mica	2.6–3.8	0.0012	10–14.5	69	117	214	Fair
Mica	2.6–3.8	0.0003–0.002	18–27			221	
Glass (general)	2.0–8.0	0.002–0.004	0.8–1.3	<34	110	697	Fair
Glass (quartz)	2.2	0.003	0.3	55		1130	Excellent

eliminated. Included in the categories above are the most widely used insulator formulations, which are described below.

A. Glasses

Most insulating applications for glasses make use of their dielectric properties, but such factors as softening temperature and thermal expansion coefficient play an important role in the selection of compositions for particular applications. Glasses are favored for many such applications because of their wide availability, low cost, and ease of fabrication. Traditional uses include ceramic glazes, wire and porcelain coatings, lamp and tube envelopes, electrical bushings, standoff insulators, fuse bodies, and fiber insulation. The more electronic applications include their use as substrates, delay lines, device passivation, capacitor dielectric, thick-film resistor components, dielectric films, and solder glass sealing [3,4].

Glass is characterized by an amorphous or random structure and wide variability in composition. Normal oxide glasses may be classified as silicates, borates, phosphates, or germanates, according to which of the glass-forming oxides (SiO_2, B_2O_3, P_2O_5, or GeO_2) make up the structure [5]. This structure is normally modified by selected additions of such intermediate oxides as Al_2O_3, Bi_2O_3, and PbO_2, which at relatively high concentrations may themselves contribute to glass formation, or by alkali (Li, Na, K, Rb, Cs) and alkaline earth (Mg, Ca, Sr, Ba) oxides. The alkali oxides in particular have the effect of opening up the structure, thereby lowering the softening point and providing increased ion mobility.

Of the glass systems above, borate glasses, although lower melting and of higher electrical resistivity and lower loss, are rarely used alone since their resistance to moisture is very poor [6]. They form, however, an important component of low-melting glasses. The phosphate glasses are also low melting and provide better electrical insulation for many device applications since they are usually free of alkali. However, moisture resistance, although an improvement on the borates, is still marginal for many applications. The silicate-based glasses, in consequence, are the most important and widely used. Of the factors that critically influence the choice of glass insulator, the electrical resistivity, temperature-viscosity relationship, thermal expansion coefficient, and chemical durability are most important. The latter property relates to the resistance of the glass to weathering and chemical attack. This is usually specified in terms of a weight change per unit area of glass surface per unit time (at 25 or 100°C) when immersed in a solution of different pH.

The temperature-viscosity behavior reflects the change in viscosity of the glass with temperature and defines a number of very important points for a glass relative to its use behavior. Figure 1 illustrates these different viscosity points for glasses of different compositions, whereas Figure 2 illustrates the relationship of some of these viscosity points to the expansion coefficient as a function of temperature. In Figure 2, T_s represents the softening point of the glass, which is related to T_m, the deformation point, and T_g is the glass transition temperature. The latter represents the transition from a glassy to a supercooled liquid state. T_{str} represents the strain point, which is the lower limit for strain relief in a reasonable time, while at the annealing temperature, T_{an}, the strain is effectively relieved in 1 hr.

In terms of composition, high alkali content (>16 wt %) degrades the insulation resistance, lowers T_s and T_g, increases the thermal

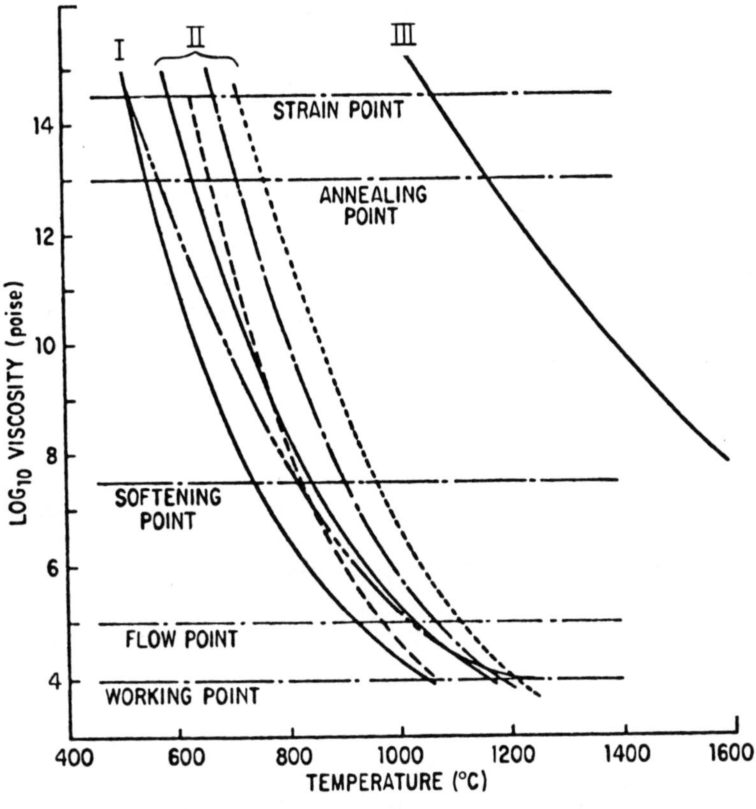

Figure 1 Plot of log viscosity versus temperature for glasses of different compositions, showing important working viscosity points. I, Soda-lime; II, borosilicate; III, fused silica glass. (From Ref. 8.)

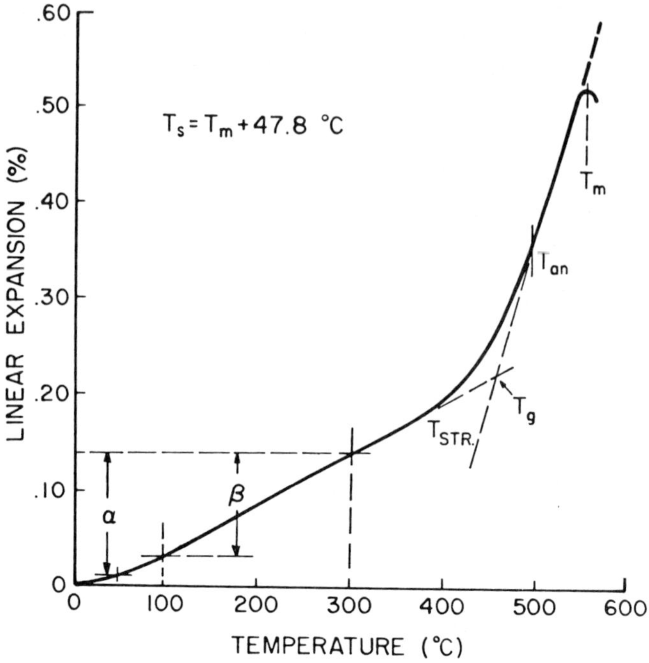

Figure 2 Plot of linear expansion (percent) versus temperature of lead-alumina-borosilicate glass, showing relationship to the viscosity points in Figure 1. $T_s = T_m + 47.8°C$.

expansion coefficient, and degrades the chemical durability. Partial replacement of the alkali content with RO- or R_2O_3-type modifiers reverses this trend. For most glasses, the dielectric constant at room temperature is linearly related to the density, which in turn can be related to the composition of the glasses, as follows [7,8]:

$$\rho_d = \Sigma_i f_i v_i \sim \Sigma_i f_i w_i \tag{3a}$$

$$k' \sim 2.2\rho_d \tag{3b}$$

where ρ_d is the density of the glass, f_i a density factor for each oxide, v_i the specific volume fraction, w_i the weight fraction, and k' the dielectric constant. Typical values of f_i for different oxides are as follows:

Oxide	SiO_2	Al_2O_3	B_2O_3	Na_2O	K_2O	CaO	MgO
f_i	2.2	2.75	1.9	2.6	3.2	3.3	3.4

PbO	ZnO	BaO
10.0	5.9	7.0

(3c)

A variety of glass compositions for different applications can be identified as detailed in Table 3. Properties for some of these glasses are listed in Tables 4 and 5. Soda-lime-silicate glasses are the most common and are used for such purposes as power and telephone line insulators, fuses, lamp envelopes, and some substrates. Lead-alkali glasses are lower melting and lower in alkali but of higher PbO content. Typical uses are for lamp stems or as graded and pinched seals. The potash-lead glasses are typically made into ribbons for capacitor use. Borosilicate glasses may be used as photomask or substrate glasses, as power line insulation, and for glass-metal sealing with Kovar, W, and Mo metals. They are typically low-alkali, low-expansion, higher-temperature glasses. Aluminosilicate glasses are used for electronic component mask substrates, for high-pressure mercury lamps (high ultraviolet transmission) and for Cu, Fe, and Ni seal glasses. Fused silica finds electrical uses delay lines and for device passivation.

Glass insulating films are widely used for the passivation of silicon and other integrated circuit devices, as crossovers or isolation layers in multilayer circuits, and for encapsulation and/or hermetic sealing of integrated and hybrid circuit packages. These various applications require very different glass compositions and indeed call into question the narrow definition of a glass, in view of the fact that such films may be partially crystalline. This definition would include both single-phase and mixed-oxide films as well as chalcogenide- and nitride-type films prepared by melting or chemical vapor deposition (CVD) techniques [9].

Glass passivation layers are required to provide good dielectric isolation, mechanical and scratch protection, act as an alkali impurity and moisture barrier, and provide low surface recombination velocity and device stability at elevated temperature under bias or operating conditions. These requirements are for the most part met by SiO_2 and PSG (phosphosilicate glass) films, sometimes in conjunction with Si_3N_4 films as an alkali diffusion barrier. Table 6 shows the use of these films for Si devices and III−V compound devices. Typically, the films are deposited by low-pressure or regular CVD or by radio-frequency (RF) sputtering techniques. Some fusible PSG films are also prepared, as are some low-melting lead-alumina-borosilicate glass films for special applications. Properties of these films are given in Table 7.

Table 3 Compositions for Typical Commercial Glasses

	wt %					
Oxide	Soda-lime	Lead-alkali	Potash-lead	Borosilicate	Aluminosilicate	Fused silica
SiO_2	71–73	56–63	34–37	70–81	54–55	99+
Na_2O	16–18	5–8.6	—	0–4	—	0–0.5
K_2O	0.8–1.0	6–7	6–8	0–1.0	—	0.02
CaO	9–12	—	—	—	13–15	—
MgO	1–3	—	—	—	—	—
BaO	—	—	—	—	3–4	—
PbO	—	21–30	57–60	0–1.5	—	—
B_2O_3	—	—	—	13–28	6–8	—
Al_2O_3	0.5–2.0	0.8–1.0	—	0–1.2	20–22	—

Table 4 Properties of Vitreous Glasses

	Soda-lime	Alkali-zinc-boro-silicate	Alkali-borosilicate	Lime-aluminosilicate
Str. point (°C)	472	506	520	670
Anneal. point (°C)	512	539	565	710
Soft. point (°C)	696	696	720	910
Thermal expansion (ppm °C)	9.2	7.2	3.25	4.6
Density (g/cm^3)	2.47	2.51	2.23	2.63
Refractive index nD	1.51	1.53	1.47	1.52
Vol. rest. (Ω-cm)	$10^{5.6}$	$10^{7.9}$	$10^{13.8}$	$10^{11.2}$
Dielectric constant (tan δ)	6.5	6.7	4.6	6.4
Young's modulus (psi)	10	10.8	9.1	12.4
Chem. durability, 5% HCl (mg/cm^2 at 100°C/24 hr)	0.02	0.03	0.005	0.4

Source: Ref. 8.

Table 5 Properties of Vitreous Glasses

	Fused silica	Alkali-lead-borosilicate	Alumino-borosilicate
Str. point (°C)	990	—	613
Anneal. point (°C)	1050	—	650
Soft. point (°C)	1580	725	820
Thermal expansion (ppm °C)	0.56	5.5–6.1	4.5
Density (g/cm^3)	2.2	—	2.76
Refractive index, nD	1.458	1.58	1.53
Vol. rest. (Ω-cm)	$10^{11.2}$	10^{12}	$10^{12.4}$
Dielectric constant (tan δ)	3.9	8.9	5.8
Young's modulus (psi)	10.5	—	9.8
Chem. durability, 5% HCl (mg/cm^2 at 100°C/24 hr)	0.17	—	0.28

Source: Ref. 8

Table 6 Dielectric Films for Device Applications

Device	Dielectric film	Application
Si	SiO	Diffusion mask
GaAs	SiO_2	Surface passivation
GaP	Si_3N_4	Impurity isolation
GaAs	PSG	MOS/MIS structures
InSb	Al_2O_3	Selected area masks
InGaAs	$ZnO\text{-}SiO_2$	Ion implantation
AlGaAs		Encapsulation

Source: Ref. 9.

In contrast to the passivation of films, the glasses used primarily for thick-film dielectric insulation or for solder glass sealing are the more conventional lead-borosilicate type of glasses. Typical composition and property ranges are given in Tables 8 and 9 for the dielectric and seal glass compositions, respectively. The dielectric glasses are used for such applications as thick-film crossover insulation, ac plasma display coatings, insulator glazes, diode encapsulation, ribbon and thick-film capacitors, and as a component of some thick-film resistor pastes.

The requirements for low softening points (T_s) and glass transition temperature (T_g), high dielectric constants (k'), and moderate thermal expansion coefficients (α) can be accommodated within the $PbO\text{-}B_2O_3\text{-}SiO_2$ composition ranges shown, with the modifier oxides (Al_2O_3, CaO, MgO, Na_2O, K_2O) being used to shift desired properties into specific directions. The wide variety of applications indicated require, for the most part, changes in one or more of these components. It would clearly be desirable to make these changes in a predictable manner so that properties could be specified in advance. For example, data analysis on more than 50 glasses using computer-aided multiple correlation/regression analysis yielded, for the composition/property ranges shown in Table 8, linear predictor equations of the form

$$P_i = r_1 w_1 + r_2 w_2 + \cdots r_7 w_7 + I_p = \Sigma r_i w_i + I_p \qquad (4)$$

where P_i is the property in question (T_s), r_i the unstandardized regression coefficient for each component i, w_i the weight percent

Table 7 Typical Properties of Dielectric Films

Dielectric	Density (g/cm^3)	Melting point (°C)	Coefficient of thermal expansion ($°C^{-1}$)	Dielectric constant	Index of refraction
SiO	2.1	—	—	3.3–5.2	1.55–2.0
SiO_2	2.2	~1600	0.55×10^{-6}	3.9	1.46
PSG	—	~900	0.87×10^{-6}	3.7–4.0	1.42
Si_3N_4	3.1	~1900 (pressure)	3.2×10^{-6}	6.1	2.1
Al_2O_3	3.6	~2020	7.58×10^{-6}	7.7	1.6
AIN	3.2	~2400	6.1×10^{-6}	9.1	2.16
TiO_2	2.6	~1850	8.22×10^{-6}	20–50	2.0
Ge_3N_4	5.0–5.4	—	—	6.5	2.06

Source: Ref. 9.

Table 8 Dielectric Glass Compositions (wt %) and Property Ranges

Component	Range	Mean	Standard deviation
PbO	50–74	59.90	5.270
B_2O_3	10–25	18.10	3.54
SiO_2	8–26	13.30	3.10
Al_2O_3	0–5	1.78	1.385
Cao	0–6	3.08	1.452
MgO	0–4	2.00	0.749
Na_2O	0–5	1.84	1.049
T_g (°C)	4[illegible]5–510	471.9	18.306
T_s (°C)	470–565	526.0	19.466
α (10^{-7} °C^{-1})	75–101	84.56	5.149
k'	12–17	14.3	0.958

Table 9 Seal Glass Compositions (wt %) and Property Ranges

PbO	B_2O_3	SiO_2	ZnO	BaO	Al_2O_3	Bi_2O_3	CuO
68–75	9–12	2–4	10–12	0–3	0–5	0–4	0–3

T_s	T_g	T_{seal}	$\alpha_{0-300°C}$
380–410°C	340–470°C	445–475°C	$80\text{-}90 \times 10^{-7}\ °C^{-1}$

associated with that component, and I_p an intercept value. Both r_i and I_p can be obtained from the computer data analysis. Thus, for the softening point T_s, values obtained at the 99% confidence level were

$$T_s(°C) = -4.67PbO - 1.14B_2O_3 + 0SiO_2 - 1.32Al_2O_3 + 1.08CaO + 5.5MgO - 9.83Na_2O + 832.4 \quad (5)$$

where the component concentration, w_i, is in weight percent. For the mean composition shown, therefore, $T_s = 526°C$. Similar analyses can be carried out for other properties for which data are available. The magnitude and sign of the regression coefficients, r_i, can be directly correlated to the effectiveness of the different oxide components in raising or lowering a particular property. These trends are illustrated in Table 10 for glasses in the $PbO\text{-}B_2O_3\text{-}SiO_2$ base system.

Solder glasses are used for sealing various combinations of glass/metal/ceramic structures in a variety of configurations [10]. This includes the bonding of magnetic (ferrite) heads for data recording (where the glass serves also as a gap material), hermetic seals for hybrid and integrated circuit packages, as well as some encapsulation functions. Typically, these glasses must be low melting and since the integrity of the bonded structures is of prime importance, expansion coefficients must be matched to minimize stresses. This is illustrated in Figure 3, which compares the stress condition for two glass films on a metal substrate on cooling to 25°C. The net compressive stress component for the lower expansion glass A is the desired condition.

Control of the thermal expansion coefficient is influenced greatly by the ZnO, Bi_2O_3, and CuO concentrations, since these oxides have the desirable effect of lowering softening points without a significant increase in thermal expansion, in contrast to the alkali oxides [11,12].

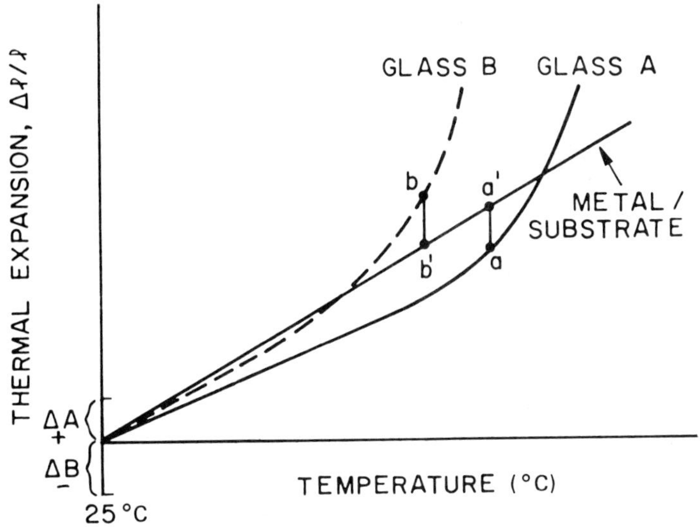

Figure 3 Thermal expansion versus temperature for two glasses A and B on a metal substrate showing residual stress components (ΔA, compressive; ΔB, tensile) on cooling.

Some seal glass compositions may be almost wholly crystalline after being formed, affording greater strength and reheat capability for the sealed structure. These compositions are subject to the same type of composition/property analysis as that discussed for the dielectric glasses. Table 11 gives composition ranges and properties for typical crystallizable solder glass compositions.

Table 10 Effects of Oxide Concentration on Glass Properties in Lead-Borosilicate Glasses

Property	Oxide decrease	Oxide increase
T_g (°C)	PbO, Na_2O	MgO
T_s (°C)	PbO, B_2O_3, Na_2O, Al_2O_3	CaO, MgO
α (10^{-7} °C^{-1})	B_2O_3, SiO_2, Al_2O_3	PbO, CaO, Na_2O
k'	B_2O_3, Al_2O_3	PbO, MgO

Table 11 Crystallizable Solder Glasses (wt %)

Component	Range	Typical
PbO	75–82	75.0
B_2O_3	6.5–12	8.0
ZnO	7–14	10.0
SiO_2	1.5–3.0	2.5
Al_2O_3	0–3.0	1.0
CuO	0–3.0	2.5
Sb_2O_3	0–1.0	0.5
Na_2O	0–0.5	—
BaO	0–4.0	—
α (0–300°C)	$80–105 \times 10^{-7}\,°C^{-1}$	85
T_s	375–395°C	385
T_{seal}	400–450°C	430 ± 10

Another important electronic application for glasses is their use as substrates for photomasks or for hybrid circuits or as porcelain enamel–coated substrates, which are finding increasing use in lower-cost thick-film circuitry [13] where the glass coating is often a lead-borosilicate glass composition. Here the properties needed are atomically smooth surfaces, chemical inertness, high volume and surface resistivity, flatness, dimensional stability, good thermal shock resistance, and low cost. Freedom from volume defects and good transparency are also needed for photomask applications and low-alkali aluminoborosilicate glasses are typically used, although soda-lime-silicate glasses may be used for less critical applications. These aluminoborosilicates are typically high-melting glasses with low to moderate expansion coefficients. A somewhat unique application for high-lead-content glasses reduced in hydrogen to yield a semiconducting surface is their widespread use in channel electron multipliers.

B. Porcelains

Porcelains can be described as polycrystalline ceramic bodies containing, typically, more than about 10 vol % of a vitreous second phase [8]. This vitreous phase essentially controls densification, porosity, and phase distribution within the porcelain and to a large extent its mechanical and dielectric properties also. Porcelain bodies can be classified as triaxial, steatite, or nonfeldspathic types, depending on the composition and amount of the vitreous phase present within the ceramic. Composition, properties, and uses of these different categories of porcelains are described below.

1. Triaxial Porcelains

Triaxial porcelains form a large base of the commonly used porcelain insulators. They exhibit fairly low mechanical strength, thermal shock resistance, and poor high-frequency characteristics (due to alkali ion migration) but are generally satisfactory for low-frequency ($<10^4$ Hz) use at ordinary temperatures. They are compounded from mixtures of feldspars [$(KNa)_2O$-$3Al_2O_3$-$6SiO_2$], kaolinite clays (Al_2O_3-$2SiO_2$-$2H_2O$), and flint or quartz (SiO_2), which places the triaxial porcelains in the phase system [$(KNa)_2O$-Al_2O_3-SiO_2)] in terms of oxide constituents [4,8].

Raw material compositions for the triaxial porcelains can be given as 40–60 wt % clays, 20–35 wt % feldspar, and 20–30 wt % flint. The raw materials each contribute different characteristics to the fired porcelain. The kaolinite clay provides the main body for the ceramic and controls the firing range and distortion during sintering. It also provides the first liquid phase in the system (∿900°C), from which a main crystalline phase, mullite, is subsequently developed. In most formulations, however, up to one-half the kaolinite content may be replaced by ball clay, which increases plasticity and strength in the formed ware but may also be the source of impurities. The feldspar provides the main flux phase, which lowers the maturing temperature and accounts for the dense nonporous structure from which some mullite phase also develops. However, the increased glassy phase and alkali content tend to degrade both the dielectric properties and mechanical strength of the porcelain. Partial dissolution of the quartz phase above 1200°C increases the liquid viscosity and helps to maintain shape during firing. An increase in the quartz content usually gives improved mechanical strength, provided that the grain size is below ∿ 10 μm, in order to minimize thermal expansion stresses. Minor ingredients (2–5 wt %) such as talc, zircon, and the alkaline earth carbonates aid in the formation and control of low-temperature eutectic phases, resulting in generally improved dielectric and mechanical properties of the porcelain [8].

Typical compositions and dielectric properties for representative porcelain bodies are given in Table 12. As indicated in Table 12, a distinction must be made between the categories of triaxial porcelains described as low tension and high tension or hard porcelains [14]. The low-tension porcelains generally have looser raw materials tolerances and are used unglazed for less exacting low-voltage and low-frequency applications, such as light sockets, fuse blocks, bushings, and hot plate and toaster insulation. They are usually sintered in the range 1260–1320°C and are characterized by moderately low dielectric strengths (∿9–11 kV/mm) and loss factors (0.05–0.10 at 1 kHz).

In contrast, the high-tension porcelains are sintered in the range 1350 to 1450°C to a high gloss finish. Lower impurity content is achieved by reduction or elimination of the ball clay content, which is replaced by bentonite clays (∿3.0 wt %) or alkaline earth additives. These porcelains can be considered as low-loss, high-dielectric-strength (∿13–18 kV/mm) materials and are therefore suitable for such high-voltage applications as transmission line insulators, high-voltage circuit breakers, and cutouts. Because of these

Table 12 Composition and Properties of Triaxial Porcelain Bodies

	wt %			
		Low tension		High tension
	Generic	a	b	c
Raw material				
Kaolinite	20–35	21	25	35
Ball clay	15–30	25	20	10
Feldspar	20–35	34	33	25
Flint	20–30	20	23	25
$(BaCa)CO_3$	0–5	—	—	5
Property				
Sintering temperature		1250–1330°C		1300–1450°C
Dielectric constant		5.5–7.0		6–7
tan δ		0.005		0.003
Loss factor		0.035		0.020

environmentally sensitive applications, the porcelain must be dense, moisture impervious, and have minimal mechanical flaws. This is aided by the use of a glaze coating, which may be a self-glaze, feldspathic glaze, or semiconducting brown glaze.

Self-glazing arises from the higher sintering temperature for the hard porcelains, resulting in a significantly higher vitreous phase and pore-free microstructure. Typical oxide constituents for a suitable high-temperature feldspathic glaze maturing in the range 1200–1250°C can be given (in wt %) as: SiO_2, >0.3; Al_2O_3, 11.9; CaO, 7.9; K_2O, 4.4; BaO, 3.6; ZnO (or MgO), 1.9. The brown glaze can be developed from this feldspathic glaze by incorporating 20–35 wt % calcined and ground coloring oxides (Fe_2O_3, TiO_2, MnO_2) into the glaze mixture and sintering in a reducing atmosphere to obtain some ions in a lower oxidation state. The juxtaposition of such ions as (Ti^{3+}, Ti^{4+}), (Fe^{2+}, Fe^{3+}), and (Mn^{2+}, Mn^{4+}) in the glaze results in partial n-type conduction and a surface resistivity (ρ_s) of $\sim 10^7$ Ω/square. This serves to reduce the corona discharge caused by ionization of moist ambient in the high potential gradient near the conductor/insulator interface.

2. Nonfeldspathic Porcelains and Oxide Insulators

Most ceramic insulators are to be found in the system $MgO\text{-}Al_2O_3\text{-}SiO_2$ [15–17]. These include the oxide end members: MgO, Al_2O_3, and SiO_2; the binary compounds: $MgO\text{-}SiO_2$ (steatite), $2MgO\text{-}SiO_2$ (forsterite), $MgO\text{-}Al_2O_3$ (spinel), and $3Al_2O_3\text{-}2SiO_2$ (mullite); and the ternary compound $MgO\text{-}Al_2O_3\text{-}2.5SiO_2$ (cordierite). Typical properties for these insulators are given in Tables 1 and 2.

Table 13 compares raw material compositions for the porcelain insulators in this system as well as zircon ($ZrO_2\text{-}SiO_2$). Raw materials used are principally talc ($3MgO\text{-}4SiO_2\text{-}2H_2O$), kaolinite clays, and alkaline earth fluxes such as $BaCO_3$ and $CaCO_3$. These porcelains are typically of higher purity with superior dielectric properties to the triaxial porcelains, but are more difficult to produce due to a shorter sintering range with these compositions.

The steatite ceramics are most commonly used for high-frequency insulation due to their low alkali content and resultant low dielectric losses. They are easy to fabricate due to the large talc content but, as indicated, control of microstructure during firing is more difficult due to tendency for excess liquid formation in the $MgO\text{-}SiO_2$ system. Steatites have fair mechanical strength but rather poor thermal shock characteristics.

The forsterite ceramics have even lower dielectric losses than the steatites and can be used for high-frequency, microwave, or high-temperature insulation. They have high expansion coefficients, are typically impervious, and in consequence, have been used for

Table 13 Raw Material Compositions for Nonfeldspathic Porcelain Insulators

Raw material	wt %			
	Steatite	Forsterite	Cordierite	Zircon
Talc	80–90	60–70	35–45	10–20
Kaolin/clay	5–7	5–10	40–50	10–20
$Mg(OH)_2$	—	20–45	—	—
$BaCO_3$	6–7	5–7	0–2	6–8
$CaCO_3$	0–1	1–3	—	0–1
Zircon	—	—	—	55–70
Al_2O_3	—	—	12–15	—
Sintering temperature (°C)	1260–1300	1275–1375	1250–1350	1300–1400

such applications as ceramic vacuum-tube envelopes. The thermal shock resistance, however, is also poor.

Cordierite ceramics possess very low thermal expansion coefficients and excellent thermal shock resistance, particularly in the porous state, and are frequently used for such applications as wire and carbon resistor cores. The porosity arises from the difficulty in controlling densification, glassy phase, and composition in the cordierite phase field due to the steep liquidus curve in the three-component system.

Zircon porcelains have the advantage of relatively low expansion coefficients, excellent dielectric properties, good thermal shock resistance, higher mechanical strength, and ease of fabrication. They represent, therefore, an attractive alternative to the other porcelains, but insulating uses are not widespread, reflecting perhaps the higher raw material and fabrication costs for the porcelain.

Of the other oxide materials, Al_2O_3 is by far the most widely used, particularly as a substrate material for microelectronic thick (and thin)-film circuitry. This derives from its ease of fabrication, excellent dielectric insulating properties, mechanical strength, surface texture, and relative chemical inertness. Alumina compositions for such applications typically range from 94 to 99 wt % Al_2O_3, with flux additives consisting of talc, clay, and MgO for sintering and grain growth control. The high thermal conduction of alumina (second only to BeO as oxides) is also a major asset for heat dissipation and thermal shock resistance in dense circuit packaging, as is the fact that it can readily be sintered to fairly exacting tolerances and surface finishes.

Of the other oxide insulators in the system, MgO and spinel ($MgO\text{-}Al_2O_3$), although excellent insulators, are not widely used due to difficulties in fabrication and cost. SiO_2 is, however, widely used as a piezoelectric transducer and it also finds usage for delay lines and high-temperature insulation. BeO oxide finds extensive use for applications requiring significant heat dissipation, but toxicity and fabrication costs limit its more general application. Other oxide materials listed in Tables 1 and 2 (zirconia, thoria, hafnia, and ceria) have mainly specialized uses except for zirconia, which has substantial insulating, oxygen sensor, fuel cell, and heater applications.

3. Special Insulators

Insulators in the $Li_2O\text{-}Al_2O_3\text{-}SiO_2$ system, including spodumene ($Li_2O\text{-}Al_2O_3\text{-}4SiO_2$), typically possess very low net shrinkage on sintering (due to significant expansion of the β-spodumene phase) and a near-zero or negative thermal expansion [15]. For these materials, dielectric properties approach those of the feldspathic porcelains, although strength values tend to be lower, but thermal shock properties are excellent. The combination of these properties would seem to be advantageous for insulator applications requiring dimensional and thermal stability. Pyroceram materials contain both crystalline and glass phases and can therefore be considered a glass ceramic. Dielectric properties of these materials tend to be dictated by the glass phase, but the cyrstalline phase imparts substantial mechanical strength [18]. These materials have been developed for substrate applications and offer advantages in controlled dimensions and ease of fabrication. Compositions fall mainly in the alkali and alkaline earth alumina silicate systems.

Silicon nitride, although being developed primarily for structural applications, nevertheless has important potential for microelectronics application, owing to its superior dielectric properties and high thermal conduction. Applications as dielectric films and as substrates are current. Boron nitride, in its hexagonal modification, has a similar structure to graphite and can readily be machined. It possess excellent dielectric and thermal shock properties and is a useful insulating material even though difficult to bond to. Mica is among the best insulating material and has the highest dielectric strength. As a layer silicate material (mainly synthetic), it is available in very thin sheets and enjoys widespread use as an insulator in electronic components [4,15].

III. THERMAL AND MECHANICAL PROPERTIES

Thermal conductivity is a relative measure of the ability of a material to dissipate heat. For ceramic materials, this encompasses a wide range of values (0.002–0.6 cal/sec-°C-cm^2/cm), as indicated in Table 2. It is determined mainly by the composition, structure, defect state, and temperature of the material. For equivalent conditions, highest thermal conduction is obtained for dense, single-phase oxides such as BeO, MgO, and Al_2O_3. The presence of complexing oxides (MgO-Al_2O_3, MgO-SiO_2) and microstructural features such as porosity, cracks, and second phases generally degrades the thermal conduction. Glasses, in particular, show low thermal conduction, which varies little with composition. Figure 4 compares the change in thermal conductivity with temperature for some typical substrate materials [19,20].

Thermal expansion coefficient can be expressed as the change in length (volume) per unit length (volume) of material for each degree rise in temperature. It is a strongly temperature dependent property, but for most ceramic insulators (Figure 5) a near-linear range can be defined between 25 and 300°C, within which the expansion

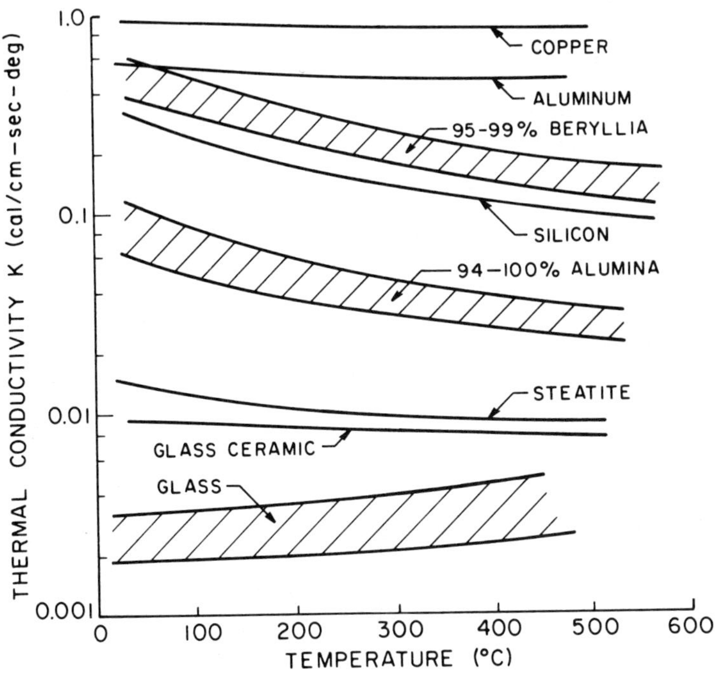

Figure 4 Plot showing thermal conductivity of various substrate materials as a function of temperature. (From Ref. 20.)

coefficient varies little. Thermal expansion coefficient is closely related to crystal structure, bond strength, and density. Consequently, changes in volume thermal expansion with temperature closely parallel changes in specific heat capacity. For glasses, the open structure and low density typically lead to low thermal expansion coefficients as with fused silica ($\sim 5.0 \times 10^{-7}$ °C^{-1}). With the addition of modifier ions this value may approach 100×10^{-7} °C^{-1} for a Na silicate glass as a result of increased density concentration and decreased bond strength of the modifier ions.

Thermal expansion coefficient is a directional or anisotropic property because of its strong dependence on crystal structure. For strongly anisotropic materials (aluminum titanate, cordierite, spodumene), this may result in negative directional expansion and overall low volume coefficient of expansion. Polycrystalline samples of these materials will, in consequence, show excessive grain boundary stresses and be inherently low in strength. Thermal shock resistance might, however, be improved.

The rupture modulus or strength measures the ability of the ceramic material to withstand the thermal and mechanical stresses to which it might be subjected during use. It is also related to the composition and structure of the material, but microstructural features such as pores, cracks, flaws, second phases, grain size, and grain boundary stresses are usually dominant. These factors are rarely quantifiable, and therefore scatter in measured data tends to be large. As with thermal conduction, dense single-phase oxides (and nitrides: Si_3N_4, BN) tend to have high rupture moduli, possibly for the same reasons, whereas complex mixed oxides and glasses have moderate to low values. The practical range of values for ceramic insulators is 40–700 MPa, but higher values can be achieved with special processing.

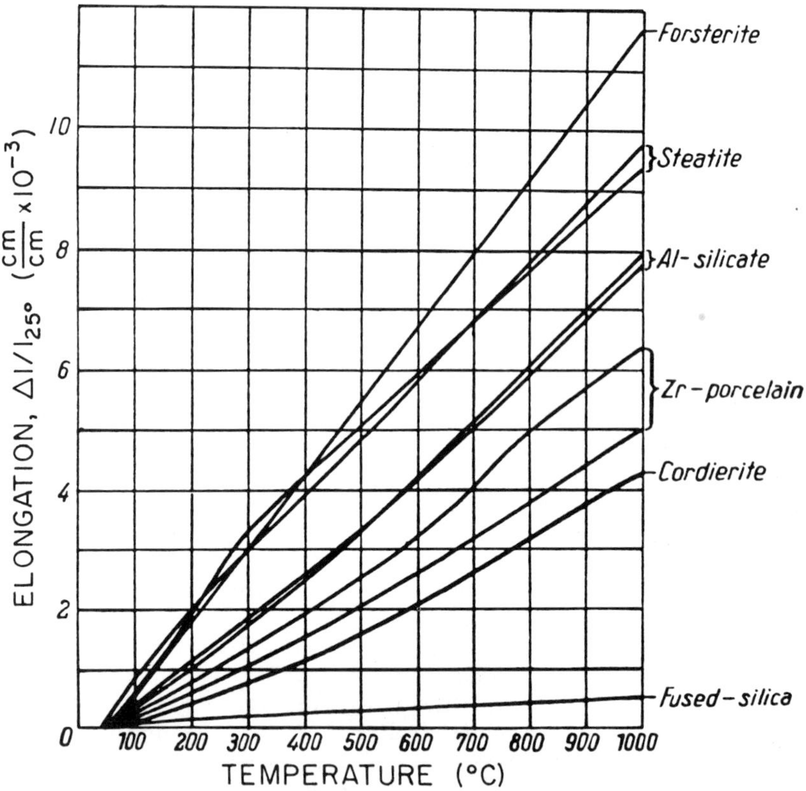

Figure 5 Thermal expansion characteristics versus temperature for some typical ceramic insulators. (From Ref. 8.)

The properties discussed above often interact where use conditions result in development of thermal stresses from the existence of thermal or expansion gradients in the ceramic. Such conditions may occur where localized hot spots develop on a substrate due to direct heating or to dissipation of heat from a resistor or active device. A combination of high specific heat and high thermal conductivity would thus serve to minimize the rise in temperature and the accompanying thermal stresses. If not relieved, such stresses could result in mechanical and dielectric degradation of the system. Similar considerations arise where there is thermal expansion mismatch across (ceramic) interfaces, as in the use of enamel or glaze coatings and passivation layers.

For ceramic insulators, possible destructive effects on heating are controlled by the thermal diffusivity, expansion coefficient, elastic modulus, and tensile strength of the material. These properties may be conveniently grouped as components of the thermal shock resistance, defined as the maximum quenching or heating rate to which the material can be subjected without undergoing physical damage or rupture. A practical measure of this quantity is the maximum temperature to which the ceramic can be heated and then quenched in water at 25°C without causing physical damage. Under these conditions, the thermal shock resistance represents simply a temperature gradient ($T - T_{25°C}$), but otherwise it may be described by a semiquantitative expression relating the temperature change from T_0 to T(°C) to the material parameters, as follows [8]:

$$\Delta T = \frac{T - T_0}{t} \sim \frac{\sigma_T}{E\alpha_v}\left(\frac{\kappa}{\rho_d C_p}\right)^{\frac{1}{2}} S \tag{6}$$

where ΔT is the thermal shock resistance (°C/sec), t the cooling time (sec), σ_T and E the tensile strength and elastic modulus (psi, MPa), α_v the volume expansion coefficient ($°C^{-1}$), and $\kappa/\rho_d C_p$ the thermal diffusivity (cm^2/sec) of the material. Elements of the thermal diffusivity are the thermal conductivity, κ (cal-cm/sec-°C-cm^2); the density, ρ_d (g/cm^3); and the specific heat capacity, C_p (cal/g-°C).

Typical values and ranges for most of these properties are given in Table 2, which, however, describes the thermal shock resistance only in qualitative terms. This is because it is not a true material property, since its magnitude depends strongly on size and shape (S) of the test specimens as well as on transient factors such as tensile strength (σ_T) and thermal diffusivity, which may be difficult to define precisely under test conditions. Thermal shock resistance is, nonetheless, an important design parameter in choosing ceramic insulators for applications where temperature changes may be severe, such as for heater and resistor cores and thick-film substrates.

Of the properties contributing to the thermal shock resistance, the modulus ratio (σ_T/E) and the thermal diffusivity tend to vary only over narrow ranges. The thermal diffusivity range for most ceramic insulators is 0.1–0.3 cm^2/sec, reflecting the fact that both κ and C_p tend to increase or decrease in response to the same materials characteristics. This is generally true also for the tensile strength and elastic modulus, which are closely related to chemical bonding and structure in the different solids. The elastic moduli for ceramic insulators fall in the narrow range 34–414 GPa, reflecting the very strong ionic or covalent bonding. Tensile strengths for ceramic insulators are considerably lower and more variable, since they are dependent on processing, test conditions, and flaws in the ceramic. Even so, high elastic moduli can usually be correlated with high tensile strengths, so that variability in the ratio of the two is small.

The dominant property in controlling the thermal shock resistance, therefore, is the linear thermal expansion coefficient, which for ceramic insulators has the range -0.4 to 145×10^{-7} cm/cm-°C according to Table 2 [α_V in Eq. (6) is approximately three times the linear expansion coefficient]. Figure 6 illustrates the sharp fall in thermal shock resistance observed for glasses of increasing thermal expansion coefficients when measured under equivalent test conditions.

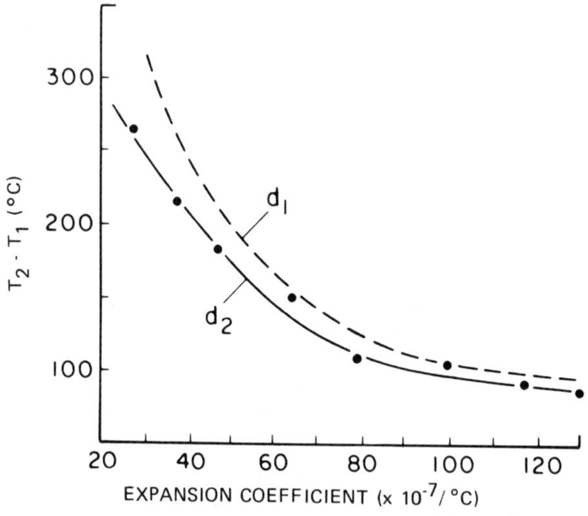

Figure 6 Variation of thermal shock resistance for glass of different thickness ($d_2 > d_1$) as a function of thermal expansion coefficient. (From Ref. 8.)

Typical ΔT values range from about 50°C for high alkali and lead (soft) glasses to ∿150°C for (hard) borosilicate glasses to ∿1000°C for a fused silica glass at equivalent cooling rates. This range of thermal shock resistance values also holds for nonvitreous ceramics as in Table 2, where high thermal conduction, high tensile strength, and low to moderate thermal expansion coefficients generally yielded good to excellent results. Examples include BeO, cordierite, Al_2O_3, zircon, and spodumene, in contrast to such high expansion ceramics as forsterite and zirconia, which give poor results.

Aside from considerations of thermal shock, thermal stresses can also develop in glass coatings at steady state due to thermal expansion mismatch with the substrate or due to temperature gradients developed across the glass. The tensile stresses developed (proportional to α, E, and ΔT) must be significantly lower than the rated tensile strength of the glass to prevent failure.

IV. DIELECTRIC STRENGTH

All dielectrics when placed in an electric field will lose their insulating properties if the field exceeds a certain critical value. This phenomenon is called dielectric breakdown and the corresponding field intensity is referred to as the dielectric breakdown strength [20–22]. Dielectric strength, therefore, may be defined as the maximum potential gradient to which a material can be subjected without insulation breakdown, that is,

$$DS = \left(\frac{dV}{dx}\right)_{max} = \frac{V_B}{d} \tag{7}$$

where DS is the dielectric strength in kV/mm, V_B the breakdown voltage, and d the thickness of the sample. Figure 7 illustrates the current-voltage characteristics up to breakdown for a typical dielectric.

Dielectric breakdown occurs in gases due to corona discharge in nonuniform fields or to photoionization and collision of electrons with gaseous atoms leading to further ionization, increased conduction, and dielectric breakdown. In liquids, breakdown is due primarily to dielectric heating and increased ionization processes. This process is greatly accelerated by the presence of impurities or high concentrations of mobile ions (Na^+, Li^+) in the liquid. For solids, the main breakdown mechanisms can be described as intrinsic, thermal, and ionization breakdown. A possible fourth mechanism, electrochemical

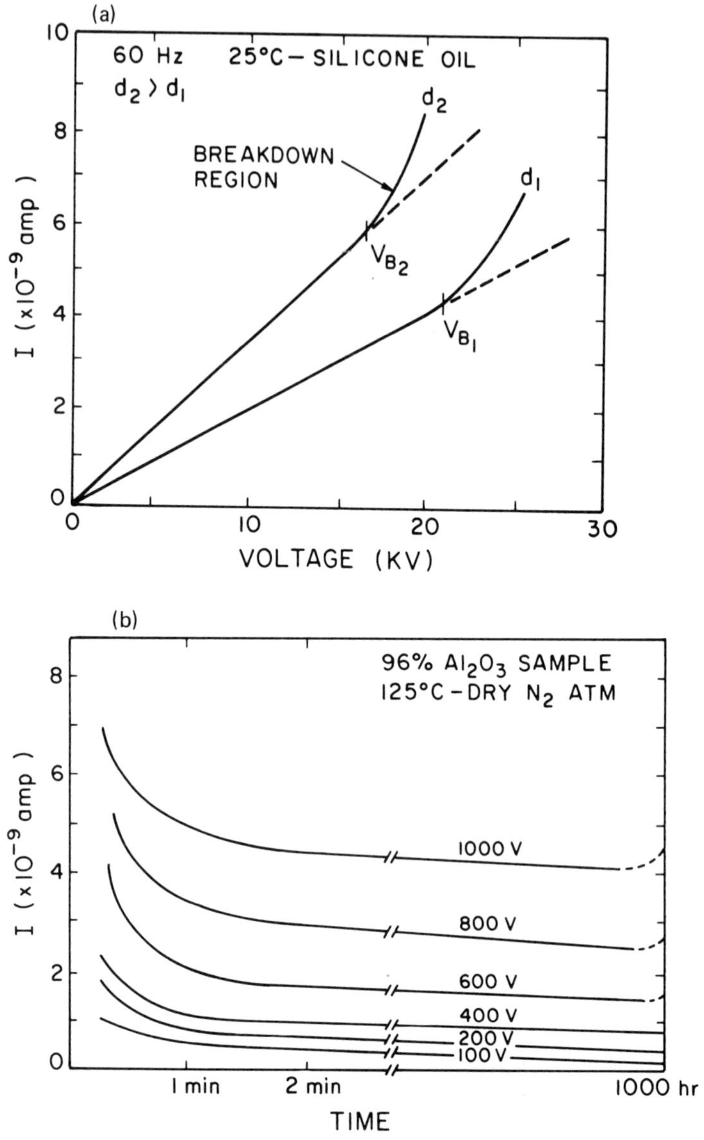

Figure 7 (a) Evaluation of breakdown voltage (V_B) from current-voltage behavior of dielectric at different thickness; (b) evaluation of electrochemical breakdown from current-voltage-time characteristics. (Dashed lines show onset of breakdown.)

degradation or breakdown, is associated with ion migration under a (dc) field gradient at elevated temperatures.

A. Intrinsic Breakdown

At high applied fields electrons may be ejected from electrode materials or may even be generated from the valence band. These electrons are then accelerated through the crystal and some undergo electron-phonon interactions, generating heat and being absorbed in the process. Other electrons collide with ions or atoms in the solid, knocking out other electrons (by elevating them to the conduction band). These newly generated electrons are in turn accelerated by the field generating other electrons, resulting finally in an "avalanche" of conducting electrons over a narrow area and a breakdown generally through a channel. Electronic or intrinsic breakdown of this type can occur in all solids but is rarely observed since the very high fields required ($\sim 10^6 - 10^7$ V/cm) usually trigger other breakdown mechanisms.

Temperature has a very pronounced effect on the electron-phonon interactions and hence on intrinsic breakdown. At very low temperatures ($<<25°C$), phonon interactions are weak, so that electrons gain more energy from the field and are able to generate many free electrons. Hence breakdown voltages are relatively low. Intrinsic breakdown voltage peaks at room temperature, where phonon interactions become important but decreases again above room temperature since thermal vibrations can now generate free electrons as well. For glasses and amorphous materials the irregular structure makes the mean free path for electrons fairly uniform with temperature. Hence the decrease in strength with temperature is more gradual and such factors as composition, pulse duration, and field strength become more important. Intrinsic breakdown is characterized by short breakdown times, $\sim 10^{-8} - 10^{-6}$ sec, as well as by increase of conductivity with increase in field strength.

B. Thermal Breakdown

This is the most common dielectric failure mode in ceramic insulators and the dielectric strengths are typically two orders of magnitude lower than intrinsic breakdown strengths. Causes for failure are generally attributed to local heating or conduction losses, which generate heat at a faster rate than it can be removed. This increases the local temperature, even to the point of melting or evaporation. Depending on the material and the thermal gradients formed, cracks may develop around the local hot spot. The combination of ionized gases and increased conduction cause dielectric breakdown. The heat generated by an (ac) field can be given as

$$W = \omega\varepsilon_0 k'V^2 \tan \delta \sim \omega V^2 C \tan \delta \tag{8}$$

where W (watts/cycle) is the energy generated, V the voltage, and C the capacitance of the sample (farads). The heat dissipated from the sample can be approximated by

$$H_D = Ah(T_S - T_A) \tag{9}$$

where H_D represents the heat dissipated (watts, cal/sec), A the surface area of the sample (cm^2), h the surface heat transfer coefficient (cal/sec-°C-cm^2), and T_S and T_A the surface and ambient temperatures, respectively. To avoid heat accumulation and breakdown, the heat dissipated must equal the heat generated. As is evident from the expressions above, heat generation is dependent on both frequency and voltage and on the material parameters k' and tan δ. These different factors can therefore be manipulated to avoid thermal breakdown, as can the size, shape, and heat transfer environment of the ceramic part. Equation (9), for example, can be used in the design of discrete capacitors, where the heat dissipated is dependent on the size (surface area), temperature gradient, and surface heat transfer coefficient (0.0002–0.002 cal/sec-°C-cm^2 for low to moderate airflow conditions).

C. Ionization Breakdown

Ionization breakdown occurs in inhomogeneous dielectric solids mainly through the mechanism of partial discharges resulting from pores or cracks being present in the ceramic. Ionization of gases inside the pores in the ceramic when subjected to a high electric field results in the local generation of heat, which is transmitted to the surrounding ceramic, resulting in a temperature gradient, the generation of thermal stresses, increased dielectric losses, and increased conduction locally. The stresses can generate cracks, leading to further ionization, and breakdown occurs via a thermoionization process similar to that described for thermal breakdown. The presence of porosity in an insulator is therefore deleterious to high dielectric strengths.

D. Electromechanical Breakdown

This is caused usually by ion migration (Na^+, Ag^+, OH^-) under a continuously applied dc field at moderately high temperatures

(50–200°C). This may lead to interelectrode shorting of closely spaced electrodes or to ion buildup or depletion at one or the other electrode, leading to high local fields and electrode corrosion. These conditions are all time dependent but can lead to thermal breakdown after prolonged exposure. To evaluate electrochemical breakdown, dc fields >3000 V/cm at 100–200°C in dry ambient (N_2) for periods of 1000 hr may be applied to the dielectric, with resulting current-voltage-time curves of the type shown in Figure 7b. For most ceramic insulators, especially glasses, an initial absorption current is followed by a more gradual approach to equilibrium, reflecting rapid initial ion migration and subsequent slowing down by the uneven potential barriers in the structure. A steady-state condition is then maintained for several hours, culminating, if breakdown occurs, in a gradual to sudden increase in current flow. The point of departure from steady state can be taken as the electrochemical breakdown for the given conditions. Such information is frequently important for insulator circuit applications such as Ag conducting lines on an Al_2O_3 substrate, where a potential difference may exist between the lines. This situation is aggravated by the presence of high humidity; hence encapsulation of such circuits is often necessary.

E. Factors Affecting Dielectric Strength

Factors that may affect dielectric strength include composition, microstructural features (porosity, cracks, flaws, second phases), and measurement parameters such as electrode configuration, specimen thickness, temperature, time, frequency, humidity, and heat transfer conditions [19].

Compositional effects relate to the amorphous or crystalline nature of the material and to the presence of mobile ions in the structure. Increased alkali concentration, for example, not only leads to higher bulk conduction, but also renders the surface more hygroscopic. The combined effects of surface porosity, surface alkali, and moisture may lead to high surface currents and breakdown; hence porcelain bodies intended for high-voltage applications are typically glazed to minimize these effects. Microstructural features such as cracks and porosity can lead to ionization of entrapped gases at high fields, particularly under high-humidity conditions. This in turn can lead to high current flow, local heating, cracking, and thermal breakdown as described. The effect of grain size and second phases depends on the details of the structure, but smaller grains will, in general, result in higher dielectric strengths, reflecting the reduced conduction in the solid.

Dielectric strength is perhaps most sensitive to changes in measurement conditions. Electrode geometry is especially important since the breakdown voltage tends to decrease with electrode area. Because of the high fields involved, point contacts can precipitate breakdown through field injection of electrons into the dielectric. To minimize this effect and provide environmental control, measurements are usually carried out with the sample immersed in a high-dielectric-grade (silicone) oil. This essentially fixes the surface heat transfer and humidity conditions.

Dielectric strength values are very sensitive to specimen thickness, which must be specified if results are to be meaningful. This is because flaw density and microstructural defects increase with the thickness (volume) of the specimen under test, leading to lower measured dielectric strengths as sample thickness is increased. Dielectric strength evaluations are typically carried out at room temperature and 60 Hz. The measured dielectric strengths are higher for shorter pulse or time duration of the applied voltage, due to the decrease in conduction and polarization losses in the dielectric at higher frequencies. For particular applications, therefore, dielectric strength evaluations should be carried out within the temperature and frequency range of interest. The effect of higher temperatures is to decrease the dielectric strength due to higher conduction in the solid.

Dielectric strength values for ceramic insulators typically range from a low of 2×10^4 V/cm for some porcelains to 1×10^7 V/cm for some micas. The single-phase oxides, such as Al_2O_3, MgO, and BeO, are intermediate at $1.0-2.0 \times 10^5$ V/cm, with equivalent values for fused silica and the alkali-free insulating glasses; high alkali glasses may range as low as 5×10^4 V/cm. Values for the typical ceramic insulators are given in Table 1.

V. DIELECTRIC PROPERTIES

A. Dielectric Constant and Loss

From Eq. (1) and the capacitive cell illustrated in Figure 8, expressions for the relative dielectric constant k', total charge Q (coulombs), and capacitance C (farads) for the cell can be developed as follows [2,22,23]:

$$k' = \frac{D}{\varepsilon_0 E} = \frac{Q/A}{\varepsilon_0 V/d} \tag{10}$$

Therefore,

$$Q = \varepsilon_0 k' \frac{A}{d} V = CV \tag{11}$$

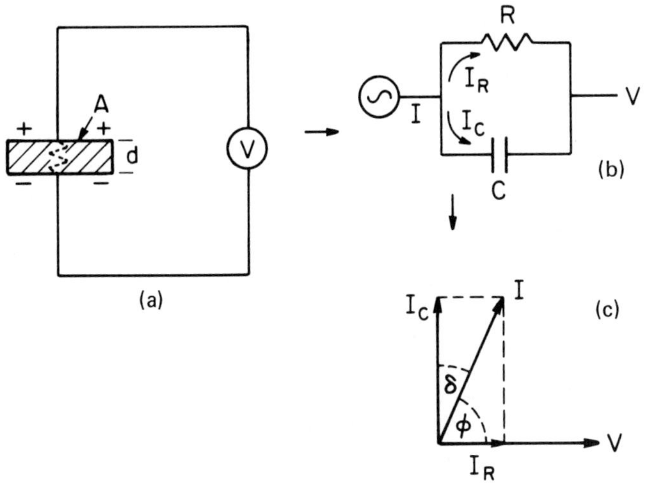

Figure 8 Equivalent circuit diagrams: (a) capacitive cell; (b) charging and loss current; (c) loss tangent for a typical dielectric.

where

$$C = \varepsilon_0 k' \frac{A}{d} \tag{12}$$

$$C_0 = \varepsilon_0 \frac{A}{d} \tag{13}$$

and

$$k' = \frac{C}{C_0} = \frac{\varepsilon}{\varepsilon_0} \tag{14}$$

In the expressions above, A represents the area of the capacitive cell, d its thickness, C_0 and C the respective air and material capacitance, V the voltage impressed across the cell, and ε the material permittivity (F/m). Thus k' also represents the ratio of the permittivities or the charge stored in the capacitive cell relative to air or vacuum as dielectric.

For the case of V being sinusoidal, Eq. (11) may be written as

$$Q = CV_0 e^{i\omega t} \tag{15}$$

Therefore,

$$I = \frac{dQ}{dt} = i\omega CV = i\omega C_0\varepsilon_0 k'V \tag{16}$$

where I represents the current flow on discharge of the capacitive cell in time t. However, for a real dielectric the current I has vector components I_C and I_R, as illustrated in Figure 8 for the condition of a lossy dielectric, represented by the circuit analog of a resistance in parallel with a capacitor. The current I_C represents a (wattless) capacitive current proportional to the charge stored in the capacitor. It is frequency dependent and leads the voltage by 90°. The current I_R is an ac conduction current in phase with the voltage V, which represents the energy loss or power dissipated in the dielectric. This condition can be represented by a complex permittivity or dielectric constant in order to deal with the loss current, as follows:

$$I = i\omega C_0\varepsilon_o\ (k' - ik'')\ V \tag{16a}$$

$$= i\omega C_0\varepsilon_0 k'V + \omega C_0\varepsilon_0 k''V \tag{16b}$$

$$= I_C + I_R \tag{16c}$$

From the magnitude of these currents, therefore, one can define a dissipation factor, tan δ, as

$$\tan\ \delta = \left|\frac{I_R}{I_C}\right| = \frac{\omega C_0\varepsilon_0 k''V}{\omega C_0\varepsilon_0 k'\ V} = \frac{k''}{k'} \tag{17}$$

These vector relationships, illustrated in Figure 8, show the phase angle ϕ as $(90 - \delta°)$ and the loss tangent, therefore representing the deviation from ideality or lossiness in the dielectric.

From Eqs. (2) and (17) the dielectric loss factor, k'', is shown to be the product of the material parameters k' and tan δ. A quality factor Q, defined as

$$Q = \frac{1}{\tan\ \delta} = \frac{\text{average energy stored}}{\text{energy dissipated per cycle}} \tag{18}$$

is frequently used as a design parameter for dielectric usage. Tables 14 to 16 define the relevant dielectric parameters and summarize the useful relationships between them for dielectric applications [1].

B. Polarization in Dielectric Solids

As indicated, the dielectric constant and dissipation factor are true material properties and can therefore be understood in terms of composition and structure, which define material behavior in a dielectric field. The macroscopic behavior can be described in terms of three vector quantities: the dielectric displacement D, field E, and polarization P:

$$D = \varepsilon_0 k'E = \varepsilon_0 E + P \tag{19}$$

Therefore,

$$P = \varepsilon_0(k' - 1)E = \varepsilon_0 x E \tag{20}$$

where x is defined as the electric susceptibility and is the proportionality factor that directly relates polarizability to field in a linear dielectric. Polarization can be further described in terms of a bound charge (surface) density related to the dielectric constant, which opposes the surface charge density σ_s created by the applied field, that is,

$$\bar{P} = \sigma\left(1 - \frac{1}{k'}\right) \tag{21}$$

or in terms of a volume charge density related to the concentration of dipoles per unit volume, N, and the local field E' in the dielectric:

$$P = \alpha N E' \tag{22}$$

and

$$E' = \frac{k' + 2}{3} E \tag{23}$$

where α is defined as the polarizability and is a true material parameter with components reflecting the different polarization mechanisms.

Table 14 Definitions for Dielectric Parameters

Units	Parameter	Definition
Farads	C_0	Capacitance of cell with air/vacuum dielectric
Farads	C	Capacitance of cell with solid dielectric
Coulombs	Q	Total charge on capacitor
C/sec (A)	I	dQ/dt = current due to charge flow
F/m	ε_0	Dielectric constant of vacuum (permittivity)
F/m	ε'	Dielectric constant of solid
F/m	k'	$\varepsilon'/\varepsilon_0$ = relative dielectric constant
F/m	ε''	Loss permittivity of solid
	k"	$\varepsilon''/\varepsilon_0$ = relative loss factor
$(\Omega\text{-m})^{-1}$	σ	Dielectric conductivity
	tan δ	D = dissipation factor = $\dfrac{\text{energy dissipated/cycle}}{\text{energy stored in solid}}$
	Q	Quality factor = 1/tan δ
(Ω)	Z	Circuit impedance = ratio of the voltage to the total current flowing in a circuit: $Z^*(\omega) = \dfrac{V^*(\omega t)}{I^*(\omega t)}$
sec	τ	RC = time constant
	ε_0	8.85×10^{-12} F/m; 8.85×10^{-14} F/cm
	C'	Real capacitance
	C"	Imaginary part (loss part) of capacitance

Table 15 Relationships Between Dielectric Parameters

Capacitance	Complex quantities
$C_0 = \varepsilon_0 \frac{A}{d}$	$\varepsilon^* = \varepsilon' - i\varepsilon''$
$C = \varepsilon' \frac{A}{d}$	also $\varepsilon^* = \varepsilon_0 k^*$
$\frac{C}{C_0} = \frac{\varepsilon'}{\varepsilon_0} = k'$	$\therefore k^* = k' - ik''$
and $\varepsilon = \varepsilon_0 k'$	and $\varepsilon'' = \varepsilon_0 k''$
	$C' = \frac{C_s}{1 + \omega^2\tau^2}$; $C'' = \frac{C_s \omega\tau}{1 + \omega^2\tau^2}$
Time const./dissipation factor	**Charge current**
$\tan\delta = \omega R_s C_s = \frac{1}{\omega R_p C_p}$	$Q = CV;\ V = V_0 e^{i\omega t}$
$\tau = RC$ (series)	$I = \frac{dQ}{dt} = i\omega CV$
$\therefore \tan\delta = \omega\tau$	$Z = \frac{V}{I} = \frac{V}{i\omega CV} = \frac{1}{i\omega C}$
$\tan\delta = \frac{\varepsilon''}{\varepsilon'} = \frac{k''}{k'} = \frac{C''}{C'}$	
$\tan\delta = \frac{\sigma}{\omega\varepsilon'} = \frac{\sigma}{\omega\varepsilon_0 k'}$	
$\tan\delta = \frac{1}{Q}$	
Dielectric power loss	**Dielectric conductivity**
$W = \omega\varepsilon_0 k' V^2 \tan\delta$ (W/cycle)	$\sigma = \omega\varepsilon'' = \varepsilon_0 \omega k'' = \varepsilon_0 \omega k' \tan\delta$

Table 16 Series and Parallel Dielectric Circuit Equations[a]

		Z	$\lvert Z \rvert$	$\tan\delta$	τ
(a)	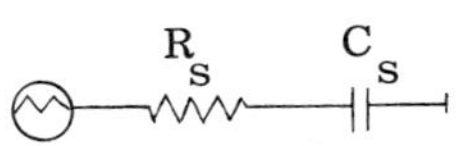	$R_s + \dfrac{1}{i\omega C_s}$	$\dfrac{1}{\omega C_s}(1 + \omega^2 C_s{}^2 R_s{}^2)^{\frac{1}{2}}$ $= \dfrac{1}{\omega C_s}(1 + \tan^2\delta)^{\frac{1}{2}}$ $= \dfrac{1}{\omega C_s}(1 + \omega^2\tau^2)^{\frac{1}{2}}$	$\omega R_s C_s$	$R_s C_s$
(b)	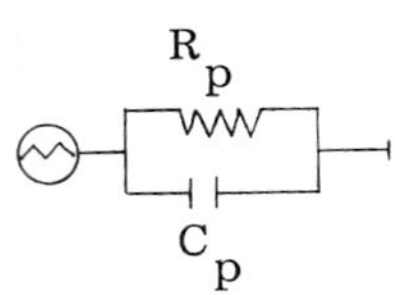	$\dfrac{R_p}{1 + i\omega R_p C_p}$	$\dfrac{R_p}{(1 + \omega^2 R_p{}^2 C_p{}^2)^{\frac{1}{2}}}$	$\dfrac{1}{\omega R_p C_p}$	$\dfrac{1}{\omega^2 R_p C_p}$

$$= \frac{R_p}{(1 + 1/\tan^2 \delta)^{\frac{1}{2}}}$$

(c)

$$R_p = R_s \; 1 + \frac{1}{\tan^2 \delta}$$

$$C_p = \frac{C_s}{1 + \tan^2 \delta}$$

$$\varepsilon'_p = \frac{\varepsilon'_s}{1 + \tan^2 \delta} \; ; \; k'_p = \frac{k'_s}{(1 + \tan^2 \delta)}$$

---

[a] Series circuits are used mainly for low-loss materials ($\tan \delta < 0.032$), which constitute most dielectrics. For $\tan \delta > 0.032$, however, the series equivalent circuit gives too high a value for k' and k" and the parallel equivalent circuit should be used [i.e., a correction should be made (using the equations above) if very accurate results are desired].

Source: Ref. 1.

From Eqs. (20) and (21) expressions can be developed (Clausius-Mosotti) relating the dielectric constant to the various polarizability mechanisms:

$$\frac{k' - 1}{k' + 2} = \frac{N\alpha}{3\varepsilon_0} = \frac{1}{3\varepsilon_0} \frac{N_0\rho_d}{M} \alpha = \frac{1}{3\varepsilon_0} \Sigma N_i\alpha_i \tag{24}$$

where N_0 is Avogadro's number with ρ_d and M the density and molecular weight of the solid, respectively. The components of the polarizability are

$$\alpha = \alpha_e + \alpha_i + \alpha_o + \alpha_s \tag{25}$$

representing susceptibilities associated with electronic, ionic, orientation, and space charge polarization, respectively.

Electronic polarization (P_e) is due to shifts or displacement of electron clouds in the dielectric field away from their equilibrium positions, resulting in a net dipolar response. It occurs in all solids up to optical frequencies $\sim 10^{16}$ Hz. Ionic polarization (P_i) results from similar ionic displacement in the field and occurs up to the infrared region of 10^{10}–10^{13} Hz. The polarization $(P_e + P_i) = P_\infty$ is mostly composition dependent, instantaneous, or nearly frequency independent for most dielectrics and only marginally affected by temperature. In contrast, orientation polarization is both frequency (time) dependent and temperature dependent, since it represents dipole orientation and ion jump polarization. The parameter $P_s = P_o + P_\infty$ thus represents total polarization from all mechanisms, including the static or low-frequency polarization.

According to Debye, P_o is time dependent, approaching its final value in an exponential manner described by

$$P_o(t) = P_o(1 - e^{-t/\tau}) \tag{26}$$

where τ is the relaxation time [period needed to diminish the polarization charge to 36.8% (1/e) of its original value]. Solution of Eq. (26) in terms of complex quantities gives the Debye equations:

$$k' = k'_\infty + \frac{k'_s + k'_\infty}{1 + \omega^2\tau^2} \tag{27a}$$

$$k'' = \frac{(k'_s - k'_\infty)\omega\tau}{1 + \omega^2\tau^2} \tag{27b}$$

$$\tan\,\delta = \frac{k''}{k'} = \frac{(k'_s - k'_\infty)\omega\tau}{k'_s + k'_\infty\omega^2\tau^2} \tag{27c}$$

where k'_s and k'_∞ represent the dielectric constants for the equivalent polarizations. The relaxation behavior representing these parameters for the ideal Debye solid (single relaxation mode) are illustrated in Figure 9. For k' and k" maximum relaxation and loss occur at $\omega\tau = 1$ and slightly higher for the tan δ peak.

Figure 10 shows the frequency dispersion or change in dielectric properties with frequency and the corresponding circuit analogs for a general solid with several relaxation modes. The dielectric constant shows an expected decrease with frequency as successive polarization modes become damped out and corresponding peaks occur in the loss spectra at the different relaxation and resonance frequencies [23,24].

The resonance losses occurring for ionic and electronic oscillations at infrared and ultraviolet frequencies are associated with absorption interactions with the incident radiation at or near the natural vibration frequencies for the ions or electrons. Relaxation losses, conversely, are associated with dipolar orientation, ion jump, or electron hopping in such solids as SiO_2, Na glasses, and semiconducting oxides, respectively, and can occur over a wide frequency range, as indicated. Space charge polarization losses typically occur at low frequency (<1 kHz) and are associated with ion migration and electrode contact losses or with the presence of grain boundary or inhomogeneous phases in the dielectric. Space charge losses in the latter case can occur up to much higher ($\sim 10^5$ Hz) frequencies.

Contributions to the overall dielectric constant from different polarization mechanisms relate to the composition, frequency, and temperature of the dielectric. For predominantly electronic solids such as diamond and molecular solids such as polymers, the dielectric constant is closely related to the refractive index (i.e., $k' \sim n^2$). Otherwise, the electronic polarizability is given by [19, 22]

$$\alpha_e = 4\pi\varepsilon_0 r_0^3 = \frac{3\varepsilon_0}{N_0} R_\infty \tag{28}$$

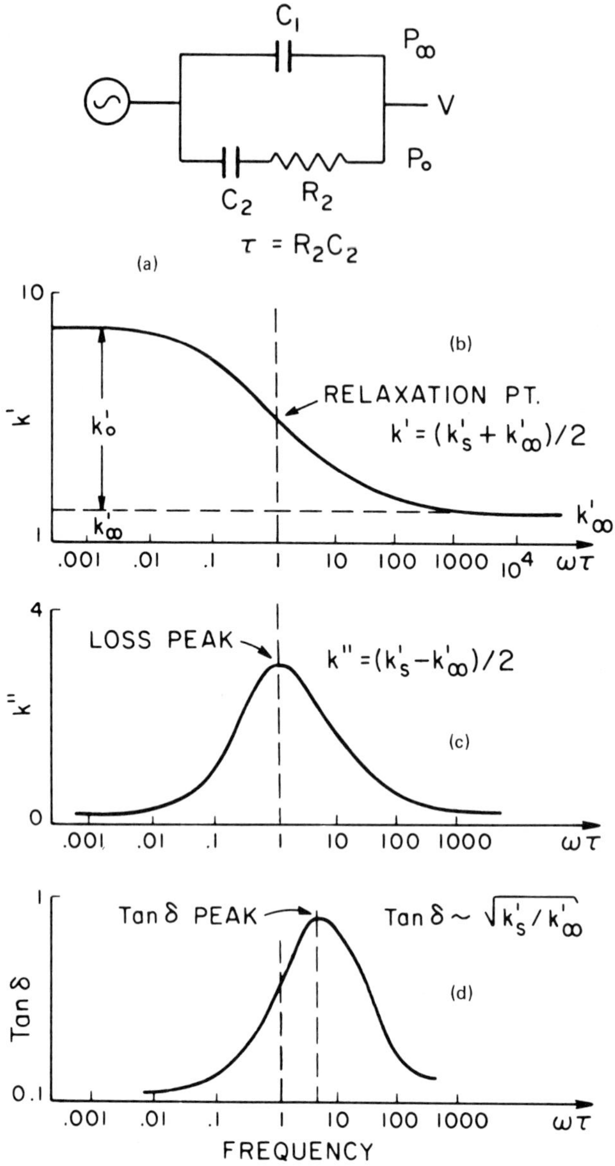

Figure 9 Representations of (a) dielectric circuit for a Debye solid, (b) dielectric constant (polarization), (c) dielectric loss factor, and (d) dissipation factor, as a function of frequency for a solid with single relaxation mode.

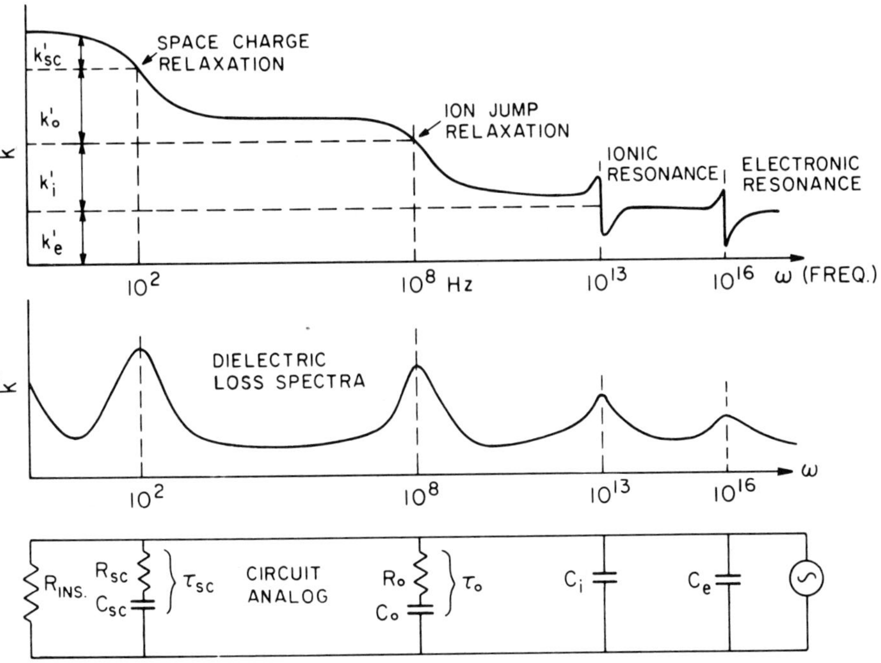

Figure 10 Frequency dispersion behavior for a general solid with multiple relaxation and resonance modes, with analog of equivalent circuits representing loss mechanisms.

where

$$R_\infty = \frac{n^2 - 1}{n^2 + 1} \frac{M}{\rho_d} \tag{29}$$

and r_0 represents the radius of the atom or ion, n the refractive index, and R_∞ the molar refractivity, values for which are available for different ions. In general, α_e increases with ion concentration, ion size, and polarizability, hence the use of such ions as Ba^{2+}, Pb^{2+}, La^{3+} and K^+ in optical glasses in order to achieve a higher dielectric constant and refractive index. For predominantly ionic solids such as MgO, Al_2O_3, and NaCl, the ionic polarizability α_i can be expressed analytically as

$$\alpha_i = \frac{(ez)^2}{2\omega_0^2 a^3}\left(\frac{1}{m_1} - \frac{1}{m_2}\right) \tag{30}$$

where ω_0 is the ionic resonance frequency ($\sim 10^{13}$ Hz), a the interionic separation, ez the charge on the ion, and m_1 and m_2 the mass of the cation and anion. Ion size and separation are seen to have a significant effect on the polarization, which in these solids generally exceeds the electronic polarization and thus contributes most to the dielectric constant.

For ceramic insulators with mobile ions, the orientation polarization can be considered equivalent to an ion jump polarization, α_j, where

$$\alpha_o = \frac{p^2}{3kT} \quad \text{and} \quad \alpha_j = \frac{(ezd)^2}{3kT} \tag{31}$$

and p = ezd is the dipole moment associated with the jump of an ion of charge ez through a distance d. As a function of frequency, the real and imaginary (loss) parts of a complex α_j may be given as

$$\alpha'_j = \frac{(ezd)^2}{3kT} \frac{1}{1 + \omega^2\tau^2} \tag{32}$$

$$\alpha''_j = \frac{(ezd)^2}{3kT} \frac{\omega\tau}{1 + \omega^2\tau^2} \tag{33}$$

$$\tau = \frac{1}{2\nu} = \tau_0 e^{\mu/kT} \tag{34}$$

and

$$n = n_0\nu = n_0 e^{-\mu/kT} \tag{35}$$

where τ is the relaxation time, ν the jump frequency, n the number of successful ion jumps per second, and μ the activation energy for ion jump ($\sim 0.4-0.9$ eV for ceramics). Since for the ceramics τ decreases with temperature, the relaxations move to higher temperature

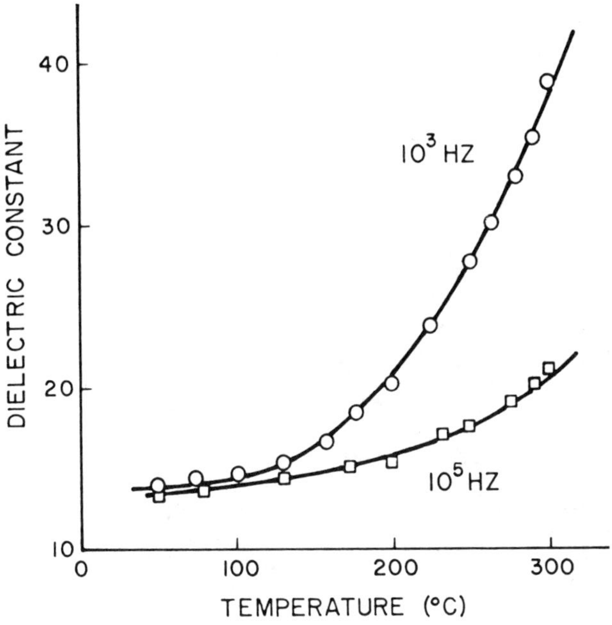

Figure 11 Plot of dielectric constant for a lead-alumina-borosilicate glass, showing change with frequency and temperature.

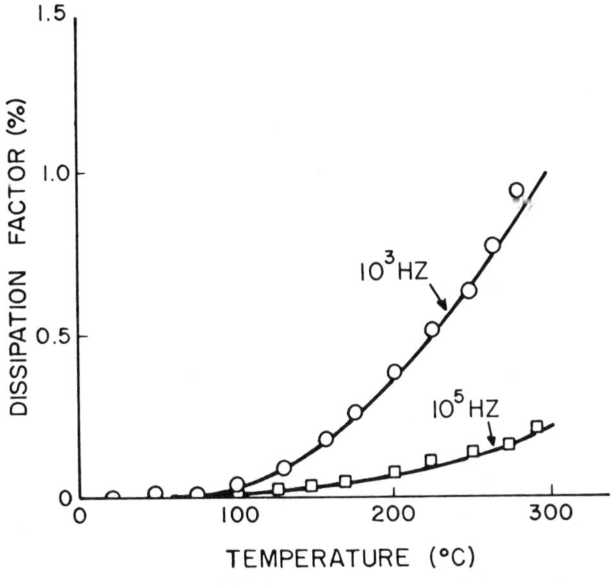

Figure 12 Plot of dissipation factor for glass in Figure 11, showing change with temperature and frequency.

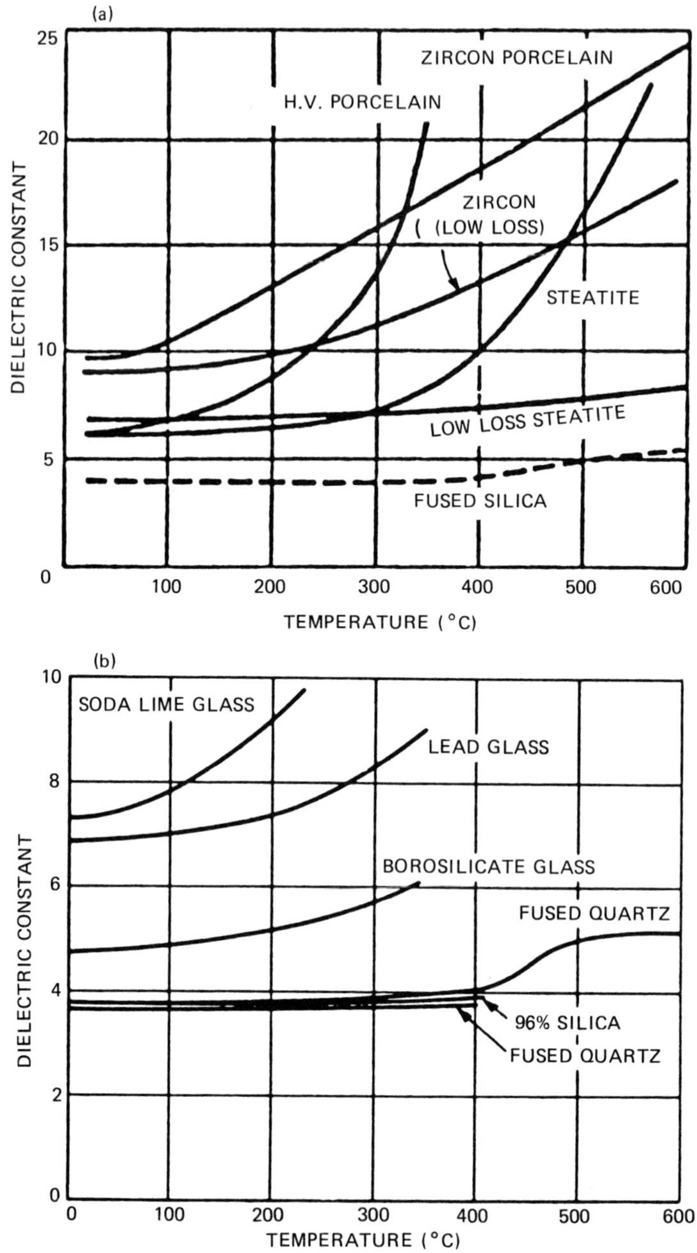

Figure 13 Plot of dielectric constant versus temperature at 10^5 Hz for (a) typical ceramic and (b) glass insulator. (From Refs. 8, 20, and 25.)

with increasing frequency. However, since n (the number of successful ion jumps per unit time) also increases constantly with temperature, a relaxation peak is not found for such materials as porcelains and glasses. Instead, the dielectric constant and loss increase continuously with temperature but decrease with frequency, as illustrated for a lead-borosilicate glass in Figures 11 and 12, where the behavior observed is due in part to the distribution of relaxation times arising from the irregular structure of the glass. Figure 13 illustrates the observed increase in dielectric constant with temperature for several glass and ceramic compositions.

VI. INSULATION RESISTANCE

A. Electrical Conduction

Electrical conduction is that property which relates the current flux in a material to the electrical field impressed across it at steady state. The current flux or density can also be defined in terms of the measurement parameters, or more fundamentally in terms of the material properties as follows [19,26]:

$$j_c = \sigma \frac{dV}{dx} = \sigma E \tag{36a}$$

$$j_c = \frac{I}{A} \tag{36b}$$

$$j_c = nq\mu F \tag{36c}$$

where j_c (A/m^2) is the current density, σ (Ω-m)$^{-1}$ the conductivity, dV/dx the voltage gradient or field E, μ(m^2/V-sec) the mobility, n the concentration of charge carriers, and q = ez (coulombs) is the charge on the carriers (e = 1.6 × 10^{-19} C and z = valence or charge on the carriers). From Eq. (10), expressions for the conductivity can be developed in terms of the measurement parameters:

$$\sigma = \frac{1}{R}\frac{d}{A} \tag{37a}$$

and

$$\frac{1}{\sigma} = \rho = R\frac{A}{d} \tag{37b}$$

or in terms of the material parameters,

$$\sigma = nq\mu \tag{37c}$$

where ρ is the electrical resistivity (Ω-m), R (ohms) the sample resistance, A its area (m^2), and d its thickness (m). For more than one type of charge carrier being present, the resultant conductivity can be defined as the sum of component conductivities (σ_i) as follows:

$$\sigma = \Sigma_i\, n_i\, (ez)_i\, \mu_i = \Sigma_i\, \sigma_i \tag{38a}$$

and

$$\sigma_i = \sigma t_i \tag{38b}$$

where t_i ($0 \leqslant t_i \leqslant 1$), the transference number, represents the fraction of the total current or conductivity attributable to the mobile charge carriers, i, which may be either electrons or holes or cations and anions. Depending on which charge carriers predominate, the solid may be classified as primarily an electronic (n or p type) or ionic conductor. However, mixed conduction does occur and this conduction is represented as

$$\sigma = \sigma_{elect} + \sigma_{ionic} \tag{39}$$

In such solids the ionic component tends to predominate at the higher temperatures, typically above 200°C. Direct determination of the transference number is difficult to achieve. In most cases the ionic component is determined by use of a coulometric-type cell or, where appropriate, by gas-phase reactions and use of the Nernst relationship, for example in measuring the oxygen ion component in ZrO_2 [17].

Where the number of charge carriers taking part in the conduction process is essentially constant with temperature [Eq. (37c)], the conduction process may be described as extrinsic or impurity controlled. This condition arises from the presence of mobile impurities in the solid, or from selective doping, for example the addition of $BaCl_2$ (Figure 14) to KCl or P to Si. The extrinsic conduction is typified by a higher conductivity and lower activation energy for conduction. The latter reflects the fact that energy need only be supplied to effect the mobility of the charge carriers. In contrast, for intrinsic conduction the fractions of charge carriers

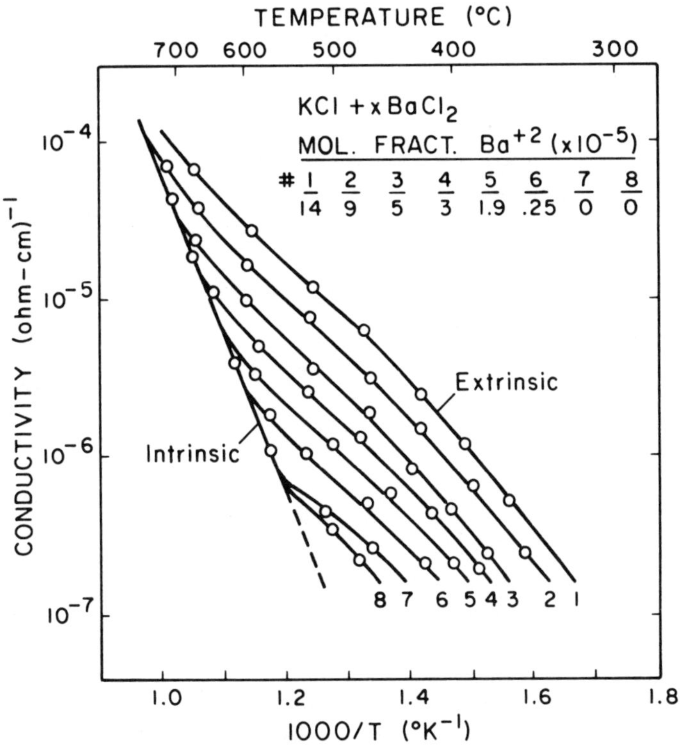

Figure 14 Plot of log conductivity versus reciprocal absolute temperature for Ba-doped KCl showing intrinsic and extrinsic conduction with increasing Ba^{2+} concentration.

(n) contributing to the conduction increases with temperature by an activated process. The combined activation energy for this process would therefore be higher, as seen in Figure 14.

The mobility term in Eq. (37c) is defined as $\mu = \bar{v}/E$, where $\bar{v}$ is the average drift velocity of the carriers under the influence of the field E. For semiconductor materials such as Si and GaAs, the mobility at all but the lowest temperatures (where impurity scattering predominates) is determined by lattice or phonon scattering of the charge carriers and decreases with temperature. This relationship may be given as

$$\mu \sim \text{const} \times T^{-3/2} \tag{40}$$

and for low-temperature impurity scattering

$$\mu \sim \text{const} \times T^{3/2} \tag{41}$$

For these materials, the mobility and conductivity are related by the Hall coefficient R_H as follows:

$$\mu = R_H \sigma \tag{42}$$

where

$$R = \pm \frac{1}{ne} \tag{43}$$

and

$$\sigma = \sigma_0 \, e^{-E_g/2kT} \quad \text{(intrinsic)} \tag{44}$$

or

$$\sigma = \sigma_0 \, e^{-E_x/kT} \quad \text{(extrinsic)} \tag{45}$$

where E_g and E_x represent the activation energies for intrinsic (band gap) and extrinsic (impurity level) conduction, respectively; k is the Boltzmann constant; and T is the absolute temperature.

For an ionic solid, mobility is related to the diffusion coefficient D (cm^2/sec) by the Einstein relationship

$$\mu = \frac{ez}{kT} D \tag{46}$$

Diffusion and conductivity are related by the Nernst-Einstein equation:

$$\sigma = \frac{n(ez)^2 D}{kT} \tag{47}$$

Since both diffusion and N (number of defects generated) are activated processes, where

$$N = ne^{-w/2kT} \tag{48}$$

and

$$D = D_0 e^{-\mu/kT} \tag{49}$$

then

$$\sigma = \sigma_0 e^{-E/kT} \tag{50}$$

and

$$E = \frac{w}{2} + \mu \tag{51}$$

where w and μ are the activation energies for defect generation and migration, respectively. For extrinsic conduction w = 0 and E = μ; that is, the ionic mobility becomes the controlling factor in the conduction.

B. Defects

The basic processes for diffusion in ionic solids require the presence of point defects in the crystal structure [27,28]. The conductivity is therefore, very closely related to the existence of these ionic defects, which are of four kinds: (1) cation vacancies, (2) cation interstitials, (3) anion vacancies, and (4) anion interstitials. Table 17 lists Kröger–Vink nomenclature for the different types of ionic defects.

Defects may be generated in a solid by (1) thermodynamic equilibrium (Schottky, Frenkel), (2) chemical substitution, (3) oxidation or reduction, and (4) energetic radiation [30]. The thermodynamic Schottky and Frenkel defects can be classified as inherent defects since one or the other is always present in an ionic solid at all temperatures above absolute zero. The formation of these defects increases the configurational entropy of the solid, thereby decreasing the overall free energy. Relationships for Schottky defect formation may be given as

$$M_M + X_X = V''_M + V^{\cdot\cdot}_X \tag{52}$$

and

$$n_s = ne^{-w_s/2kT} \tag{53}$$

where n_s is the concentration of Schottky defects and w_s is the activation energy for defect formation. Schottky defects are typically

Table 17 Kröger–Vink Nomenclature for Ionic-Type Defects

Symbol	Definition[a]
M_M	Cation on regular lattice site (Ni_{Ni})
X_X	Anion on regular lattice site (O_O)
V_M	Cation vacancy
V''_M	Effective charge on cation vacancy
M_i	Cation on interstitial site
$M^{\cdot\cdot}_i$	Effective charge on interstitial cation
X_i	Anion on interstitial site
X''_i	Effective charge on interstitial anion
V_X	Anion vacancy
$V^{\cdot\cdot}_X$	Effective charge on anion vacancy

[a]Referenced to MO lattice.
Source: Ref. 29.

formed in close-packed (face-centered cubic, hexagonal close packed) structures such as the alkali halides, MO oxides, and corundum (Al_2O_3) since such structures do not readily accommodate interstitial ions. The cation and anion vacancies are considered to form separately and migrate to the surface of the crystal, thereby increasing the volume and decreasing the density. Relationships for Frenkel defects may be written as

$$M_M = M^{\cdot\cdot}_i + V''_M \tag{54}$$

or

$$X_X = X''_i + V^{\cdot\cdot}_X \tag{55}$$

and

$$n_f = (nn^*)^{\frac{1}{2}} e^{-w_f/2kT} \tag{56}$$

where n_f is the concentration of Frenkel defects, n* the number of interstitial sites, and w_f the activation energy for defect formation. Frenkel defects typically form in more open structures (CaF_2), where there is a size and charge disparity between the cations and anions. This includes such solids as BaF_2, ZrO_2, UO_2, and the Ag and Cu halides. Frenkel solids are often electronic conductors at lower temperatures, due to the prevalence of covalent bonding, for example, in the Cu halides.

Chemical substitution can result in any of the four defects indicated being formed in the crystal lattice, as follows:

1. Cation vacancy:

$$x(CaCl_2) \xrightarrow{NaCl} x(Ca^{\bullet}_{Na} + 2Cl_{Cl} + V'_{Na}) \tag{57}$$

2. Anion vacancy:

$$x(CaO) \xrightarrow{ZrO_2} x(Ca''_{Zr} + O_O + V^{\bullet\bullet}_O) \tag{58}$$

3. Cation interstitial:

$$x(AlNa) \xrightarrow{SiO_2} x(Al'_{Si} + Na^{\bullet}_i) \tag{59}$$

4. Anion interstitial:

$$x(NdF_3) \xrightarrow{CaF_2} x(Nd^{\bullet}_{Ca} + 2F_F + F'_i) \tag{60}$$

Reaction (59) would represent the classic weathering of silica to form a feldspar mineral. Constraints on the substitution are that the impacted ions should be within ±15% in ionic size, of not too great disparity in valence state, and of similar coordination structure. Any other condition leads to only limited substitution of the ion in the host lattice. Chemical substitution may also be used to induce valence changes in the material, as follows:

$$x(La_2O_3) \xrightarrow{BaTiO_3} x(2La^{\bullet}_{Ba} + 3O_O + 2Ti'_{Ti}) \tag{61}$$

where, as indicated, a reduction of Ti^{4+} ions to the Ti^{3+} state takes place to offset the increased charge on the A site in $BaTiO_3$. The juxtaposition of Ti^{3+} and Ti^{4+} ions on equivalent sites facilitates electron transfer or hopping from the Ti^{3+} to the Ti^{4+} ions with increased n-type conduction, depending on the Ti^{3+} concentration.

This mechanism defines another type of point defect, the polaron, which is an electron or hole trapped on a particular atom, ion, or vacancy. The electron or hole and its associated polarization field may be considered as a particle since it can propagate throughout the crystal, usually by hopping from one ion to another as above. If the dimension of the polarization field is small compared to the lattice dimension of the crystal, this condition is described as a small polaron.

C. Defects in Nonstoichiometric Oxides

Point defects that are formed in oxides as a result of equilibrium in a gas-phase ambient may occur in two different ways: (1) by oxygen deficiency (excess metal) or (2) by metal deficiency (excess oxygen) with respect to a stoichiometric composition, as follows.

1. Oxygen-Deficient Oxides

For such an oxide, the overall nonstoichiometric reaction can be written as

$$MO = MO_{1-x} + \frac{x}{2} O_2 \tag{62}$$

The oxygen vacancy is formed by the transfer of an oxygen atom on a normal lattice site to the gaseous state:

$$O_O = \tfrac{1}{2}O_2(g) + V_O^{\cdot\cdot} + 2e' \tag{63}$$

Since the vacancy has two trapped electrons, it can act as a donor and become singly or doubly ionized. This would lead to n-type conduction at high temperature in such fixed valence oxides as Al_2O_3, MgO, and CaO. The free electrons may also become associated with variable valence cations and the following reactions can occur:

$$M_M + V_O = V_O^{\cdot} + M_M' \tag{64}$$

$$M_M + V_O^{\bullet} = V_O^{\bullet\bullet} + M''_M \tag{65}$$

Equation (64) is the more likely reaction and leads to n-type hopping conduction in such oxides as TiO_2, Fe_2O_3, CeO_2, and Nb_2O_5.

In the case of excess metal atoms, such atoms may be found in the interstices:

$$M_M + O_O = M_i + \tfrac{1}{2}O_2(g) \tag{66}$$

The neutral M_i atom can subsequently be ionized:

$$M_i = M_i^{\bullet} + e' \tag{67}$$

$$M_i^{\bullet} = M_i^{\bullet\bullet} + e' \tag{68}$$

Again, the singly ionized state is the more likely and leads to n-type conduction in oxides such as ZnO.

2. Metal-Deficient Oxides

In an oxide (MO) cation vacancies can be formed through the reaction of ambient oxygen with the oxide:

$$\tfrac{1}{2}O_2(g) = V_M + O_O + 2h \tag{69}$$

The holes present in the vacancy may be excited and transferred to other parts of the crystal:

$$V_M = V'_M + h^{\bullet} \tag{70}$$

This condition leads to p-type conduction in such oxides as MnO, NiO, FeO, and CoO. As a valence defect, Eq. (70) can be written as

$$M_M + V_M = V'_M + M_M^{\bullet} \tag{71}$$

With excess oxygen, neutral interstitial oxygen atoms may be formed thus:

$$\tfrac{1}{2}O_2(g) \longrightarrow O_i \tag{72}$$

but this condition is not common in oxide lattices.

The relationship of oxygen partial pressure to conduction in metal deficient oxides can be illustrated as follows:

$$2M_M + \tfrac{1}{2}O_2(g) = V''_M + 2M^{\bullet}_M + O_O \tag{73}$$

The equilibrium constant, K, for this reaction is

$$K_T = \frac{[M^{\bullet}_M]^2 \, [V''_M] \, [O_O]}{(P_{O_2})^{\frac{1}{2}} \, (M_M)^2} \tag{74}$$

Where $[O_O]$ and $[M_M] \sim 1$ and $V''_M = \frac{1}{2}[M^{\bullet}_M]$. Since σ, the conductivity, is proportional to $[M^{\bullet}_M]$ (e.g., $[Ni^{3+}]$), we can derive a relationship:

$$\sigma = KP_{O_2}^{1/6} \tag{75}$$

Thus, if the P_{O_2} is increased, the conductivity also increases. In oxygen-deficient oxides it can be shown, similarly, that the effect of increase in P_{O_2} is just the opposite, that is,

$$\sigma = KP_{O_2}^{-1/6} \tag{76}$$

Depending on the oxygen partial pressure, therefore, n- or p-type conduction may occur at very low and high P_{O_2}, respectively. For P_{O_2} ranges where the oxygen and metal vacancy concentrations are of similar magnitude, the oxide is essentially stoichiometric and ionic conduction predominates. All three types of conductions are exhibited by such oxides as ZrO_2 over a wide range of oxygen partial pressures.

D. Band Conduction

The energy-band model for the transition metal oxides is unique in that the energy bands that are important for semiconduction are related to the inner electron levels on the atoms in the structure rather than to outermost valence electron levels. If in an oxide, the 3d levels are only partially filled, the magnitude and type of conduction will be related to the extent of overlap of the 3d orbitals [31,32]. If the overlap is barely sufficient to develop an energy band, conduction can occur only within this band, and this is generally described as narrow-band conduction. Oxides such as Ti_2O_3, TiO, and V_2O_3 exhibit this type of conduction. In certain other oxides, the 3d wavefunction overlap is so small that a 3d band does not develop. The 3d electrons are localized on the individual ions and can move from ion to ion only by the hopping process described. Examples of materials exhibiting such conduction are Li-doped NiO, La-doped $BaTiO_3$, and Fe_3O_4.

E. Diffusion

As indicated previously, lattice diffusion in ionic solids is effected by the presence of point defects in the structure [30,33,34]. The various diffusion mechanisms are described below.

1. Vacancy Mechanism

Diffusion takes place by vacancy mechanism when an atom or ion on a normal lattice site jumps into an equivalent lattice site or vacancy. This moves the vacancy to the site exited by the atom; thus the atom migrates in a direction opposite to the vacancy. Vacancy migration is important in such materials as NaCl, ZrO_2, UO_2, and MgO. It should be noted that the vacancy carries a charge opposite to that of the ion.

2. Interstitial Mechanism

If an atom on an interstitial site moves to a neighboring interstitial site, the diffusion process is described as interstitial migration. Such a movement or jump of the interstitial atom involves a considerable distortion of the lattice and this mechanism is probable only when the interstitial atom is smaller than the atoms on the normal lattice positions or in glasses where the open structure permits ready migration of alkali and hydroxide ions via this mechanism.

3. Interstitialcy Mechanism

If the distortion becomes too large to make the interstitial mechanism probable, interstitial atoms may move by another mechanism. In the interstitialcy mechanism, an interstitial ion pushes one of its nearest-neighbor atoms located on a normal lattice site into another interstitial position and itself occupies the lattice site of the displaced atom. In the interstitialcy mechanism one may distinguish between two types of atom movements. If the atom on the normal site is pushed in the same direction as that of the interstitial atom, the jump is termed collinear. If the atom is pushed to one of the other neighboring sites so that the jump direction is different from that of the interstitial ion, the jump is termed noncollinear. This mechanism is important in the Ag halides, for example, where the migration of the large Ag^+ ion interstitially would cause severe lattice distortions.

4. Divacancy Mechanism

Cation and anion vacancies can be coupled and migrate in tandem through the solid. The migration can be considered as dissociative if it temporarily breaks up and reforms or nondissociative if no breakup occurs. The latter mechanism requires less energy for propagation. The divacancy mechanism can account for material transport without accompanying charge transport since the charges on the vacancies offset each other. Discrepancies between measured conductivities and diffusion coefficients and the Nernst–Einstein relationship, discussed previously, are often attributable to this cause. In any event, the Nernst–Einstein equation must be modified to reflect differences in the various diffusion mechanisms, as follows [27]:

$$\sigma_i = \sigma t_i = \frac{n_i(ez_i)^2 D_i}{kT} B \qquad (77)$$

where the mobility parameter, B, is related to the various diffusion mechanisms as follows: B = 1.27 (vacancy), 1.0 (interstitial), 1.38 (interstialcy-collinear), 3.0 (interstialcy-noncollinear), and B << 1.0 (divacancy). These empirical factors are intended to normalize differences between the diffusion mechanisms related to number of available jump sites and jump distances. For glasses, where interstitial migration predominates, agreement between the Nernst–Einstein equation and measured data is good [19,27].

F. Electrical Conduction in Glasses

Electrical conduction in glasses is mainly attributed to the migration of mobile ions such as Li^+, Na^+, K^+, OH^-, and H^+ under the influence of an applied field. At temperatures above 250°C, divalent ions such as Ca^{2+}, Mg^{2+}, and O^{2-} can also contribute to the conduction process, but their mobility is considerably lower than that of the alkali ions [35–37].

The current carriers are considered to be located in potential wells and undergo thermal vibrations of magnitude proportional to the temperature. In the absence of an applied field the ions make random jumps according to the probability of surmounting the surrounding energy barriers but migrate preferentially in the direction of the applied field, which effectively lowers the energy barriers in the forward direction. The number of conducting ions therefore increases with both temperature and electric field.

Early work indicated that oxide glasses containing large amounts of alkali ions were essentially electrolytic in nature. Under a dc field electrolysis was found to occur and Faraday's laws obeyed. Electronic conduction was found to take place only in transition metal–oxide-containing glasses. Otherwise, the transference number for the alkali ions is close to unity, the mobility of divalent and higher valence ions being considerably lower. However, ionic conduction in oxide glasses is strongly influenced by such factors as composition, temperature, and thermal history as described below.

1. Composition

The basic ingredients of oxide glasses may be classified as network formers (SiO_2, B_2O_3, P_2O_5), modifiers (Na_2O, K_2O, CaO, MgO), and intermediates (Al_2O_3, PbO). During fusion of the glass mixture a structural readjustment takes place such that the very strong Si-O-Si bonds are broken and the bridging oxygens are converted to nonbridging oxygens. Such a reaction can be represented as

$$-\overset{|}{\underset{|}{Si}}-O-\overset{|}{\underset{|}{Si}}- + Na_2O \longrightarrow -\overset{|}{\underset{|}{Si}}-O^{\;-Na^+-}\;O-\overset{|}{\underset{|}{Si}}- \qquad (78)$$

where the alkali ions or alkali earth ions become electrostatically attached to the nonbridging oxygens and can migrate freely. As more alkalis are added to the glass, fewer and fewer bridging oxygens remain. The network "coherence" decreases and it becomes easier for the ions to move [38]. Table 18 lists the values of E for the various alkali and alkali earth glasses. Stevels describes the

Table 18 Activation Energies of Alkali and Alkali Earth Glasses

Glass composition (mol %)	Activation energy (eV)
15 K_2O–85 SiO_2	0.83
40 Na_2O–60 SiO_2	0.51
40 K_2O–60 SiO_2	0.63
20 Na_2O–20 K_2O–60 SiO_2	1.03
20 K_2O–20 Li_2O–60 SiO_2	1.05
10 Na_2O–20 PbO–70 SiO_2	1.04
30 Na_2O–20 PbO–50 SiO_2	0.82
30 Na_2O–20 BaO–50 SiO_2	0.86
20 Na_2O–30 BaO–50 SiO_2	1.10
10 Na_2O–10 CaO–80 B_2O_3	1.17
10 Na_2O–30 MgO–60 B_2O_3	1.13
40 CaO–60 SiO_2	1.10
45 CaO–55 SiO_2	1.35
50 CaO–50 SiO_2	1.30
55 CaO–45 SiO_2	1.25

Source: Ref. 40.

structure change in terms of a coherence parameter Y, defined for silicate glasses as [39]

$$Y = 8 - 2R \tag{79}$$

where $3.2 \leqslant Y \leqslant 4$, typically, and is related to the number of non-bridging oxygens per Si^{4+} polyhedra. The parameter R may be defined as

$$R = \frac{2 - x}{1 - x} \tag{80}$$

where x is the mole fraction of modifier ion (Li^+, Na^+, K^+). R may also be related to the average number of oxygen ions per Si^{4+} (or other glass-forming) ion.

Schwartz and MacKenzie [40] studied the conductivity of calcium silicate glasses with Na_2O impurities in the temperature range 250–500°C. They found that the electrical resistivity was insensitive to the Na_2O content but decreased with increasing CaO concentration. The activation energy for conduction ranged from 1.4 to 1.25 eV as the CaO concentration increased from 40 mol % to 55 mol %. They concluded that the detailed structure of glass, relating to the coordination number of the cations and not simply to their random distribution, played an important role in ionic conduction.

Badger and White [41] measured the electrical resistivity of some glasses in the soda-lime-silica system. They performed measurements on glasses containing equivalent amounts of various oxides incorporated in a parent soda-lime glass. The specific resistance at 400°C ranged from 0.06 to 5.0×10^5 Ω-cm. They found that an increase in resistance was produced by additions of MnO, ZnO, B_2O_3, Fe_2O_3, BaO, PbO, TiO_2, and K_2O, the resistance increasing in this order. A decrease in resistance was produced by Na_2O, CaO, and Al_2O_3.

When a second alkali is added to an alkali-silicate glass, the conductivity sharply decreases. The simple concept of a random silicon network cannot explain this anomalous behavior and many theories have been proposed to explain it [42–44]. One such theory assumes that the glassy structure is composed of two regions: polar regions, where the alkali ions preferentially aggregate, and nonpolar regions, consisting of the glass-forming oxide. When two ions are present there will be two different types of polar regions, with each ion associated only with one type of polar region. The decrease in conductivity was attributed to the inability of ions to exchange between the polar regions and to the obstacles caused by one polar region interacting with another ion. It was also found that for a fixed concentration of charge carriers, addition of MgO, PbO, BaO, and CaO tended to decrease the conductivity. This behavior was attributed to the "blocking" of the conducting ions by the heavier divalent atoms.

2. Temperature Dependence

For most oxide glasses, the temperature dependence of the electrical resistivity over a fairly wide range (but below the glass transition temperature) is adequately described by the Rasch–Hinrichsen equation [8,45]:

$$\log \rho = A + \frac{B}{T} \tag{81a}$$

or

$$\rho = Ae^{B/T} \tag{81b}$$

where, for most classes, $A = -0.45$ to 1.5, $B = 3000–6000\ K^{-1}$, $\rho = 10^{10}–10^{15}$ (Ω-cm) at 25°C, and $E = 0.2 \times 10^{-3}B$ (eV) (E is the activation energy for conduction). Figure 15 shows a plot of log resistivity versus reciprocal temperature for various types of

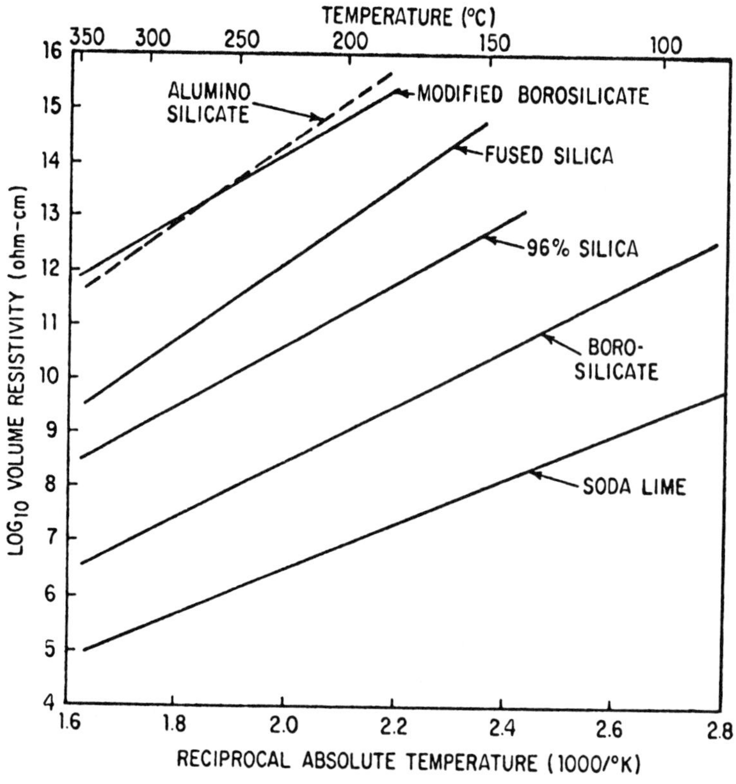

Figure 15 Plot of log resistivity versus reciprocal absolute temperature for commercial glasses, showing wide range in values. (From Ref. 8.)

commercial glasses, where, as indicated, the presence of mobile cations generally leads to much lower resistivity values. The Rasch–Hinrichsen relationship is useful but gives no direct information related to the structure of the glass or the conduction mechanisms.

Several resistivity-temperature relationships relating to structure have been developed, but that due to Stevels is perhaps most widely used:

$$\rho = \frac{6kT}{\lambda \upsilon \alpha (ez)n} e^{E/kT} = \rho_0 e^{E/kT} \tag{82}$$

where ρ_0 defines the preexponential terms, υ is the natural vibration frequency (10^{12}–10^{13} sec^{-1}), λ (8–9 Å) the average jump distance, α ($\sim$3) the number of adjacent jump sites, and n the concentration of charge carriers.

As the mole fraction and concentration (n) of modifier ion increases, the effect is to loosen the structure, resulting in freer migration of the conducting ions and a lower activation energy, E. Although α does not vary much with modifier content, the jump distance λ would increase somewhat, since the network expands, but this is offset by the increase in mobile ion concentration. However, both λ and particularly α are greatly affected by the presence of blocking cations in the glass structure, such as occur when two or more mobile species are present simultaneously. This condition leads to the so-called mixed-alkali effect, as illustrated in Figure 16 for lead-alkali-borosilicate glasses with different total Na and K concentrations.

G. Oxide Conduction

1. Porcelains

The resistivity of all dense porcelain bodies is typically greater than 10^{11} Ω-cm at 25°C and could be higher except for the ever-present impurities in the porcelains [46]. The type and quantity of glassy phase in the ceramic is most important in determining the value of ρ at elevated temperatures and in fact for all the other electrical properties. Since in the glassy state the alkali ions are more weakly bonded than in the crystalline state, large amounts of an alkali-containing glassy phase in a ceramic body produce, as a rule, lower resistivities at high temperatures and also higher dielectric losses. This group of ceramics includes the porcelains, cordierites, and steatites. In the case of certain high-fired cordierite, forsterite, low-loss steatite, zircon, and high-alumina porcelains, the glassy

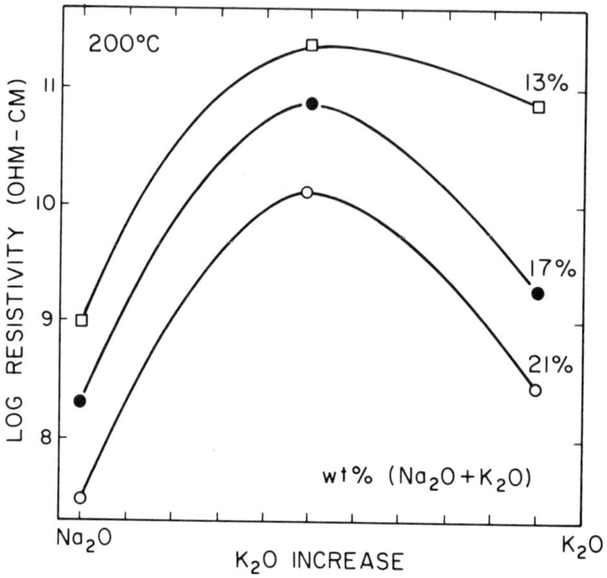

Figure 16 Plot of electrical resistivity (at 200°C) versus total alkali (Na and K) in alkali-lead-borosilicate glass, showing mixed-alkali effect.

phase consists more of BaO and other alkaline earth oxides so that the electrical losses are lower than in the alkali-containing porcelains. Pure crystalline, glass-free ceramics (Al_2O_3, BeO, MgO, SiO_2) exhibit the lowest losses, compared to glassy-phase ceramics, and they also possess higher resistivities and lower losses at elevated temperatures. Figure 17 shows log volume resistivity data versus reciprocal absolute temperature for common porcelains and oxide insulators.

2. Alumina (Al_2O_3)

The subject of electrical conductivity of alumina has been dealt with in several articles [47]. While it is assumed that the oxide is an electronic conductor, ionic conductivity may be an important part of the total conductivity. Earlier studies indicated Al_2O_3 to be an electronic conductor above 1300°C, while it was a mixed ionic and electronic conductor at lower temperatures. Cohen [48] found that the conductivity mechanism in alumina was sensitively dependent on purity. The "pure" (polycrystalline) specimens were found to be substantially electronic conductors. Experiments in which samples were heated in air and in vacuo showed alumina to be an n-type semiconductor at temperatures below 450°C. "Impure" specimens appeared to be mixed conductors at high temperatures (i.e., both electronic and ionic).

From measurements of the temperature dependence of conductivity the band gap in alumina was estimated to be 5.0 eV. Pappis and Kingery [49] measured the electrical conductivity of both the polycrystalline and single-crystal aluminum oxide at 1300–1750°C and at oxygen partial pressures of 10^0–10^{-10} atm. At 10^0 atm the activation energy was 2.97 eV and the conductivity was p type; at 10^{-5} atm the activation energy was 2.62 eV below 1627°C and 5.5 eV above with intermediate-type conductivity; and at 10^{-10} atm the activation energy was 2.48 eV below 1627°C and 5.8 eV above and the conductivity was n type. They concluded that the conductivity of α–Al_2O_3 did not result from any single simple process over wide ranges of temperature and oxygen partial pressure studied.

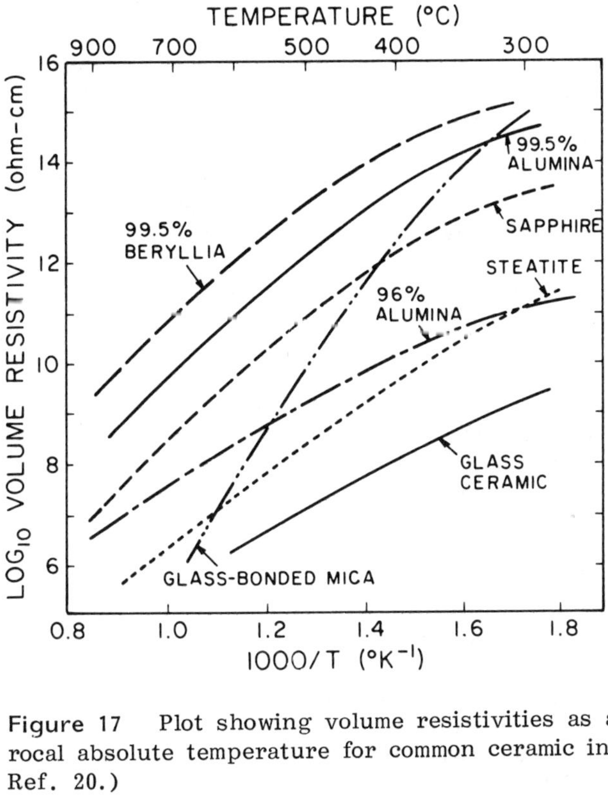

Figure 17 Plot showing volume resistivities as a function of reciprocal absolute temperature for common ceramic insulators. (From Ref. 20.)

3. Spinels

The conductivity of the spinels has been determined to be mainly electronic, the ionic contribution, if it occurs at all, being very small [50,51]. A perfect lattice of the spinel ($X^{2+}Y_2^{3+}O_4$) structure will consist of X^{2+}, Y^{3+}, and O^{2-} ions. For current to pass through this lattice, movement of electrons must occur. In the perfect spinel there are no free electrons, since they are all in close combination with the various ions. Electrons could be set free from the O^{2-} ions by the addition of thermal energy, but the amount of energy needed is high and would probably occur only at very high temperatures.

When an electron is introduced into the lattice, it might associate with a cation and convert it to the X^+ or Y^{2+} state. These valence states are unstable and the electron would be readily released. Such a process is believed to require no activation energy. As indicated, such electron excess can be produced by oxygen deficiency, and vice versa. Similar to this type of electronic interaction, there is also the possibility of hole interaction, where, for example, an O^{2-} ion becomes devoid of one electron (O^{1-}), or where the X^{2+} or Y^{3+} cations exist in a higher valence state. The electron-deficient ion can appear to propagate through the lattice by movement of electrons in the opposite direction. In the inverse spinels [$Y^{3+}(X^{2+}Y^{3+})O_4$], if the Y^{3+} ion can exist in the Y^{2+} state, it is possible for an X^{2+} ion to release an electron, becoming X^{3+}, and for a Y^{3+} ion to take up an electron to become a Y^{2+} ion. These changes in valence states result in hopping conduction in the spinel lattice, which may be either electron or hole type.

There is no striking difference in the conductivities or methods of conduction in spinels of the normal type compared to those of the inverse type. According to their conductivities, the spinels can be divided into four groups: (1) aluminates having low conductivities, (2) chromites with moderate conductivities, (3) ferrites with fairly high conductivities, and (4) magnetites with an extremely high conductivity approaching metallic conduction. In the aluminates the method of conduction is controlled by the divalent cation, with magnesium and zinc aluminates being electron conductors while the cobalt and nickel aluminates show hole conduction. With the exception of ferrous compounds, the conductivity of the chromites is controlled by the trivalent cation. The chromites are all hole conductors, while ferrites are electron conductors. Magnetite is an intrinsic semiconductor due to the presence of ions of the same metal (Fe^{2+}, Fe^{3+}) but with differing valencies in equivalent octahedral lattice sites. Most spinels obey the Arrhenius-type resistivity-temperature relationship to some degree. In particular, the aluminates of Co, Ni, and Zn, zinc and cobalt chromites, nickel and zinc ferrites, and magnetite conform to the law over the temperature range 650–1000°C. Table 19 sums up the electrical conductivity data for various spinels.

Table 19 Properties Relating to Resistivity of Spinels

Specimen	E (eV)	Temperature range (°C)	σ at 900°C (mhos)	Method of conduction
Co aluminate	1.55	700–1020	8.55×10^{-6}	Hole
Mg aluminate	1.14	740–1010	1.33×10^{-6}	Electron
Ni aluminate	1.82	780–1020	2.75×10^{-6}	Hole
Zn aluminate	0.86	640–980	5.38×10^{-6}	Electron
Co chromite	1.05	500–1010	2.49×10^{-3}	Hole
Mg chromite	0.95	535–1000	9.71×10^{-4}	Hole
Zn chromite	1.27	800–1020	1.03×10^{-4}	Hole
Zn chromite	1.09	480–800		Hole
Mg ferrite	1.16	610–830	3.98×10^{-2}	Electron
Ni ferrite	1.01	500–1000	1.95×10^{-2}	Electron
Ni ferrite	0.77	225–500		Electron
Zn ferrite	0.96	500–1020	1.28×10^{-2}	Electron
Zn ferrite	0.83	290–500		Electron
Ferrous chromite	0.76	655–1000	1.83×10^{-3}	Hole
Magnetite	0.106	90–400	1.85×10^{-1} @300°C	Intrinsic semiconductor

4. Magnesia (MgO)

MgO has been determined to be a mixed electronic and ionic conductor, the relative importance of the ionic and electronic conduction being a function of temperature, oxygen partial pressure, and impurity content [52,53]. Extensive studies made of these aspects have generally found that the total conductivity decreases with decreasing oxygen partial pressure at high oxygen pressures, then goes through a minimum at about 10^{-3} atm and increases with decreasing pressure at low pressures. Electronic conduction is favored at low or high P_{O_2} and at higher temperatures (>1300°C), while ionic conductivity is favored at low temperatures and at intermediate oxygen pressures. The minimum in conductivity reflects a roughly equal balance between Mg vacancies which predominate at high P_{O_2} and oxygen vacancies at low P_{O_2}. The conductivity in impure oxides was attributed to Mg vacancies. Other oxides such as BeO are essentially similar to Al_2O_3 and MgO in their conduction behavior.

5. Zirconia (ZrO_2)

The magnitude and type of conductivity in ZrO_2 depend on the structure since ZrO_2 has three distinct phase modifications:

$$\text{monoclinic} \xleftrightarrow{1000-1200°C} \text{tetragonal} \xleftrightarrow{2000-2400°C} \text{cubic}$$

In addition to the change in structure brought about by the temperature changes, addition of stabilizers such as CaO, MgO, and Y_2O_3 results in the formation of the stable cubic phase and also affects the conductivity [54].

a. Monoclinic ZrO_2. Vest, et al. [55] studied the electical conductivity as a function of partial pressure of oxygen at 1000°C using ac techniques. The conductivity was found to go through a minimum at about 10^{-16} atm O_2, with n- and p-type conduction being observed at lower and higher oxygen pressures, respectively, the ionic transport number being less than 0.01. The oxygen pressure dependence of the p-type conduction was interpreted to be proportional to $P_{O_2}{}^{1/5}$. The monoclinic ZrO_2 was determined to be stoichiometric at 10^{-16} atm O_2 with the p-type conduction and the $P_{O_2}{}^{1/5}$ pressure dependence attributed to the predominance of fully charged zirconium vacancies. Monoclinic ZrO_2 under these conditions is, therefore, a metal-deficient rather than an oxygen-deficient oxide conductor. The hole mobility at 1000°C was found to be 1.4×10^{-6} cm^2/V-sec, which indicates a thermally activated hopping mechanism for charge transport since this mobility is too small for hole conduction to occur in any simple band.

Kumar et al. [56] studied the defect structure of monoclinic ZrO_2 by measuring the transference number and electrical conductivity as a function of O_2 pressure and temperature. The data suggested the existence of doubly ionized oxygen vacancies at low pressures (i.e., $<10^{-19}$ atm) and singly ionized oxygen interstitials at pressures $>10^{-19}$ atm consistent with an anti-Frenkel disorder. The zirconia was found to be primarily an ionic conductor below ∿700°C and an electronic conductor at 700–1000°C for $10^{-22} \leqslant P_{O_2} \leqslant 1$ atm.

b. Tetragonal ZrO_2. Vest and Tallan [57] found that tetragonal ZrO_2 is a mixed electronic and ionic conductor with a significant ionic contribution except at very high temperatures or very low oxygen partial pressures. The n-type electronic conductivity observed at high temperatures (1400°C) and low P_{O_2} ($<10^{-13}$ atm) was interpreted as arising from completely ionized oxygen vacancies. At the monoclinic-to-tetragonal transition (∿1150°C), there was an abrupt isothermal change, with the tetragonal form showing the lower conductivity.

c. Stabilized ZrO_2. Kingery et al. [58] found that the electrical conductivity was independent of P_{O_2} for cubic $Ca_{.15}Zr_{.85}O_{1.85}$, indicating that the conductivity is wholly ionic. The conductivity could be represented over the temperature range 700–1725°C by the expression

$$\sigma = 1.5 \times 10^3 \exp\left(\frac{-1.26}{kT}\right) \tag{83}$$

From a calculation of the electrical conductivity resulting from oxygen ion diffusion it was concluded that the conductivity is completely ionic in calcia-stabilized ZrO_2 and the transference number for oxygen ions is unity. Similar observations can be made regarding yttria-stabilized zirconia, which shows even higher ionic conduction with a maximum at approximately 9.0 mol % Y_2O_3. The transport mechanism has been confirmed to be oxygen ion conduction via oxygen vacancies, the concentration of which is increased by the Y^{3+} or Ca^{2+} substitution for Zr^{4+}. Figure 18 compares the conductivity as a function of reciprocal absolute temperature for a calcia (CSZ-13.3 mol % CaO)- and yttria (YSZ-4.5 mol % Y_2O_3)-stabilized zirconia body. Both had average grain sizes of ∿1 μm and were sintered at 1300–1350°C for 4 hr.

Kröger [59] explained the promotion of n-type electronic conduction by relatively small concentrations of donors (vanadium,

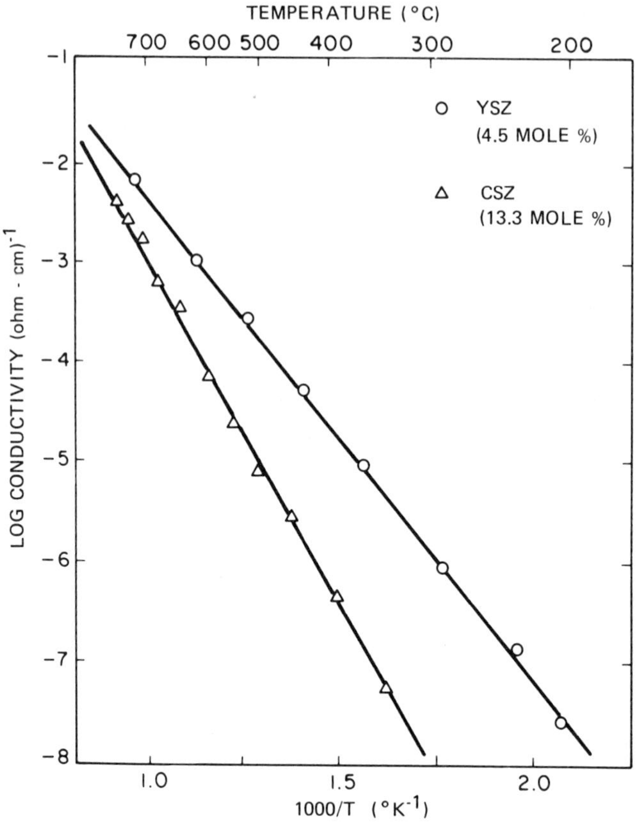

Figure 18 Plot of log conductivity versus reciprocal absolute temperature for calcia (CSZ-13.3 mol %)- and yttria (YSZ-4.5 mol %)-stabilized zirconia ceramics.

tantalum, etc.) in the presence of calcium, which is an acceptor, by assuming that most of the calcium was present as neutral calcium ion–oxygen vacancy pairs or clusters.

To summarize, ceramic materials can be classified into insulators and semiconductors. Conduction can be ionic or electronic depending on composition, structure, and defect state, defects being essential to keep the process of conduction going. Several methods, including chemical substitution, nonstoichiometry created by oxygen atmospheres, and temperature activation can be adopted to create such defects. Conduction in glasses and porcelains is mainly ionic, with oxide materials such as alumina, spinels, zirconia, and magnesia tending more toward electronic conduction.

REFERENCES

1. V. V. Daniel, *Dielectric Relaxation*, Academic Press, New York, 1967, Chaps. 1, 2, 5, 6, 12, 13, and 16.
2. A. A. Zaky and R. Hawley, *Dielectric Solids*, Dover, New York, 1970, Chaps. 1, 2, and 6.
3. M. Spector, *Insul./Circuits*, *25*(1), (1979).
4. *Insul./Circuits (Directory/Encyclopedia)*, *19*(7), 194–208, (1973).
5. H. Doremus, *Glass Science*, Wiley, New York, 1973, Chap. 2.
6. B. Locsei, *Molten Silicates and Their Properties*, Chemical Publishing Company, New York, 1970, Chaps. 3–6.
7. J. M. Stevels, *Progress in the Theory of the Physical Properties of Glass*, Elsevier, London, 1964.
8. W. Espe, *Materials of High Vacuum Technology*, Vol. 2: *Silicates*, Pergamon Press, Oxford, 1968, Chaps. 10–12.
9. B. C. Sharma, *J. Solid State Technol.*, *21*(2), 48–53 (1978); Part II: *21*(4), 122–126 (1978).
10. D. W. A. Forbes, *Glass Technol.*, *8*(2), 32–42 (1967).
11. J. Gallup and A. G. F. Dingwall, *Am. Ceram. Soc. Bull.*, *36*(2), 47–51 (1957).
12. R. H. Dalton, *J. Am. Ceram. Soc.*, *39*(3), 109–112 (1956).
13. W. H. Class, *Solid State Technol.* *17*(8), 32–36 (1974).
14. S. P. Chaudhuri, *J. Br. Ceram. Soc.*, *73*(2), 37–41 (1974).
15. I. J. McColm, *Ceramic Science for Materials Technologist*, Leonard Hill, Glasgow, 1983.
16. R. W. Sillars, *Electrical Insulating Materials and Their Applications*, Peter Peregrinus, London, 1963, Chaps. 4, 6, 11.
17. L. L. Hench and D. B. Dove, *Physics of Electrical Ceramics*, Marcel Dekker, New York, 1971, Chaps. 17–20.
18. D. G. Grossman and J. O. Isard, *J. Phys. D: Appl. Phys.*, *3*(7), 1058–1067 (1970).
19. W. D. Kingery, H. K. Bowen, and D. R. Uhlmann, *Introduction to Ceramics*, 2nd ed., Wiley, New York, 1976.
20. F. M. Clark, *Insulating Materials for Design and Engineering Practice*, Wiley, New York, 1962, Chap. 16.
21. P. J. Harrop, *Dielectrics*, Wiley, New York, 1972, Chaps. 1–6.
22. J. C. Anderson, *Dielectrics*, Reinhold, New York, 1964, Chaps. 1–8.
23. B. U. P. Sceife, *Complex Permittivity (Theory and Measurement)*, English Universities Press, London, 1971, Chap. 3.
24. N. P. Bogoroditskii and V. V. Pasynkov, *Properties of Electronic Materials*, Part 1: *Dielectrics*, Boston Technical Publishers, Cambridge, Mass., 1967.
25. M. D. Rigterink, R. O. Grisdale, and S. O. Morgan, *J. Am. Ceram. Soc.*, *25*(15), 439–443 (1942).

26. L. V. Azàroff and J. J. Brophy, *Electronic Processes in Materials*, McGraw–Hill, New York, 1963, Chaps. 8–9.
27. Per Kofsted, *Nonstoichiometry, Diffusion, and Electrical Conductivity in Binary Metal Oxides*, Wiley–Interscience, New York, 1972.
28. A. B. Yidiard, *Handbuch der Physik*, Vol. 20, Springer–Verlag, Berlin, 1957, pp. 246–349.
29. F. A. Kröger, *The Chemistry of Imperfect Crystals*, North-Holland, Amsterdam/Wiley, New York, 1964.
30. N. N. Greenwood, *Ionic Crystals, Lattice Defects and Nonstoichiometry*, Chemical Publishing Company, New York, 1970.
31. J. H. DeBoer and E. J. W. Verwey, *Proc. Phys. Soc. (Lond.)*, *49*(274), 59–73 (1937).
32. G. L. Swell, *Phys. Rev.*, *129*, 597–608 (1963).
33. P. Süptitz and J. Tetlow, *Phys. Status Solidi*, *23*, 9–56 (1967).
34. R. J. Brook, in *Electrical Conductivity in Ceramics and Glass*, N. M. Tallan, ed., Marcel Dekker, New York, 1974, Chap. 3.
35. A. E. Owen, in *Progress in Ceramic Science*, Vol. 3, Pergamon Press, Elmsford, N.Y., 1963.
36. P. L. Kirby, *Br. J. Appl. Phys.*, *1*(8), 193–202 (1950).
37. J. Vermeer, *Physica*, *22*, 1247–1253, 1269–1278 (1956).
38. R. J. Charles, *J. Am. Ceram. Soc.*, *46*(5), 235–243 (1963).
39. J. M. Stevels, *Handbuch der Physik*, Vol. 20, Springer–Verlag, Berlin, 1957, pp. 353–391.
40. M. Schwartz and J. D. Mackenzie, "Ionic Conductivity in Calcium Silicate Glasses," *J. Am. Ceram. Soc.*, *49*(11), 582–585 (1966).
41. A. E. Badger and J. F. White, *J. Am. Ceram. Soc.*, *23*(9), 271–274 (1940).
42. D. E. Day, *J. Non-Cryst. Solids*, *21*, 343–372 (1976).
43. J. O. Isard, *J. Non-Cryst. Solids*, *1*, 235–261 (1969).
44. R. M. Hakim and D. R. Uhlmann, *Phys. Chem. Glasses*, *8*(5), 174–177 (1977).
45. H. Beyersdorfer, *Silikatech*, *5*(11), 459–462 (1954).
46. H. E. Schwiete and H. F. Eberle, *Ber. Dsch. Keram. Ges.*, *47*(10), 609–612 (1970).
47. D. W. Peters, L. Feinstein, and C. Peltzer, *J. Chem. Phys.*, *42*(7), 2345–2346 (1965).
48. J. Cohen, *Am. Ceram. Soc. Bull.*, *38*(9), 441–446 (1959).
49. J. Pappis and W. D. Kingery, *J. Am. Ceram. Soc.*, *44*(9), 459–464 (1961).
50. V. R. K. Murthy and J. Sobhanadri, *Phys. Status Solidi A*, *38*, 647–651 (1976).
51. R. A. Weeks and E. Sonder, *J. Am. Ceram. Soc.*, *63*(1–2), 92–95 (1980).
52. T. J. Lewis and A. J. Wright, *Br. J. Appl. Phys.*, *1*(4), 441–447 (1968).

53. S. P. Mitoff, *J. Chem. Phys.*, *36*(5), 1383–1389 (1962).
54. A. V. Ramana Rao and V. B. Tare, *Ser. Met.*, *6*(2), 141–148 (1972).
55. R. W. Vest, N. M. Tallan, and W. C. Tripp, *J. Am. Ceram. Soc.*, *47*(12), 635–640 (1964).
56. Arun Kumar, Dilip Rajdev, and D. L. Douglass, *J. Am. Ceram. Soc.*, *55*(9), 439–445 (1972).
57. R. W. Vest and N. M. Tallan, *J. Am. Ceram. Soc.*, *48*(9), 472–475 (1965).
58. W. D. Kingery, J. Pappis, M. E. Doty, and D. C. Hill, *J. Am. Ceram. Soc.*, *42*(8), 393–398 (1959).
59. F. A. Kröger, *J. Am. Ceram. Soc.*, *49*(4), 215–218 (1966).

2

Ceramic Capacitor Materials

GILBERT GOODMAN
Johnson Controls, Inc., Milwaukee, Wisconsin

I. INTRODUCTION

The phenomenal growth of integrated circuit technology is common knowledge. It tends to obscure the fact that the consumption of discrete circuit components such as resistors and capacitors has never ceased to climb. Ceramic capacitors account for a major segment of this volume. In 1981, not a bountiful year for electronics, 7.3 billion discrete ceramic capacitors of domestic and foreign origin were consumed in the United States. Their gross value was $490 million. (The capacitors contained in imported electronic equipment would be in addition to these numbers.)

This vigorous component technology is traceable to the Leyden jar of 1745, the first capacitor design to have enjoyed widespread use. In its most advanced, dry, design with metal foil electrodes in contact with glass dielectric, this early device had volumetric efficiency and loss characteristics that are poor by today's standards. Nevertheless, it figured in pioneering experiments on the nature of static electricity, some, like Franklin's lightning investigations, hazardous and others of questionable value.*

The capacitor's basic function, energy storage, has seen broadened application since the early experiments. Through it a capacitor is called upon to block direct current, or to couple ac circuits. In bypassing applications the capacitor separates the ac and dc portions of a mixed signal. Alternating currents are also separated by capacitors according to frequency, and the charge/discharge characteristics of resistance-capacitance combinations are applied in timing. Physically large-scale tasks such as high-energy storage and power factor correction fall to larger and different types of capacitors from those to be discussed here.

---

*In a demonstration before members of the court of Louis XV of France, 700 monks with joined hands allegedly were caused to jump simultaneously by a Leyden jar discharge [1].

The properties of capacitor dielectric materials which determine in detail the manner in which the energy storage function is performed are:

1. Dielectric constant relative to that of vacuum, commonly known as dielectric constant or permittivity: the measure of the response of the dielectric's polarization mechanisms to an electric field
2. Dissipation factor, power factor, and loss factor: measures of the net inefficiency of the polarization process(es)
3. Insulation resistance: measure of the efficiency of dc blocking
4. Temperature dependencies, frequency dependencies, and field strength dependencies of the parameters above

This chapter is confined to capacitors whose dielectrics are ceramic materials.

II. HISTORICAL BACKGROUND

In a sense, the very glass material that served as the Leyden jar's dielectric persisted in use as a dielectric into the twentieth century. That is, the properties of electrical porcelain are largely determined by the composition and content of glass in its microstructure. Porcelain insulators figured importantly in early electrical technology as isolating and structural members. Steatite ceramics were then developed for the same applications. Containing feldspar as a fluxing constituent, they were still subject, although to a lesser degree, to similar high-loss behavior as that of porcelain. For capacitors for the first radio circuitry in the early twentieth century the premium dielectric materials were paper and mica. Besides having the best available loss characteristics, these materials could be fabricated in thin sheets to provide high capacitance in spite of their modest dielectric constants.

Thurnauer [2,3] described the finding in Germany around 1920 that steatite ceramics that were fabricated with alkaline earth oxide fluxes instead of alkaline oxides (feldspar) possessed improved low-loss behavior. It was a natural direction to try to employ such materials as substitutes for mica and paper. A challenge to the ceramic technology, therefore, was to develop ways of fabricating the materials in very thin cross section. At this time the extruded tubular steatite ceramic capacitor made its appearance as a mica capacitor replacement. Such has been the commercial hardiness of the mica capacitor, however, that it is still in use, accounting for about 5% of the electronic capacitors consumed in the United States. Many mica applications were preempted by ceramics during World War II when the overseas mica sources became uncertain, and the product never regained its prewar share of the market.

Another mineral displaced by a ceramic because of wartime shortages was soapstone or block talc. Easily machined and fired into vacuum-tube insulators, the mineral became a strategic material and was largely replaced by ceramic process steatite. As has been described, it was basically this same ceramic which then found application in fixed tubular, trimmer, and other capacitor types.

Thus displacement of natural dielectric materials by ceramics, under the technical demands of a growing electronics industry and of wartime shortage situations, has been a key pattern in ceramic capacitor development. Another is that the development of ceramic dielectrics technology has been influenced by, and in turn has influenced, the miniaturization of the equipment in which capacitors are used. This cycle has historically been initiated by the military and then been followed by civilian application. The increase in available dielectric constant from less than 10 for porcelain to, effectively, 100,000, in grain boundary barrier layer capacitors has been instrumental in this size reduction. Additionally, suitable capacitors for thin- and thick-film circuits have been developed.

Where minimization of the dielectric loss is of foremost importance, as in high-frequency circuits, ceramic capacitors have been developed with exceptionally low dissipation factors and large time constants.

Finally, the technology of fabricating ceramics has kept pace with the range of material properties, providing geometries that capitalize on the properties and are appropriate to the electronic equipment assembly methods used. Hybrid circuits, for example, employ chip capacitors with closely defined electrical characteristics and physical geometries, which are installed in place by automated assembly methods.

III. FERROELECTRICITY IN CAPACITOR TECHNOLOGY

The discovery of ferroelectricity in barium titanate during World War II made available for ceramic capacitor design dielectric constants up to two orders of magnitude greater than theretofore known. This class of ceramic materials consequently has dominated capacitor technology to the degree that the standards of performance incorporate ferroelectricity-related idiosyncrasies. It is appropriate to mention here these behaviors and their origins.

In ferroelectricity, preexisting electric dipoles, whose presence in a material is predictable from the crystalline symmetry, interact at a distance, generally not predictably, to spontaneously polarize

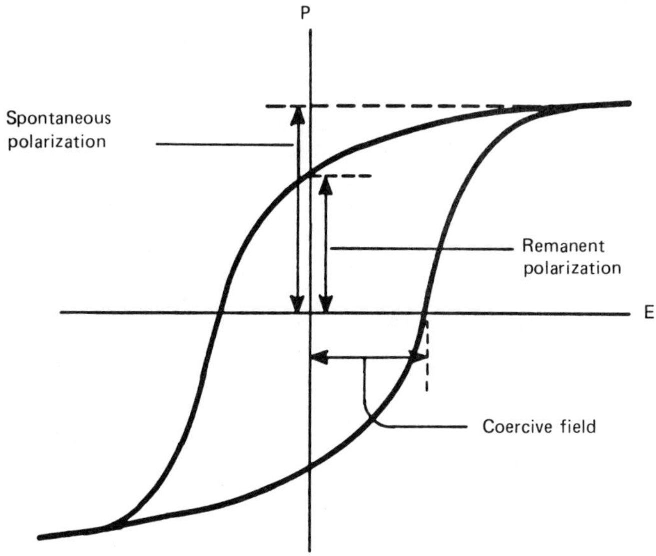

Figure 1 Ferroelectric hysteresis loop.

subvolumes of a crystal or ceramic mass. The spontaneously polarized regions with a single direction of polarization are called domains, after ferromagnetics terminology. Orientational relationships of domains to one another are governed by the crystal symmetry. In the barium titanate family of materials on which most ceramic capacitors are based,* the crystal structure is the cubic or pseudocubic perovskite structure. The spontaneous polarization can orient parallel to any of the pseudocube unit cell edges, so that adjacent domains have 180° or near-90° relationships to one another. The domain structure of a titanate ceramic as fabricated can be quite complex, a consequence of the random, sintered microstructure and strain patterns due to anisotropic dimensional changes developing on cooling through the Curie temperature.

The domains are also orientable by an externally applied electric field whose effect is to increase the component of polarization in the field direction. At the Curie temperature the spontaneous polarization disappears, but significant effects of the applied field on permittivity persist for at least 50°C higher. If the applied field is

*Relaxator-type dielectrics, to be discussed, form in the tungsten bronze structure.

lifted, some of the regions that were oriented retain the new orientation; others revert. Much of the behavior of a capacitor is describable in relation to the spontaneously polar character or the associated ferroelectric hysteresis loop (Figure 1).

With increasingly high ac field the dielectric constant ($\delta P/\delta E$) first increases, then decreases as the dipolar contribution increases, then saturates. When increasingly dc biased, a dielectric shows decreasing low-signal ac dielectric constant since the bias serves increasingly to repress domain reversibility. Increasing frequency also leads to a decrease of dielectric constant in the form of a relaxation, centered in the gigahertz range, when the spontaneous polarization lags behind the applied frequency. Associated with this, the dissipation factor passes through a maximum value.

IV. DIELECTRIC PROPERTIES OF MULTIPHASE SYSTEMS

Except in research studies, orientation of microstructure to enhance electrical characteristics has not generally been practiced on ceramic dielectrics. Two major exceptions are the orientation of ferrite permanent-magnet materials and the poling of ceramics to produce piezoelectric transducers. In these cases there are accompanying effects in the dielectric properties, but not of magnitudes useful for application commercially in capacitors.

However, efforts to rationalize the characteristics of dielectric systems in terms of the constituent phase properties and their microstructural arrangement are numerous. The case of microstructures in which there are large differences of conductivity among the phases is most important to capacitors and is taken up in the section on grain boundary barrier layer capacitors.

The major other microstructural case is that in which permittivities of the constituent phases differ significantly but their conductivities are the same. Actually, the prevalence of minor low-frequency relaxation phenomena is evidence that this pure condition does not usually apply.

A simplification is to discuss this subject in relation to a system of only two phases. Maxwell's "Treatise" [4] addresses the problem of the dielectric constant of a mixture of two isotropic phases, given their dielectric constants and the geometry of the mixture. The analysis deals first with planar parallel arrangements of two phases oriented either normal to or parallel to the specimen electrodes. More realistic, however, is the case where phase 1 is continuous and has dispersed in it phase 2 in the form of spheres. The spheres are assumed to be separated enough that they do not disturb the lines of flux around one another. Although this is strictly true only at infinite dilution, Maxwell's relations have some applicability at up to 10–50% volume concentration of phase 2. For $k'_2 > k'_1$,

$$k'_T = k'_2 \frac{k'_1 + 2k'_2 - 2(1 - v_2)(k'_2 - k'_1)}{k'_1 + 2k'_2 + (1 - v_2)(k'_2 - k'_1)} \quad \text{for phase 2 continuous}$$

and

$$k'_T = k'_1 \frac{k'_2 + 2k'_1 - 2(1 - v_2)(k'_1 - k'_2)}{k'_2 + 2k'_1 + v_2(k'_1 - k'_2)} \quad \text{for phase 1 continuous}$$

where v is the volume fraction of a phase. The shape and dimensions of the disperse phase do not enter into this calculation. The relationships have had application in such systems as resins loaded with a high-k' titanate powder, or a partially devitrified glass. Mitoff [5] has provided a useful discussion of the extent to which these Maxwell relationships can be modified for applicability where there are disparate values of permittivity of the two phases, deviations from sphericity of the disperse phase, and where both phases are continuous instead of only one. In general, the most accurate predictions can be made for the most dilute dispersions.

Figure 2 is a graphical representation of these behaviors when the two phases have widely differing dielectric constants ($k'_1/k'_2 \sim 1000$). The extreme curves represent laminated structures arrayed parallel and normal to the electrodes. Within these extremes fall the cases of dilute dispersions of spheres of one phase in a matrix of the other. Curves for the latter terminate in the dashed sections, where the condition of dilute dispersion no longer applies.

The diagonal straight line of Figure 2 represents a widely used empirical relationship due to Lichtenecker [6]:

$$\log k'_T = v_1 \log k'_1 + v_2 \log k'_2 + \cdots$$

In use of this relationship there is no concern at all for the physical geometry of the system.

A summary of the extensive literature on modeling of heterogeneous dielectric systems is given by Van Beek [7]. The many physical models that have been proposed and analyzed have had mainly academic interest and have rarely played a role in the design or realization of a ceramic capacitor. Lichtenecker's empirical mixing rule, in fact, seems to have been the most frequently used.

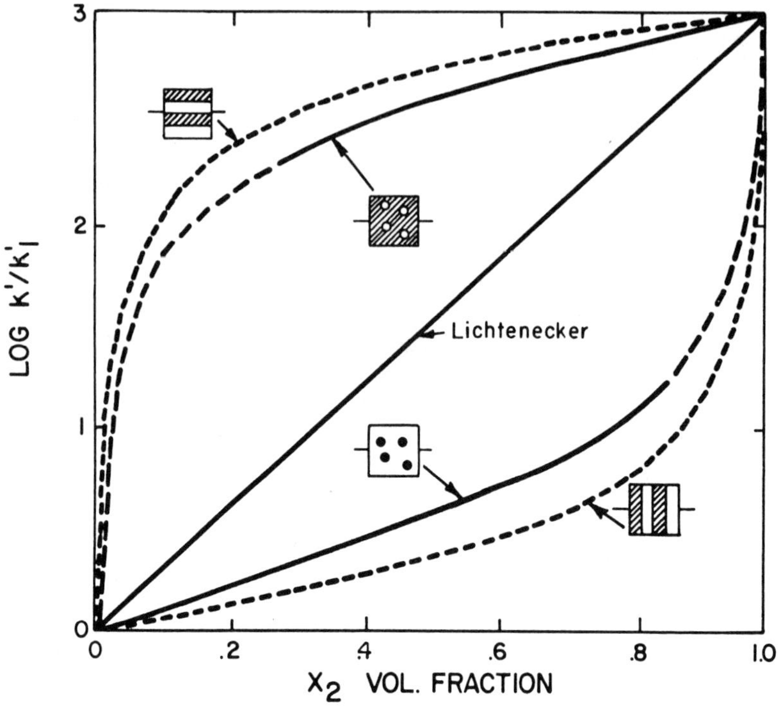

Figure 2 Plot of Maxwell equation for dielectric mixtures with $k'_1/k'_2 = 1000$. (After Ref. 5.)

V. BASIC CERAMIC DIELECTRIC MATERIALS

A. Porcelain and Steatite

Classical porcelain and steatite have little remaining place as the dielectrics of modern electronic circuit capacitors, but they leave some heritages:

1. The dogma that alkaline oxide—free compositions are essential for high resistivity and low dissipation factor came out of the development of low-loss porcelains and steatites, and is adhered to in specifying the raw materials of dielectrics of other types, as well as of metallizations, soldering fluxes, coatings, and encapsulants.
2. The use of steatite as a diluent in dielectric formulations with graded dielectric constants, to enable a series of capacitors with fixed dimensions but graduated capacitance values.

B. Rutile

Electrical porcelains and steatite ceramics have dielectric constants in the range 5–7 and positive temperature coefficients of capacitance. The move to blends of steatite with rutile was thus a natural one—it opened the way to 10-fold higher dielectric constant and to the possibility of lowered positive temperature coefficients or negative ones. Single-crystal rutile's properties had been known since 1902: dielectric constant of about 170 along the tetragonal c axis and 80 in the a directions. The ceramic mixtures, typified by steatite-TiO_2, have near-zero temperature coefficients of capacitance for mixtures whose dielectric constants are 15–20. The dissipation factor of rutile is typically two to five times that of low-loss steatites, and their blends have intermediate loss values.

C. Barium Titanate

The discovery of ferroelectric barium titanate opened the present era of ceramic dielectrics; predecessor materials are scarcely represented in the components manufactured today. The discovery was a World War II occurrence. The high dielectric constant was first described by American Lava Co. personnel in the United States in 1942 [2]. Recognition of $BaTiO_3$ as a new ferroelectric compound followed, possibly independently, in several countries [8–10]. A question of historical interest is posed by the lapse of a decade or more between actively ongoing development of titania-based dielectrics and the discovery of $BaTiO_3$. A case can be made that there may have been earlier synthesis of the material without recognition of its dielectric character.

Barium titanate has become the basic ceramic capacitor dielectric material in use today. As earlier with rutile, the main mode of its use has been in combination with other materials. Primary application requisites are a high capacitance and, to the degree possible, stable capacitance over the temperature range of use of the component. Practical temperature ranges of use have been specified as from −55 to 125°C, or segments within that range. Since pure, crystalline $BaTiO_3$ has a peak permittivity at about 130°C and another, secondary peak at 10°C [11], corrective modification is required for practical application. The formulation of a titanate-based dielectric is an "arty" matter and compositions in use are based on several principles.

1. Solid Solutions

Solid solution with an isostructural compound broadens the Curie peak [12]. This was first perceived in $BaTiO_3$-$SrTiO_3$ solid solutions (Figure 3) and many such combinations are now known. The amount of peak broadening accompanying solid solution does not of itself enable a useful capacitance versus temperature characteristic. However, limited solubility or nonequilibrium processing can promote the formation of two or more solid solutions, each with its Curie peak. When a phase with a high permittivity peak coexists with a lower-permittivity material, their summed permittivity is less sharply peaked. Further contributing to flattening of permittivity versus temperature for $BaTiO_3$ is that the smaller peak associated with the 10°C tetragonal-to-orthorhombic phase transition tends to be shifted upward in temperature as the main, cubic-to-tetragonal peak shifts downward.

2. Fine-Grained Materials

Reducing the grain size of the barium titanate ceramic below about 1 μm in diameter has a flattening effect on capacitance versus temperature. The origin of this effect is in the nature of the symmetry transition associated with the 130°C permittivity peak of $BaTiO_3$. The microstructure of a large-grained ceramic contains "90°" twins formed to relieve the strains generated when the c axis elongates

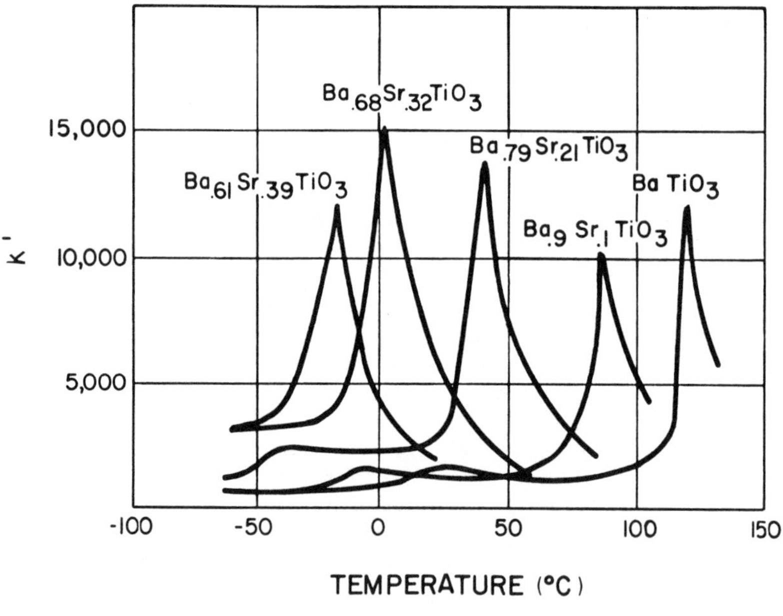

Figure 3 Effect of solid solution on k' versus temperature characteristic. (From Ref. 12.)

on passing from the cubic to the tetragonal symmetry; in the strained state the highest permittivity is displayed.

The domain boundary planes have surface energy proportional to the square of the grain diameter; the strain energy responsible for domain formation, a volume effect, is related to the cube of the grain diameter. As indicated schematically in Figure 4, with decreasing grain size a critical size, D_c, is reached where it is energetically less costly to support an elastic strain than to relieve it by twin formation. The ceramic in which this situation exists will have average grain size of the order of 1 μm or less and be untwinned; that is, each grain will comprise a single domain and have one orientation of spontaneous polarization, be stressed, and tend to cubic symmetry [13]. The permittivity of a ceramic in this condition will be greater than when unstrained, having a room-temperature value of about 2000, compared with 1400 for large-grained, freely twinned barium titanate. To take advantage of this effect, $BaTiO_3$ starting material of very fine grain size and processing that minimizes grain growth are required.

3. Additives

A number of additives have been associated with the ability to flatten the permittivity versus temperature characteristic of barium titanate ceramic. Nickel oxide and bismuth stannate have been described to function in this way [14,15]. The additives are aliostructural and

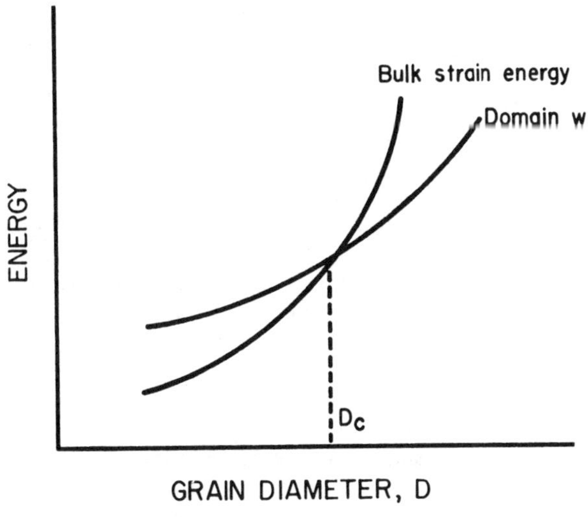

Figure 4 Limiting grain size for ferroelectric domain formation.

are effective in small quantities. Their function is attributed to grain growth inhibition accomplished by lodging in the grain boundaries. The terminology "peak shifting" and "peak depressing" has been applied to the functions just described and is due to Coffeen [15].

Plessner and West [16] early enumerated materials combinations that provide usefully flattened, high permittivity versus temperature characteristics, and others have extended the list and demonstrated also the utility of nonstoichiometric substitutions:

$BaTiO_3$ and $MgSnO_3$

$BaTiO_3$, $SrTiO_3$, and $CaTiO_3$

$BaTiO_3$, $CaSnO_3$, and CaO

$BaTiO_3$ and $CaZrO_3$

$BaTiO_3$ and $CaSnO_3$

$BaTiO_3$ and $Bi_2(SnO_3)_3$

The latter combination, although still widely used, has been a victim of precious-metal economics. The cost-reducing change to palladium-based alloys in multilayer capacitor electrodes from platinum-gold, and the fact of intermetallic compound formation between bismuth and palladium, have largely reduced the use of bismuth-containing Curie peak depressors (see Section VII.C.3.a).

4. Relaxator Dielectrics

A class of ceramic compositions originally described by Smolenskii and Isupov [19] is characterized by high dielectric constants and a broad peak in the permittivity versus temperature characteristic, even in the absence of additives and even in single crystals. This behavior is tentatively attributed to a microstructure of small (∿100 Å), ordered regions, too small to yield the sharp phase transition of normal ferroelectrics [20]. As a result, some spontaneous polarization and associated ferroelectric behavior are retained over a very broad temperature range. Compositions in the tungsten bronze structure system $Pb(Mg_{1/3}Nb_{2/3})O_3$-$PbTiO_3$-$Ba(Zn_{1/3}Nb_{2/3})O_3$ have diffuse peaks with heights corresponding to permittivities of 25,000 [21]. A large number of specific dielectric compositions of this type are described in the patent literature. An attractive secondary characteristic is that with some of them dense ceramics are achievable at sintering temperatures low enough for applicability with silver-palladium alloy electrodes in multilayer capacitors.

In brief, high-permittivity dielectrics for modern ceramic capacitors are based on (1) ferroelectric barium titanate and its isomorphs and their solid solutions, and (2) ceramic processes that lead to nonequilibrium mixtures of such phases. The resultant locally inhomogeneous microstructure possesses a broadened, flattened permittivity-temperature characteristic, which becomes further flattened and elevated if grain growth during sintering can be sufficiently inhibited.

VI. PERFORMANCE CATEGORIES OF CERAMIC CAPACITORS

In the United States, the Electronic Industries Association has defined three categories of ceramic capacitors [22] having the following characteristics:

Class 1: Relatively low capacitance with low dielectric loss, low temperature coefficient of capacitance, and low rate of aging of the capacitance value. Their specified temperature range of use is −55 to 85°C and temperature coefficient of capacitance values range from "zero" to +75,000 or −75,000 ppm/°C, with tolerances on the temperature coefficient ranging from ±30 ppm/°C to ±2500 ppm/°C. This class of capacitors uses relatively low permittivity dielectrics ($k' \leqslant 100$). The most commonly specified class 1 temperature coefficient is 0 ± 30 ppm/°C.

Class 2: Based on high-permittivity dielectrics ($k' \leqslant 15{,}000$), primarily ferroelectric barium titanate. Because of the prevalence of ferroelectric character, important specified parameters are the voltage coefficient of capacitance, voltage coefficient of series resistance, voltage coefficient of dissipation factor, and rate of capacitance decrease upon aging.

Subclasses of class 2 are based on the temperature coefficient of capacitance and the temperature range of use. Temperature ranges of use are specified according to limits: as low as +10, −30, or −55°C and as high as +45, +65, +85, +105, or +125°C. Within these spans, maximum specified deviations from the room-temperature capacitance value may range from ±1 to ±22% and from +22, −33% to +22, −82%. Some commonly specified temperature coefficients are:

Z5U: less than +22% and −56% deviation from the room-temperature capacitance over the range +10 to +85°C

Y5U: ditto, except over the temperature range −30 to +85°C

X7R: less than ±15% change between −55 and +125°C

Class 3: These capacitors are characterized by very high capacitance values derived from electrode barrier layers or grain boundary barrier layers, and low-voltage-withstanding ability.

VII. VARIETIES OF CERAMIC CAPACITORS

A. Film Capacitors

Wafer or disk discrete ceramic capacitors have dielectrics at least 100 μm thick bearing electrodes with thicknesses of about 25 μm. In thick-film capacitors, both the electrode material and the dielectric have thicknesses similar to that of the electrodes in discrete capacitors. Their formulations include low-melting glass binders which enable adherence to a substrate to be developed at well under 1000°C. Usually the electrode and dielectric thick-film materials are formulated as paints and deposited in the desired pattern by the silk screen process.

In thin-film capacitors all layers, dielectric and metal, are deposited on a substrate by vacuum evaporation or cathode sputtering in thicknesses about two orders of magnitude less than those of thick films. The dielectric layer may be initially deposited as a metal and require an oxidation or anodization treatment to produce an oxide film.

MOS or MNOS capacitors are discrete-component versions of the type of capacitors incorporated into integrated circuits. A single-crystal silicon wafer is gold-diffused on one face for conductivity and provided with an oxide film (or oxide-nitride film) on the other. The oxide is then coated with a 1 μm-thick evaporated film of aluminum. These are not strictly ceramic materials or components and will not be taken up further.

1. Thin-Film Capacitors

Thin-film capacitors do not figure greatly in the discrete-component market. Most commonly they are seen in hybrid circuits which employ resistors and conductors also made by thin- and/or thick-film methods, and active device chips. The nature of thin-film processes dictates batch processing, so the product tends to be more costly than ceramic wafer or thick-film components. Consequently, thin-film capacitors are most likely to be found in military and industrial products where compactness and reliability are important, operating voltage is low, and the limited range of capacitance values and higher cost are acceptable.

Silicon monoxide, a common vacuum-deposited thin-film dielectric, has room-temperature permittivity around 6 and is typically deposited in 5- to 30-μm thicknesses. The deposited material can contain silicon and silica as secondary phases. Deposition conditions govern the proportions of the three phases and the net capacitor characteristics. Evaporated aluminum on a glass or glazed alumina ceramic substrate makes the most common base electrode, and is also used as a counterelectrode in approximately 5-μm thicknesses. Typical characteristics of such a structure would be 0.01 μF/cm^2, 25 V working voltage, and 25 Ω-F RC product. The temperature coefficient of capacitance (TCC) is 100 ppm/°C or less over the capacitor operating temperature range. Details of the procedure for depositing smooth SiO for dielectric use tend to be customized and to vary from facility to facility. Evaporation source temperatures can be up to 1600°C. As deposited, the layered device may be significantly stressed, and annealing below 700°C is resorted to reduce this and stabilize the electrical characteristics. Pinholes are a cause of failure by shorting, and therefore areas larger than 1 cm^2, which might be too defect susceptible, are not commonly made. SiO films as produced are bipolar (i.e., function as capacitors in either polarity).

Sputtered, or evaporated and anodized, tantalum oxide, on the other hand, provides a unipolar capacitor, necessitating for some applications two capacitors to do the work of a single bipolar. Glass or glazed alumina substrates provide the necessary smoothness and high-temperature resistance.

In one published procedure [23] a 4-μm-thick tantalum metal film is first deposited by cathode sputtering on a lapped or glazed 99% alumina substrate. The geometry of the sputtered device is defined by use of a metal mask or by removing undesired material in a photoresist mask and etch operation. The tantalum film is then made the anode in an electrolyte solution containing a platinum cathode. Approximately 30 min of anodization is conducted, including a period of reversed current to improve the quality of the Ta_2O_5 film. A counterelectrode of gold or palladium-gold is applied in the same manner as the base electrode. Such a capacitor, formed at 200 V, has a capacitance density of 0.06 μF/cm^2. TCC is 150–250 ppm/°C. The sixfold increased capacitance density over that mentioned earlier for silicon monoxide is due to the higher permittivity (k' = 21.2) of anodic tantalum oxide. Numerous possibilities exist for modifying the capacitor construction and the dielectric material composition. For example, either of two crystalline forms of tantalum can be produced, which react differently to anodization and to carbon and oxygen additives.

A major rationale for the use of these thin-film capacitors is that their positive TCC is opposite in sign to and of about the same magnitude as the temperature coefficient of resistance of thin-film resistors. In combination they can then be used to construct precise, stable RC circuits.

Titanium oxide, silicon dioxide, aluminum oxide, and aluminum silicate make alternative films suitable for capacitors. Sometimes the high dielectric constant of tantalum oxide is tempered by applying a second film, such as SiO, to give a duplex dielectric. This is done to provide precise lower capacitance values without having to deviate from an adopted geometry. Also, a duplex film is less vulnerable to pinhole defects than a single layer. This type of capacitor would have an evaporated gold counterelectrode and capacitance density of about 0.01 $\mu F/cm^2$.

Thin films of titanate ceramics have been prepared by evaporation or reactive sputtering, using both metal and oxide precursor materials. There has been no commercial-scale application.

2. Thick-Film Capacitors

Designers of thick-film hybrid circuits have the option of using prefabricated discrete capacitors or of constructing them in place on the circuit substrate using the same screen printing process and equipment that are used for the resistors and conductors of thick-film hybrids. The convenience and low cost of thick-film capacitors are offset, however, by their limited range of capacitance values and working characteristics. This is due to constraints on process conditions imposed by the nature of thick-film processing. A thick-film capacitor is comprised of substrate, screened base electrode, dielectric layer, and counterelectrode. Ideally, each layer is processed at a lower temperature than the one beneath, limiting interaction and preserving the dimensions and compositional identity of each. With some materials systems one attempts to avoid interaction of the layers despite the use of a single firing for the entire capacitor, or for the dielectric and counterelectrode. Process temperatures are in the range 600–900°C. Dielectric pastes maturable within this range incorporate, as do the conductive and resistive pastes, ingredients of low-melting glasses. When the glass is nonreactive and has a typically low dielectric constant, its volume fraction must be minimized so that the high-permittivity phase is continuous and a maximum capacitance level is maintained. The reason for this constraint is discussed in Section IV. Yet, for cohesion of the dielectric and a strong bond to the substrate, at least 10 vol % of the bonding material is required. Another approach is to employ the glassy material in higher proportion but of a composition that will devitrify in firing. Then, if the crystalline product is itself a high-permittivity material, a fairly high net permittivity may be realizable without sacrifice of the mechanical bond.

As an example of the interaction problem, Biggers et al. [24] describe thick-film capacitors fabricated from a $BaTiO_3$-based k'-9000 ceramic Y5U dielectric with 10% of a glass frit of molar composition 60 $CdO-20B_2O_3-20\ SiO_2$. Fired at 900°C in discrete disks the mixture has permittivity of about 40. The same mixture was used as a thick-film capacitor dielectric in the following manner: A silver paint bottom electrode was screened and fired on 96% alumina substrates. The dielectric paste and a counterelectrode were then applied in successive screenings and fired in one operation under the same conditions as before. The capacitance of the resultant capacitors corresponded to about 40% lower effective permittivity of the dielectric (i.e., $k' \sim 25$), as a thick-film capacitor than as a sintered, subsequently electroded disk. This is attributable to reaction during cofiring of the dielectric and the counterelectrode. The temperature coefficient of capacitance was less than ±20% over −55 to +125°C, and room-temperature insulation resistance was about $10^{12}\ \Omega$.

The patent literature is replete with examples of thick-film capacitor materials compositions and processing [25–28]. Current art is typified by U.S. patent 3,666,505 [26]: A calcined mixture of $BaTiO_3$ and 1–5 mol % of such oxides as ZrO_2, Al_2O_3, SiO_2, or Nb_2O_5 is mixed with 1–5% of Fe_2O_3 and 1–5% of lead bismuth borosilicate glass. A double application of the paste is applied to a prefired platinum-gold electrode on an alumina substrate. The dielectric and a Pt-Au counterelectrode are fired together at 1050°C. A dielectric constant of 1400 and dissipation factor of 0.025 are typical. The k' value varies less than 20% over −25 to +150°C.

B. Single-Layer Discrete Capacitors

Figure 5 illustrates the common ceramic disk capacitor. Disks, rectangular plates, and tubes with Electronic Industries Association class 1 and class 2 performance are manufactured by substantially classical ceramic methods, highly automated. The powdered raw materials are batched, and mixed, then pressed, cast, or extruded into the desired shape. Binders and plasticizers of types and in concentrations appropriate to the forming method are added to provide formability and green strength. Sintering to a dense ceramic is a final dielectric fabrication step, followed by application of electrodes and leads and packaging or protective coating. Class 3 capacitors receive somewhat modified handling and will be taken up separately in Section VII.D.

The dielectric powder materials can be purchased ready for batching and forming or prepared from the constituent oxides or their precursor compounds. In the latter case it is customary to provide a lower-temperature calcination heat treatment followed by crushing and/or ball milling before the final forming step. Where the final dielectric properties depend on a particular, nonequilibrium mixture of phases in the final product, the selection of starting raw

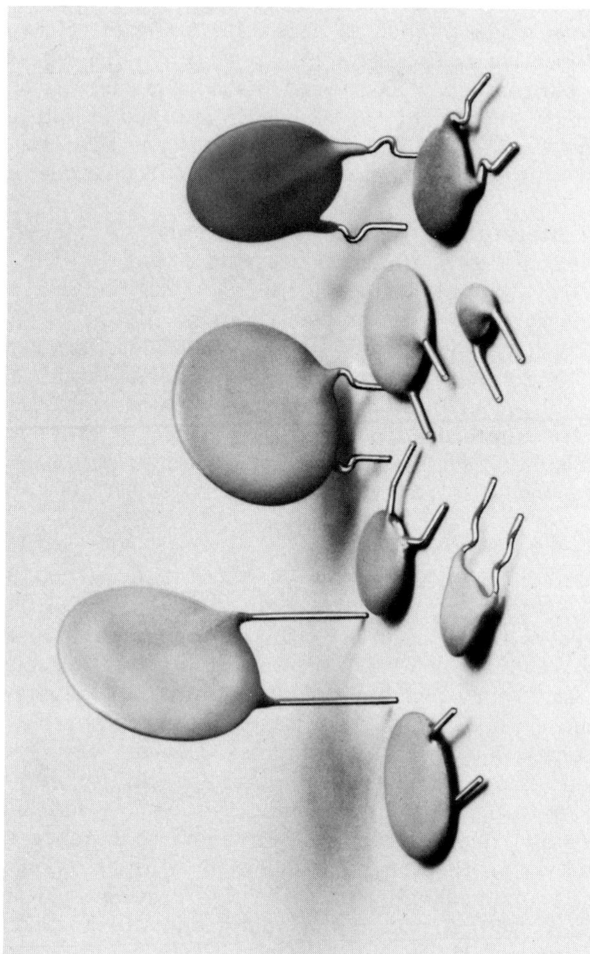

Figure 5 Ceramic disk capacitors. (Courtesy of Centralab Inc.)

materials (i.e., whether titanate compounds or the elemental oxides or carbonates, etc.) and the specific conditions of this part of the process are quite critical.

An important current trend in fine ceramic technology is to employ special powder preparation methods based on chemical synthesis which produce fine, unagglomerated material with narrow particle size distribution [29]. Such powders lend themselves to sintering at lower temperatures than are normally required for densification and produce a ceramic of very uniform grain structure. The implications of this capability with respect to dielectric materials are not yet plain, but the ability to produce a controlled and reproducible powder, alone, should be of great importance.

When disks and plates are to be pressed, an organic binder/lubricant is included in the pressing powder. For very large quantities of material, spray drying produces powder suitable for pressing without an excess of fine fractions. The compaction into disks or wafers is done in steel or carbide dies in high-speed automatic presses.

In a less common form of powder compaction the powder is sprinkled continuously into the "pinch" of a pair of steel rolls and a continuous sheet of compacted material emerges. This process is best suited to thicker slabs and disks.

Extrusion of an appropriately plasticized powder mass into strip, tubes, or rods is another continuous process, which is then followed by baking to a hard condition and cutting into the desired lengths or slices.

Because multilayer ceramics are most commonly formed by continuous casting of ceramic sheet, that process has been diverted, to a degree, also to the production of capacitors comprised of a single thickness of dielectric. The cast sheet is then cut or punched into wafers or disks. These processes are all based on a ceramic slurry which contains, in addition to the powdered dielectric material, organic binder(s), a solvent, and plasticizers. These ingredients are mixed and dispersed by ball milling. Acrylic resin systems are commonly used. The suspension, deaired, is cast continuously onto a moving substrate and drawn under a doctor blade, or the doctor blade is drawn over the slurry, in either case defining a precise thickness of the cast suspension. The most common substrate is a silicone-treated paper strip. An early design of one such piece of equipment is described by Howatt et al. [30]. Modern equipment is of precision construction, casting tape automatically at rates of 1–10 ft/min and removing the solvent rapidly so that the tape can be wound onto reels.

Subsequently, the ceramic tape is separated from the paper substrate and disks or wafers are punched or cut from the sheet, which is described as being in a "green" or "leather-hard" condition. An interesting, although little-used variation is to wind a strip of the green dielectric on a mandrel into a tube which ultimately serves as a cylindrical capacitor [31].

Ceramic firing of dielectrics of the type described is done in air atmosphere at temperatures of 900–1300°C; an arrest in the heating schedule at 300–400°C permits elimination of the organic contents.

Electrodes are applied to single-thickness capacitors in a subsequent heat treatment at a lower temperature so that there is no interaction with the dielectric. Thereby the sintering requirements of the electrode material and dielectric material can be addressed separately. Electrode materials will be taken up later in this chapter.

C. Multilayer Capacitors

The largest class of ceramic capacitors produced in the United States, in numbers and in value, is the multilayer type. Relative to a capacitor with a single thickness of dielectric, multilayer construction affords greater capacitance density at the cost of lowered operating voltage. Small multilayer units or chips are physically compatible with solid-state semiconductor components in hybrid circuit construction, where the low-voltage rating is not of consequence. Figure 6 is a cutaway view of a multilayer capacitor and Figure 7 shows an assortment of multilayer capacitor types.

1. Basic Multilayer Fabrication Methods

The multilayer ceramic capacitor was initially the result of efforts to replace mica with ceramic dielectrics. In mica capacitors thin mica splittings are electroded, stacked, and electrically connected

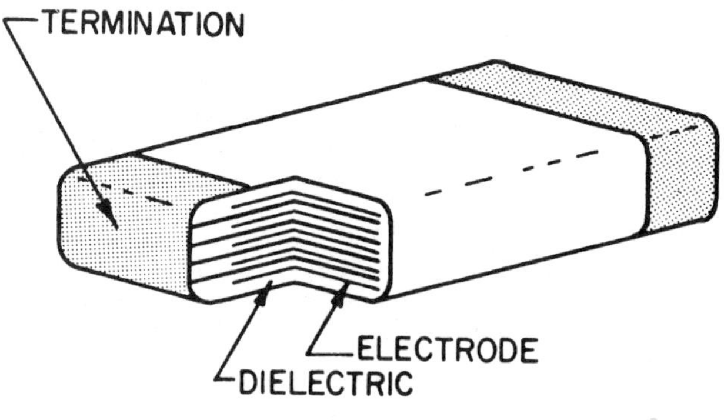

Figure 6 Cutaway view of multilayer ceramic chip capacitor. (Courtesy of Vitramon, Inc.)

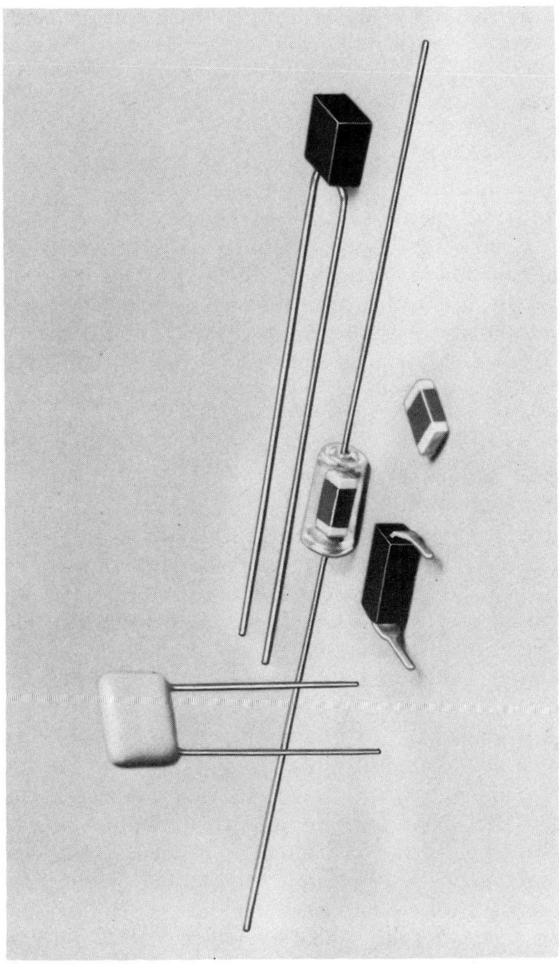

Figure 7 Varieties of multilayer ceramic capacitors. (Courtesy of Centralab Inc.)

in parallel. The first mica substitute materials to be made into multilayers were low-melting and devitrifying glasses developed by DuPont during World War II [32,33].

Two general fabrication procedures were employed in this early work: (1) Alternate layers of slurries of ceramic dielectric powder and of silver powder are deposited on a substrate by successively spraying or casting the dielectric slurry and screen printing the electrode. Following drying and sintering, a monolithic laminated structure is obtained whose laminae can be connected in parallel to form a capacitor. (2) In a related process, the dielectric is cast into individual sheets as described in Section VII.2.B, dried, and imprinted on one side with silver electrode paint. Following drying, the electroded green strips are stacked, compacted into a slab, fired, and parallel connected.

The low-melting dielectric glasses for these components derived compositionally from vitrifiable enamels used in decorating bottles and other glassware. These materials consist of inorganic colored pigments dispersed in and ultimately bonded by low-melting glasses based on lead borosilicate. Melting in the low range 500–800°C, the vitrifiable colors can be matured without deformation of the glass articles to which they are applied. More important from the monolithic capacitor standpoint is that silver powder makes a compatible electrode for use with these glasses, remaining solid while the glass dielectric fuses, but being close enough to its 960°C melting point to sinter readily into conductive layers.

A 1946 patent of Deyrup [34] provides the following example of the composition (in mol %) of one of these vitreous materials: 23.2 PbO, 44.0 SiO_2, 2.7 K_2O, 2.5 Na_2O, 2.3 Li_2O, 10.4 NaF, 8.7 MgO, and 6.2 SrO. The composition is formulated from oxides, carbonates, and fluorides, is melted at 1000–1200°C and quenched in water, is pulverized into fine powder, and is sintered into its finished form at 740°C. The material is described as having a 1-MHz loss tangent of 2×10^{-4} and a permittivity of about 10. The glass crystallizes to a large degree during the 740°C firing.

An interesting aspect of this dielectric material is emphasized in the patent—the low dielectric loss despite the substantial content of alkali-metal ions. The latter, traditionally, are regarded as mobile and agents of dielectric loss. In the light of current understanding the contradiction is attributable to the lead ion, or to pairing of alkali modifier ions so that their usual mobility in the glass network is greatly reduced.

Modern versions of this 35-year-old dielectric composition are still in use. The capacitors are, in fact, mica replacement units characterized by permittivity around 10, low dielectric loss, high insulation resistance, small temperature coefficient of capacitance (90 ppm/°C), and maintenance of these characteristics to 100 MHz or higher. The partially devitrified glass is white and is sometimes referred to as a porcelain dielectric. Actually, in composition, microstructure, and electrical properties, such materials are far removed from the classical clay-flint-feldspar porcelain.

Starting from this development, the evolution of the multilayer capacitor has been in response to two needs: (1) dielectric materials, which enable greater capacitance density and reduce component size; and (2) lower-cost electrodes. Thus, starting with the glass-based dielectrics, which were low melting and could be cosintered with silver electrodes, the trend was to higher-permittivity titanates, which, however, necessitated higher-sintering electrode material combinations of platinum, palladium, and gold. As the costs of these metals mounted during the 1970s and 1980s, ceramic fabrication stratagems were devised to reduce the electrode materials costs: (1) simple minimization of the quantity of metal employed in an electrode by optimizing its composition, paint characteristics, and application technique; (2) use of glass-derived dielectrics which fire at sufficiently low temperatures to be compatible with a silver-palladium alloy of as much as 70 wt % silver content; (3) use of base metal electrodes, enabled either by employing firing atmospheres reducing to the electrode metal or by infiltrating the capacitor's electrode regions, after ceramic sintering, with the molten base metal; (4) use of relaxator-type dielectrics, many of whose compositions have firing temperatures which are low enough for use with silver-palladium; and (5) further development of the grain boundary barrier layer capacitor to provide multilayer-equivalent capacitance density, or greater, in single-layer ceramics with silver or base metal electrodes. Further, there is effort to combine multilayer and barrier layer technologies. These variations on the multilayer capacitor concept are now taken up individually.

a. Lamination Stacking. Thurnauer [2] describes the observation by U.S. technical teams in Germany after the close of World War II of a pilot plant layout for producting thin ceramic sheet for capacitors by continuous casting and doctor blading. A similar method was under development in the United States during the war and was patented in 1949 [30]. The method was used to produce thin sheets of ceramic barium titanate and of a number of related materials: TiO_2, $MgTiO_3$, $CaTiO_3$, $SrTiO_3$, and $BaTiO_3$ + $MgZrO_3$. The process described in the patent and in the 1947 Howatt et al. paper [30] contains the basic elements of modern titanate multilayer technology.

The solvent-resin system employed in this process, which has enjoyed remarkable longevity and had many offspring in the capacitor technology, was comprised as follows: (1) A polymer mixture consisting of 78.9 wt % Albalyn polyester (methyl abietate), 15.8% Ethocel 50 (ethyl cellulose), and 5.3% DuPont M95 resin, a natural varnish. (The reasons behind this complex resin mixture may have been to maximize mutual solubility, to provide the necessary strength and flexibility to the film, and to spread pyrolysis during firing over a wide range of temperature. (2) Diethyl oxalate plasticizer—13.2% of the total resin content. (3) Staybelite (triethylene glycol dihydroabietate), a cross-linking agent for the Albalyn used in equivalent amount as the polyester. (4) Toluene as the system solvent.

These organics were mutually dissolved, with heating, and ball milled with the dielectric powder at an appropriate milling fluidity. For titania, the ratio of organic medium to dielectric powder was 34 wt %. Following milling and deairing, the suspension was permitted to flow through a slit onto a moving stainless steel belt which carried it under infrared drying lamps. Firing of lengths of the strip (100–800 μm thick) was at 1350°C for 4 hr. In general it was the practice to apply electrodes by firing on a commercial silver paint at 600°C. However, a momentous step into multilayer technology is described as follows:

> It was also found possible to combine the firing of the ceramic pieces and the electrodes. . . . The dry sheets (4 × 4 cm squares) were coated on both sides with a gold-platinum paste . . . and stacked several . . . squares high. . . . These were fired successfully into one dense block in a single operation. . . . The objection to the one-fire process was that the high firing temperature of the ceramic body necessitated using the expensive noble metal paste.

This experiment in effect teaches a current manufacturing process. In it layers of green dielectric are individually electroded, stacked, and pressed with or without heat into a slab or block. The baked or calcined blocks are cut into individual multilayer capacitors ready for final firing. Accurate alignment of the electrodes of the stacked laminae is necessary for maximum capacitance and can be problematical when the thin, deformable green ceramic sheets are handled. This basically is the multilayer capacitor process in major current use. Similar processing is also used for the construction of electronic modules in which as many as 50 layers of alumina ceramic substrate bearing thick-film resistors and conductors are laminated into a block with conductive "vias" linking the layers.

b. Spray Deposition. The limiting thinness of dielectric elements in a capacitor made as just described is imposed by the tendency of the resin-bonded dielectric sheets to deform and misalign the stacks. Alternative processes avoid this problem by casting, spraying, or dip-coating layers of dielectric slurry onto a platen or substrate.

The spray process was first described in connection with the devitrifying glass-type dielectric [36]. Subsequently, at least one U.S. manufacturer has applied it with titanate ceramics and noble metal electrodes. In the process, platens pass, successively, under a spray head which deposits dielectric layers, through a dryer, to a silk-screening station where a multiple-capacitor electrode pattern is applied, through a dryer, and back for the next spray coat.

c. *Build-Up Process*. In place of spraying, a process successively casting thin layers of ceramic slurry onto a platen with intermediate silk screening of electrodes is also practiced today [37]. The dielectric slurry has flow characteristics that permit casting a 250-μm-thick layer repeatedly. Following more or less superficial drying, the cast surface is smooth and receptive of the silk-screened electrode paint pattern.

In modern monolithic capacitor production, methods of the types described produce hard, dry, green dielectric and metal laminate stacks measuring 50,000–100,000 mm^2 in area. From these are cut or sawn individual capacitors ranging from approximately 150 to 2 mm^2 in area. The potential for producing large numbers of parts is obvious. Depending on the electrode configuration printed into a stack, individual capacitors can be cut out with exposed electrodes at the ends, to be subsequently metallized and soldered, or with a margin of dielectric material all around so that the electrodes are entirely internal. In the latter case it is then necessary to drill lead holes or otherwise gain access to the opposing sets of electrodes.

2. Infiltrated Electrodes

Rutt [38–40] developed an alternative method which makes it possible to have in one component, multilayer construction, a high-sintering "conventional" barium titanate–based dielectric, and low-melting base metal electrodes. This is done by first accomplishing ceramic maturation, and subsequently putting the electrodes in place. In the laminate assembly step, laminae of the dielectric material are alternated with, instead of electrode material, layers of the dielectric material precursor compounds. When these are cofired, the dielectric sinters to a high density while the precursor layer materials combine into chemically similar compounds but in a very porous state. In addition, the organic binder content of the porous layer is made greater, to enhance its ultimate porosity, and additional pyrolyzables such as carbon can be included for the same purpose. Following firing, the multilayer structures are evacuated, then immersed in a molten low-melting solderlike alloy under 10–15 atm pressure. This causes the alloy to intrude into the porous-layer regions to provide electrodes.

In a variation of the process, the multilayers are first impregnated with silver nitrate solution. This is thermally decomposed in place to provide a silver electrode prestructure, which is then more readily wetted by the solderlike alloy.

The following example composition is from the second-cited Rutt patent [39]:

> *Composition of the dielectric strata*: 93 w/o $BaTiO_3$ and 7 w/o $Bi_2O_3 \cdot 3\ ZrO_2$, mixed and uncalcined, both materials about 1.5 μm average particle size. 100 g of this powder is mixed with 4 g ethyl cellulose binder, 3 g butyl benzyl phthalate, 2 ml acetic acid defoamer, and 30 ml dichlorethane solvent.
> *Composition of the porous strata*: 66.9 w/o $BaCO_3$, 27.1 TiO_2, 3.32 Bi_2O_3, 2.64 ZrO_2. The TiO_2 has an average particle size of 5–10 μm, the other solids 1–2 μm. One hundred grams of this dielectric powder is mixed with a medium containing 14.5 g acrylic resin, 1.3 g ethyl cellulose, 1.6 g lecithin dispersant, and 83 ml pine oil. The unreacted percursors, the high binder content, and the relatively large particle size all lead to the high porosity of these layers following firing. The two different solvent and binder systems help maintain the individualities of the two layers. The dielectric layer slurry is cast into 50 μm thick sheets, and an electrode pattern is silk-screened onto them using the porous layer paint. The layers are assembled into a multilayer stack and pressed in a die at 85°C and 28 kg/cm^2 to consolidate the stack. The resultant slab is then cut into individual capacitors with the porous layer edges exposed on at least two sides. The high content of binder material is eliminated in an 86 hour heating to 420°C; the temperature is then raised more rapidly for final firing at 1260°C.

Example electrode alloy compositions range from pure lead to alloys of lead, aluminum, copper, zinc, bismuth, tin, cadmium, and others. The capacitors and electrode metal are maintained apart from one another in a closed vessel heated above the melting temperature of the metal. A rough vacuum is applied, the capacitors are immersed, and the pressure is then raised to about 14 kg/m^2 to force the metal into the porous ceramic. At least one U.S. component manufacturer produces multilayer capacitors by a process of the type just described.

3. Low-Sintering-Temperature Dielectrics

Another approach to less costly construction materials in multilayer capacitors is to use silver-based electrodes in conjunction with dielectric ceramics having intermediate sintering temperatures. Such compositions require higher firing than do the fusible glasses already described, but lower than do conventional titanate materials. That is, they are sinterable in the range 900–1150°C.

a. Electrode Alloys. As introduction to dielectrics with lower sintering temperatures, it is appropriate to digress to the subject of electrode materials for ceramic dielectrics in general. The original application of the classes of materials that now serve in electrodes for ceramic capacitors was for decorating dinnerware and for bonding metals to ceramics or glasses. Gold, platinum, silver, and palladium, in the form of finely divided powders, in admixture with low-melting glasses have been used individually and in various combinations for these purposes. For multilayer capacitor electrodes the glass content is generally omitted and the metal powders must be heated to the range where atomic mobility is sufficient to produce sintering and densification in a reasonable time. Sintering temperatures of platinum, palladium, gold, and silver are about 1250, 1100, 750, and 650°C, respectively. The densification of platinum or combinations of platinum-palladium or platinum-gold therefore takes place compatibly with the firing of titanate ceramics, and these were, in fact, the original electrode metal combinations of choice. Silver, or its alloys with gold, palladium, or platinum, would be appropriate with lower-firing ceramics or, with very little or no alloying, with low-melting crystallizing glasses.

Because sintering shrinkages of the metal and ceramic powders differ, delamination of the multilayer structure would be commonplace in the absence of any but mechanical bonding. Fortunately, the noble metal powders, especially Pt and Pd, experience surface oxidation and reaction with the ceramic to provide a chemical bonding mechanism. This is particularly effective in titanate capacitors with nickel electrodes, due to the latter's partial oxidation and solution in the titanate (see Section VII.C4).

The reactivity of dielectric with electrode necessitates empiricism in the formulation of dielectric compositions. The dielectric properties are altered by the interaction. The magnitude of the effect depends on the materials, the ratio of electrode thickness to dielectric thickness, and the conditions under which they are sintered. Whereas the base metal case has been quite thoroughly documented, the effect in noble-electroded capacitors, where not cataclysmic, has tended to be discounted. Figure 8 illustrates, however, the effect of reaction during air firing of various noble metal electrodes, screen printed, with a 50-μm-thick BX dielectric. The intrinsic dielectric behavior is represented by the curve for the dielectric with sputtered silver electrodes.

In practice electrode metals are used in the form of fine, generally submicron-diameter powders produced by precipitation with

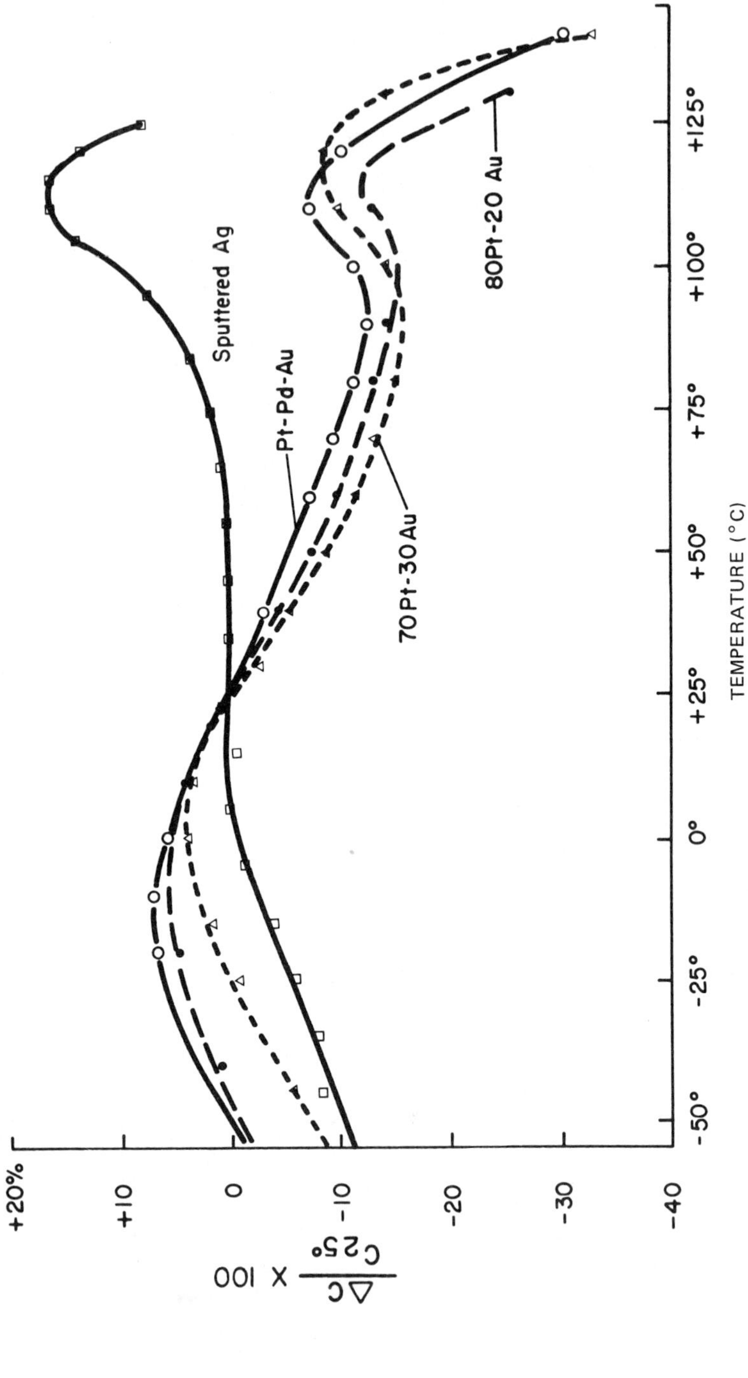

Figure 8 Effect of electrode-dielectric interaction on capacitor characteristics. (From Ref. 41.)

reductants from aqueous solution. All binaries in the system Pt-Au-Pd-Ag form continuous solid solutions, so that binary combinations are available with sintering characteristics to suit a range of ceramic processing. The binary metal combinations can be simply powder mixtures, but if coprecipitated under suitable conditions can be indistinguishable from, and are described as, solid solutions [42]. The powders are suspended in a solution of one or more organic resins with a net rheology suitable for silk screening.

In the Ag-Pd binary system, compositions with palladium as the minority alloying element provide a desirable combination of low cost and high liquidus temperatures [43]. Such electrode compositions are always formulated with concern for "silver migration," an electrolytic phenomenon associated with the conjunction of silver, electric field, and moisture [69]. In this process, silver goes into water solution at the anode, migrates to the cathode, and plates out in dendritic form. Similar, related mechanisms operate whether the silver is initially present as exposed metal, is encapsulated in a glassy mitrix, or is dissolved in a borosilicate glass. According to Miller [43], the rate of silver migration is one-tenth as fast in an 80/20 silver-palladium alloy as with silver.

An effect of the trend to use of palladium in electrode alloys has been to displace bismuth from many dielectric formulations. Compounds such as bismuth stannate or bismuth zirconate have had wide use as Curie peak suppressors [15]. However, reaction of the bismuth with palladium to form Bi-Pd intermetallic compounds occurs below the titanate capacitor sintering range [44]. The reaction is exothermic and disruptive of the not yet coherent multilayer structure. When the upper temperature stability limit of PdO is depressed by a low ambient oxygen partial pressure, the Bi-Pd interaction effects in a capacitor are ameliorated. This is believed due to preferential solution of the palladium in silver [45] and suggests a way of circumventing this electrode-dielectric materials incompatibility.

An obvious step further in the direction of low-cost electrode materials is to make use of base metals. Developments in this direction are taken up below. Attempts to employ combinations of base and noble metals have not been followed by significant use [46].

b. End Termination Materials. In addition to the material that comprises the electrodes of a multilayer capacitor, another metallization is used to connect its two sets of electrodes in parallel. This material is applied to the exposed electrodes at each end of the capacitor. It may consist of the electrode material itself, cofired with the capacitor, but most commonly is a low-cost silver-frit combination applied at a lower temperature, 600–900°C, than the electrode. Such material is vulnerable, in subsequent handling, to dissolution by molten solder. Protective platings of copper or nickel or coatings of silver-bearing solders are sometimes employed to prevent this [47].

c. Glass-Reacted Capacitors. To permit substitution of low-cost silver-based electrode alloys for platinum or palladium-based alloys, a dielectric densifying at about the sintering temperature of the alloy is needed. For example, an alloy of composition 70 at % Ag–30 at % Pd, melting at 1225°C, would sinter in the range 700–900°C. Maher [48–50] and others have developed titanate compositions that meet this requirement through the use of glass formers which foster liquid-phase sintering. The novelty of these compositions is that ceramic sintering at the desired low temperature is accomplished with a small enough content of the glass-forming constituents that the high permittivity of the titanate constituent(s) is not substantially compromised. An example of the compositions used for a high-permittivity dielectric is as follows:

1. A low-melting glass of composition (in wt %) 36 CdO, 23 Bi_2O_3, 25 PbO, 5 ZnO, 1 Al_2O_3, 5 B_2O_3, 5 SiO_2 is prepared by mixing the constituents, melting in a platinum crucible at 980°C, crushing, and ball milling in water to less than 1 μm average particle size.
2. A "barium titanate" (presumably a zirconate- and strontia-modified $BaTiO_3$, to shift and broaden the permittivity versus temperature characteristic) is mixed and reacted at 1350–1400°C, then crushed and milled.
3. The products of compositions 1 and 2 are combined in proportions up to 10 wt % glass, fabricated into thin sheets, and fired at 1100–1150°C for 1 hr.

The resultant ceramic has a k' of 1550 and a dissipation factor of 0.015–0.017 at 1 kHz using a 40-V/mm measuring field. The temperature coefficient of capacitance is ±10% over −55 to 125°C.

Maher [50] describes compositions in which the same glass phase is used with dielectric combinations from the system BaO-TiO_2-Nd_2O_3 to provide a series of dielectrics with graduated positive and negative temperature coefficients of capacitance. An NOP composition in this family has a k' value of 68. Again, the process sequence is such that the titanate portion is well reacted before the glass phase is introduced. Capacitor firing temperatures are no higher than 1150°C, permitting the use of Ag-Pd electrodes in the composition range 15–40 at % Pd.

Guiding principles with the flux-sintered compositions are to minimize the content of flux to begin with, and to select for fluxing constituents those that can be molecularly assimilated by the high k' phase(s), thereby lessening the dilution effect due to the glass. Burn [51] has examined a variety of types of glass compositions for this purpose in the system CdO-Bi_2O_3-PbO-B_2O_3, measuring their efficacy by the ability to cause grain growth during sintering and in the process to be homogeneously assimilated so that they do not remain to function as diluents. SiO_2 and GeO_2 were found to be useful alternative glass formers for this purpose; ZnO, Cu_2O, and Li_2O are alternative glass modifiers also readily taken up by the perovskite lattice, while Nb_2O_5, Y_2O_3, Sb_2O_3, or WO_3 are glass modifiers which are less perovskite compatible.

Many of the resultant reactive glass additives, in amounts well under 5 wt %, enable the titanate ceramics to be densified. An example (in wt %):

Glass: 1.5 ZnO, 0.56 $BO_{1.5}$, 1.5 $BiO_{1.5}$

Ceramic: 96.44 $BaTi_{0.903}Zr_{0.097}O_3$

This combination is sinterable at 1100°C in multilayer structures with 30 wt % Pd–70 wt % Ag electrodes. The reacted glass flux dielectric in combination with silver-palladium electrodes in a multilayer capacitor is commercially important and in use by several manufacturers.

A somewhat differently oriented use of low-melting additives is described by Payne and Park [52]. The simple combination of $BaTiO_3$ and $Pb_5Ge_3O_{11}$, for example, is said to provide in a 950°C firing a dielectric with permittivity of 1210 and dissipation factor 0.012. The lead germanate flux is described as being, in amounts up to 10 vol %, segregated entirely in the grain boundaries. This is unexpected; in view of the level of permittivity, significant assimilation of the flux phase would have been more usual. A broadened composition of fluxing material is given as $(PbO)_x$ $(GeO_2)_{y-z}$ $(SiO_2)_z$ [53], where $1 < x \leqslant 6$, $1 < y < 3$, $3 \leqslant (y + z)$. This type of composition is more chemically resistant than $Pb_5Ge_3O_{11}$.

d. *Glass Ribbon Capacitors*. Several methods have been advanced for producing multilayer ceramic capacitors by stacking laminae of glass ribbon and metal foil and subjecting the stack to hot pressing. In the latter step the layers are plastically compacted into a dense, coherent body and the glass is substantially crystallized into a high-permittivity crystalline dielectric.

The attraction of such a process is that technology exists for fast, continuous production of thin, uniform glass ribbon. In addition, a sufficiently low hot-pressing temperature would permit the use of low-cost aluminum foil electrodes.

The success of such a process hinges on the ability to formulate a glass with a minimum of network-forming constituents so that conversion to the crystalline dielectric approaches 100%. Compositions have been designed for this purpose [54,55], based on barium titanate, such as 30–40 (in mol %) BaO, 15–40 TiO_2, 9.5–26 SiO_2, 7–25 $AlO_{1.5}$. The high content of residual glass-forming constituents presumably accounts for the limited success of this concept. However, there is current commercial production of multilayer capacitors by a closely related method. The dielectric is, instead, a noncrystallizing glass ribbon with a permittivity of about 20. The electrodes are of aluminum foil and the entire unit is encapsulated in glass with the leads emerging. This product has performance characteristics in a class with the mica replacement, glass-derived capacitors with silver electrodes discussed in Section VII.C.1.

e. *Relaxator Dielectrics.* Another group of dielectric materials with which it is possible to fabricate multilayer structures at low enough temperatures for compatibility with silver-palladium electrodes is the relaxator class of materials. Relaxator, or relaxor dielectrics, attracted interest over 20 years ago as a special class, phenomenologically, of the perovskite structure ferroelectrics [56]. Only in the last few years, with the increased costs of noble metals. have they assumed some commercial interest. As dielectrics, the materials are distinguished by a ferroelectric-to-paraelectric transition with a diffuse character attributed to micro-inhomogeneities of composition. These stem from disorder of the ions occupying octahedral positions in the unit cells. The broad peak of dielectric constant versus temperature that results is of interest as the basis for a capacitor dielectric. Figure 9 shows such a characteristic.

Relaxator behavior was early recognized as a widespread phenomenon among ternary oxide perovskites. A number of the latter are characterized by being sinterable at temperatures as low as 1000°C. Mixtures of two or more relaxator materials offer the opportunity to place their dielectric constant peaks, which individually are broad, so that they are influential over the entire temperature range of use of the intended capacitor. Iizawa et al. [58] describe compositions based on $Pb(Fe_{2/3}W_{1/3})O_3$-$PbZrO_3$ plus MnO_2 or CeO_2, which sinter at 950°C, have room-temperature k' of 5000–8000, tan δ of 0.013–0.019, insulation resistance of 10^{10} Ω-cm, and TCC over 20–85°C of less than 30%.

Tanei et al. [59] list 99.5–99.95 wt % $Pb(Fe_{2/3}W_{1/3})_{1-x}Ti_xO_3$ plus 0.05–5% MnO_2. In these compositions the purpose of $PbTiO_3$ is to provide an elevated Curie point and MnO_2 is described as an acceptor dopant to decrease tan δ and increase resistivity. The possibility of peak suppression as an additional function of the MnO_2 is not addressed. This ceramic, following calcination at 650–850°C is sintered at 1000°C or lower. Room-temperature dielectric properties are k' $\geq$ 2000, tan $\delta \leq 0.05$, and volume resistivity $\geq 10^9$ Ω-cm.

Jang et al. [21] have combined relaxator compounds to provide in one dielectric two broad k' versus temperature peaks possibly useful for capacitors:

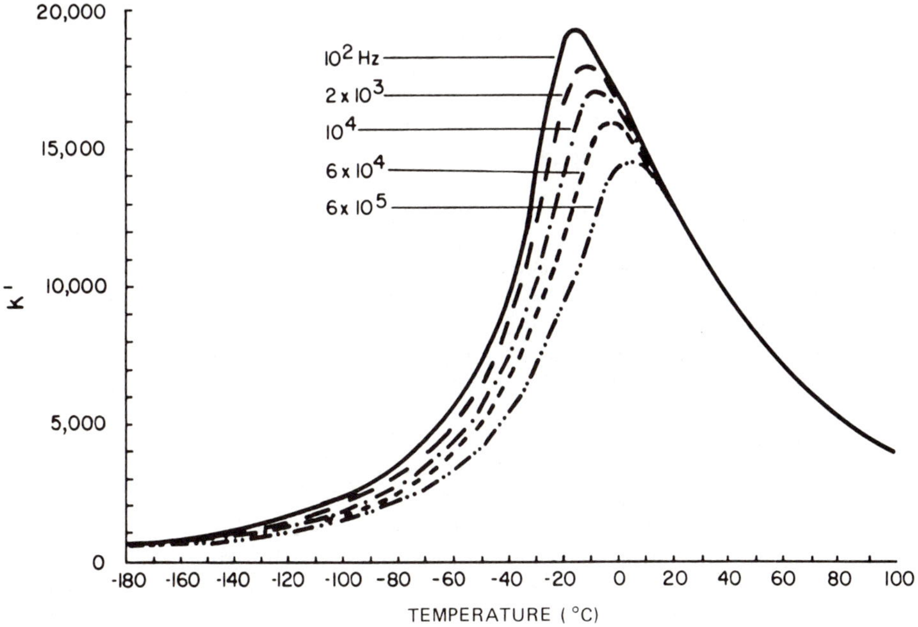

Figure 9 Relaxator behavior in single-crystal lead magnesium niobate. (From Ref. 57.)

$$Pb(Mg_{1/3}Nb_{2/3})O_3\text{-}PbTiO_3\text{-}Ba(Zn_{1/3}Nb_{2/3})O_3$$

The heights of the maxima of k' versus temperature are controllable by the composition. However, the firing temperatures indicated for these materials are 1280°C or higher, necessitating the higher-cost electrode compositions.

Negative aspects of the relaxator dielectrics may account for their small application to date. Large electrostrictive effects present in some compositions may be reflected in instability of the capacitance value. Tungsten and niobium oxides are more costly constituents than zirconates and titanates. Also, by definition, the relaxator character is associated with strong frequency dependencies of k' and tan δ at the diffuse phase transitions. With a conventionally ferroelectric-based material, relaxations tend to be smaller and to occur at higher frequencies.

4. Base-Metal-Electrode Multilayer Capacitors

The concept of ceramic dielectric compositions and processing compatible with the fields of stability of base metal electrodes (BMEs), for cofiring, is over 25 years old [60,61]. The original intent was to eliminate the need for noble metals in multilayer capacitors, but in several ways the concept was premature: Noble metal prices were still far below the levels to come, commercial titanate raw materials had not yet arrived at the purity and degree of consistency available in the industry today, and, above, all, the present-day understanding of the technology was lacking.

The Herbert invention was a multilayer capacitor with nickel electrodes. To prevent oxidation of the metal, the capacitors were fired in a reducing atmosphere. To prevent formation of an oxygen-deficiency semiconductive dielectric under these conditions, the titanate was modified with manganese oxide. The total system is complex and its understanding, though increased, is not yet complete. The equilibria of nickel and manganese with their oxides are shown in Figure 10.

When a multilayer capacitor with a titanate-based dielectric and nickel electrodes is fired at a temperature and oxygen partial pressure reducing to NiO, the Curie peak is determined by the titanate dielectric. Burn and Maher [63] presented analytical data, for example, indicating essentially no solubility of NiO in $(BaTiO_3)_{0.865}(CaZrO_3)_{0.135}$ fired at 1350°C with an oxygen pressure of 10^{-11} atm. With increasing oxygen pressure to 10^{-7} atm, NiO solubility up to 0.30 mol % was found. Depression of the Curie peak and shift of the Curie peak to a lower temperature accompanied the increased oxidation. It is assumed that in commercial BME systems, similarly, some degree of oxidation of the electrodes and solution in the dielectric take place.

Under the effect of the substantially reducing sintering conditions, electron-donor oxygen vacancies form on the titanate-zirconate solid solution lattice, leading to n-type conductivity [64]. To regain the volume resistivity essential in a practical capacitor, acceptor dopants are used [65] whose ionic radii permit their substitution for Ba^{2+} or Ti^{4+} but which have lower valence than the ion replaced. Herbert's finding that manganese serves this function has been followed by analytical studies showing trivalent or quadrivalent manganese substituting for Ti^{4+} in air firing. In lower oxygen pressures the function of manganese is probably based on the substitution of Mn^{2+} for Ti^{4+}, in which form it shifts the Curie temperature; as Mn^{4+} in air firing it has no such effect.

The stoichiometry, or ratio of large (e.g., Ba^{2+}, Ca^{2+}, Pb^{2+}) to small (e.g., Ti^{4+}, Zr^{4+}) cations, has been identified by Eror et al. [66] as basic to the successful fabrication of BME multilayer capacitors. The critical lower limit of this large-to-small cation ratio is defined as 0.95. Values much greater than 1 make for refractory, difficultly sintered ceramics. Since raw material impurities include these ions, detailed knowledge of material composition is vital. Similarly, ions such as Al^{3+} and Sn^{4+} can be introduced in processing

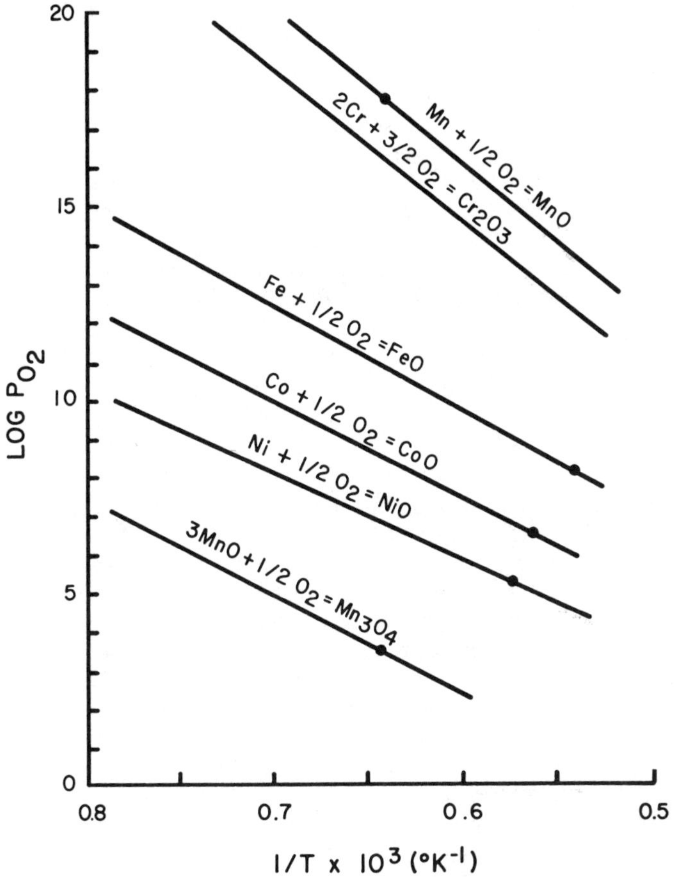

Figure 10 Redox characteristics of base metal electrode material. (From Ref. 62.)

and need to be corrected for. The Eror et al. patent [66] defines 10^{-7} atm as the maximum oxygen pressure permissible with firing temperatures up to 1400°C.

Several U.S. manufacturers produce nickel electrode multilayer capacitors today. These tend to be characterized by lower insulation resistance and higher temperature coefficients of capacitance than counterparts with silver-based or noble metal electrodes. Presumably, the greater demands of the base metal process on raw materials purity and on process control are accountable. It is of interest that the greater chemical reactivity between electrode and dielectric makes BME capacitors exceptionally strong and delamination resistant.

D. Barrier Layer Capacitors

Ceramic capacitors of the types discussed up to this point store energy by means of electronic, ionic, and orientational polarization, with a possible inadvertent and minor contribution from interfacial polarization. The latter is evident as an increase of energy dissipation on descending through power frequencies toward dc. Increasingly, however, components are becoming available whose function is based *primarily* on polarization at interfaces between grains and/or between electrode and dielectric. By their nature these capacitors are limited to low operating voltages. The growth of interest in them derives largely from the growth of solid-state circuitry with compatible low operating voltages.

The theory of the dielectric properties of multiphased systems as developed by Maxwell [4] has already been discussed. This was expanded by Wagner [67] to cover the ac case and has been simplified for applicability to ceramic systems by Volger [68]. When the phases of a ceramic have differing conductivities, imposition of an electric field causes more charge to be removed from the more conductive phase than can be taken up by the less conductive phase. The phase interfaces thereby become charged or polarized. Outwardly, this is evidenced by the flow of a decreasing current when dc voltage is applied or an opposite current when the voltage is lifted.

If one makes assumptions as follows regarding a ceramic dielectric material:

One species of grain
Uniform grain size
Continuous grain boundary
Grain boundary thickness << grain diameter
Grain boundary permittivity = grain permittivity
Grain boundary resistance >> grain resistance

$d_g/d_{gb} << \rho_{gb}/\rho_g$, where d_g and d_{gb} are the grain and grain boundary dimensions, respectively, and ρ_{gb} and ρ_g their resistivities

then a simple two-layer model is applicable (Figure 11). This predicts the following dispersion behavior of the grain boundary barrier layer system below its relaxation frequency:

1. $k'_{effective} = k'_{intrinsic} (d_g/d_{gb})$
2. $\rho_{effective} = \rho_{gb} (d_{gb}/d_g)$
3. $RC_{effective} = RC_{intrinsic} (d_g/d_{gb})^{\frac{1}{2}}$

Above the relaxation frequency, all characteristics assume the intrinsic values.

The first commercial barrier-layer capacitors were described in a 1950 patent [70] as a layer of insulating titanate ceramic bearing on one side an electrode layer of reduced, conductive titanate and on the other a metal electrode. The structure was produced from a barium strontium titanate material containing 0.5 wt % of didymium oxide as a "reduction catalyst." A ceramic disk of this composition is first sintered in air, then refired at a lower temperature in air with one face in contact with a carbon slab. Effective dielectric constants of 20,000 were claimed, with resistance over 20 MΩ.

The above-described component has developed into a widely manufactured type of disk or wafer capacitor. Present practices are generally as follows. The green ceramic is fired in air to full density with insulating character. It is then heated in hydrogen or forming gas and reduced homogeneously to a semiconductive body. A thin layer at the surfaces is then reoxidized to insulating condition by heating in air. Finally, silver paste (metal plus glass) electrodes are applied to the two faces. It is possible to combine the sintering and reducing steps into one; similarly, the reoxidation and electrode firing steps are combinable into one.

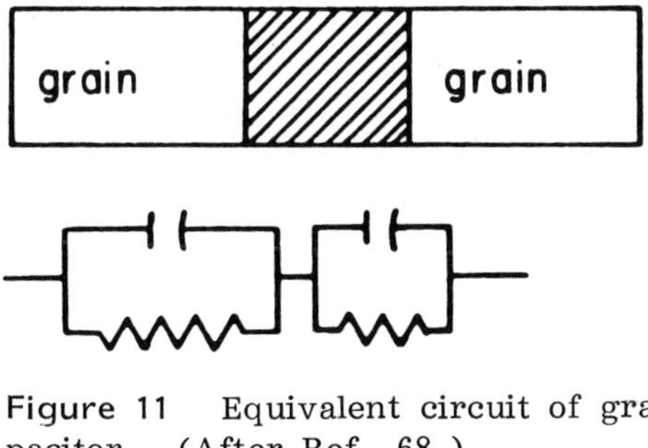

Figure 11 Equivalent circuit of grain boundary barrier layer capacitor. (After Ref. 68.)

Commercial capacitor products of this technology are characterized by low RC, about 0.1 to 1 MΩ-μF depending on size, rather high dielectric loss (tan δ of 0.05–0.10 at room temperature, 1 kHz), and low to moderately high voltage ratings. In spite of their 20-year history of use, the physical character of these components seems not to have been definitively described as yet. Schottky barrier–like behavior has been described when one of the two junctions is removed by grinding [71]. However, dielectric layers produced in more extensive reoxidation appear to have normal dielectric characteristics. Furthermore, a reaction of the glass frit and/or of the silver of the counterelectrodes with the dielectric layer seems to have beneficial effects on the latter's voltage withstanding ability and possibly on other dielectric characteristics.

A subsequent development was the combination of atmosphere reoxidation of the reduced titanate with counterdoping by electron acceptor ions [72,73]. These treatments were intended to provide electrode barrier layer p-n junction capacitors, but to a greater degree, it turned out, they provide active junctions at the contacts between grains. Specific experiments with n-type $BaTiO_3$ counterdoped with in-diffused copper oxide show no contribution to capacitance by an electrode barrier effect [74]. A detailed description of the production process in the first commercialization of grain boundary barrier layer (GBBL) capacitors is given by Waku [74]. To a commercial-grade barium titanate is added up to 0.5 m/o Dy_2O_3 to help reduce it to a semiconducting state and small quantities of Al_2O_3 and SiO_2 intended to promote grain growth. (For high capacitance from series-connected grain boundary barrier layers, the grain size needs to be large; for high voltage-withstanding ability the applied field needs to be distributed over the many boundaries of a fine-grained ceramic). The green ceramic disks are fired in nitrogen. A slurry of copper oxide is then painted onto each disk face and fired in air at 1000–1100°C. Copper ions diffuse throughout the specimen, mainly via a liquid phase. Then, with silver paste applied, capacitance corresponding to a room-temperature permittivity of about 20,000 is measured in a $BaTiO_3$ material whose intrinsic k' value is 1000. For grains of 20–40 μm, the boundary layer thickness is thereby indicated to be (20–40 μm) × (1000/20,000) or about 1–2 μm.

A succeeding development in GBBL process technology has been to incorporate both n and p dopants in the titanate initially so that they can accomplish their functions in a single firing [75,76]. A typical composition is the following [77]:

$$Ba(Ti_{0.9}Sn_{0.1})O_3 + 0.1 \text{ mol \% } Dy_2O_3 + 0.4 \text{ wt \% } SiO_2$$

$$+ 0.5 \text{ mol \% } BaCO_3 + 0.6 \text{ mol \% } CuO$$

Specimens of this mixture are fired in hydrogen and cooled in air. The electroding procedure is not specified. The n and p dopants, Dy^{3+} and Cu^{2+}, are both batch constituents. The dysprosium presumably takes up Ti^{3+} sites, while copper confines itself to the grain boundary regions, counterdoping the n-type grains to produce insulating $Ba(Ti,Sn)O_3$. The importance of large grains is stressed, and the silica additive and barium nonstoichiometry are for the purpose of promoting grain growth.

Heywang [78] has accounted for the characteristics of GBBL capacitors in terms of a Schottky barrier between n-type conductive grains. The p-dopant content must be high enough to give a thick grain boundary region. The p-type semiconductive grain boundary has low conductivity because of the depth of the acceptor energy states available for conduction. A Philips group have refined the Heywang model to one where n-i-n, rather than n-p-n, junctions determine the capacitor characteristics [79–84].

Most recently, strontium titanate has become a commercially significant GBBL material. As a linear dielectric, $SrTiO_3$ is free of the shortcomings of ferroelectric $BaTiO_3$-based materials: the capacitance value does not age, is not voltage dependent, is not highly temperature dependent, and the dissipation factor is low. Also, $SrTiO_3$ is intrinsically capable of higher conductivity than $BaTiO_3$ [85,86]. Thus its Maxwell–Wagner dispersion comes at about 1 GHz, a higher frequency than for $BaTiO_3$. A composition and procedure from the patent literature are as follows [87]:

Base material: $SrTiO_3$ + 0.8 wt % Nb_2O_5 (or Ta_2O_5) + 0.2 wt % SiO_2 + 3.0 wt % ZrO_2. Sinter in 99% N_2 + 1% H_2 for 2 hr at 1450°C.

Fire silver paint containing a frit of 50 wt % Bi_2O_3, 10 wt % B_2O_3, 40 wt % PbO for 2 hr at 1180°C.

Nb_2O_5 is the n dopant. The prolonged and exceptionally high temperature electrode firing indicates that an electrode paint constituent, presumably bismuth, serves as counterdopant. The capacitor has an effective k' value of 23,000, tan δ value of 0.008, insulation resistance of 5×10^{10} Ω, and 2.4% capacitance decrease between room temperature and 80°C.

A model of this $SrTiO_3$ ceramic with semiconductive grains, resistive or insulating grain surfaces, and an insulating phase in the grain boundaries has been developed whose results are in good agreement with device properties [88,89]. Burn and Neirman [88] attribute the grain semiconductivity to donor electrons produced when oxygen is lost during low pO_2 firing as well as to those due to donor impurities. A room-temperature grain resistivity of $<6 \times 10^4$ Ω-cm was estimated. Impregnation of the grain boundaries by 5 wt % of a mixture of 90 wt % Bi_2O_3 and 10 wt % Cu_2O added to the silver metallizing paint imparted overall apparent resistivity around 10^{11} Ω-cm. Figure 12 demonstrates the coincidence of maxima of grain size and apparent dielectric constant, a GBBL characteristic.

E. Multilayer GBBL Capacitors

The concept of combining multilayer geometry with GBBL microstructure to obtain exceptionally high capacitance density is initially attractive but contains several inherent problems: (1) GBBL capacitor microstructures are large-grained to maximize capacitance. But a multilayer device with excessively thick laminations dictated by the

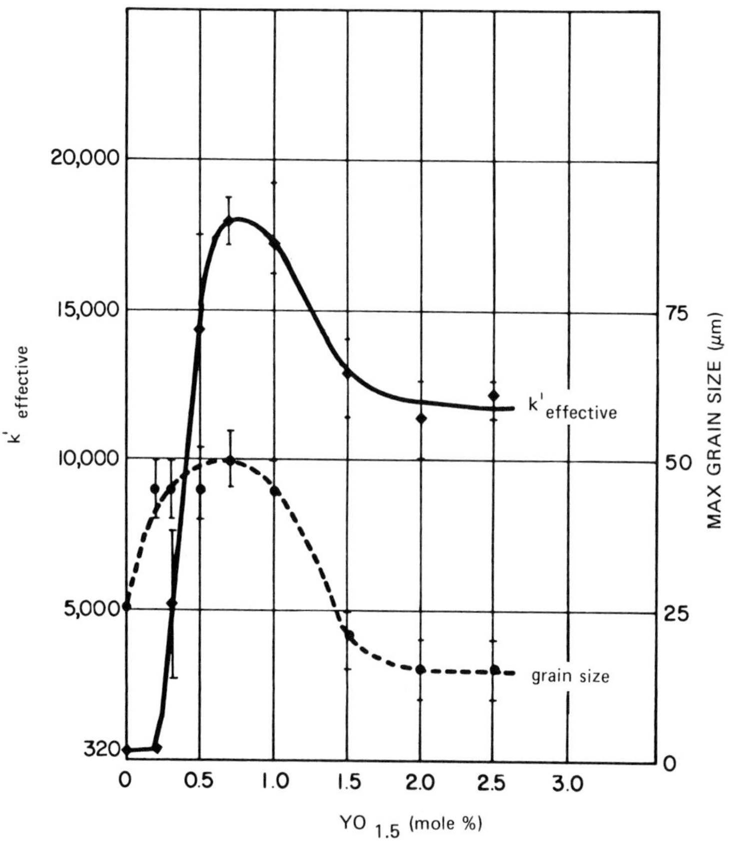

Figure 12 Dependence of $k'_{effective}$ on grain size in a GBBL capacitor. (From Ref. 88.)

grain size is limited in bulk capacitance. (2) The chemistry of a multilayer GBBL capacitor becomes even more complex due to interaction of the cofired dielectric and electrodes. Schmelz [90] has reported that Pd^{2+} or Pt^{2+} counterdope n-type $BaTiO_3$, an effect that would reduce an intended GBBL device to an ordinary multilayer one. Infiltrated electrodes or the low-temperature electroding methods discussed earlier might solve this problem. There has been preliminary work along these lines [91] in which the GBBL chemistry is first built into the dielectric powder and a low-temperature multilayer process is then practiced. None of these concepts has been demonstrated commercially, however.

VIII. CAPACITOR PERFORMANCE PARAMETERS

The parameters whose values together describe the performance of a capacitor are:

1. Capacitance and its dependence on temperature, voltage, and frequency
2. Dissipation factor and its dependence on temperature, voltage, and frequency
3. Insulation resistance and its short-term dependence on temperature and voltage
4. Response of insulation resistance to long-time testing under dc field and elevated temperature
5. Maximum permissible operating voltage
6. Stability of dielectric characteristics with aging in the absence of an electric field

In the following section these parameters are discussed in greater detail.

A. Capacitance

Dielectric materials differ from one another in the numbers and types of microscopic charged entities available to respond to an applied electric field. Total response to the field is the polarization P, proportional to the number of charged entities N, the sum of their individual polarizabilities α_T, and the electric field(s) they experience:

$$P = N\alpha_T E_l$$

P in ceramic dielectric materials can be made up of four components: electronic, atomic, orientational, and interfacial polarizations.

Electronic polarization P_e occurs within atoms when the electric field slightly displaces the electron clouds of the atoms relative to the nuclei. The electronic polarizability α_e of an atom is given by

$$\alpha_e = \varepsilon_0 4\pi R^3$$

where R is the atomic radius. Materials with large atoms therefore have higher electronic polarization. There is no temperature dependence except that due to thermal expansion effect on density. This would make for a slight negative temperature coefficient of dielectric constant:

$$\frac{1}{k'}\frac{dk'}{dT} = -\frac{1}{3}\frac{(k' - 1)(k' + 2)}{\rho}\frac{d\rho}{dT}$$

where $1/3\ d\rho/dT$ is the linear thermal expansion coefficient. Electronic polarization can respond to frequencies into the optical and ultraviolet range (i.e., beyond the frequency range of interest for capacitors). The electronic contribution to k' is given by Maxwell's rule from electromagnetic theory as the square of the refractive index. Dielectrics having electronic polarization only are linear dielectrics. Only organic compounds such as benezene are in this category, with permittivity values less than 5.

Atomic or ionic polarization is molecular in scale but can also be considered instantaneous in comparison with the periods of power and radio frequencies. In ionic polarization crystal lattice ions of opposite sign separate under the influence of the external field. The oscillation periods are greater than those of electrons because of atomic inertia. The dielectric constant of a dielectric material with atomic polarizability begins to decrease with increasing frequency in the infrared region. The temperature dependence of atomic polarization is positive; for alkali halides its magnitude is about the same as for electronic polarization.

Dipolar or orientational polarization P_o occurs when molecules possess a permanent electric moment which is alignable by the applied electric field. This mechanism is most pronounced in liquids and gases, but is also important in polymers with long polar molecules and in the special cases of electrets and ferroelectric materials. The limiting frequency for orientational polarization in liquids and gases is in the microwave region, about 10^{10} Hz. In ceramic dielectrics such moments would not contribute to k' beyond radio frequencies. Under dc potential long-period polarizations can be built up. Temperature is a strong influence, having an inverse relationship to relaxation time.

Ferroelectricity [92,93] can be considered a special case of orientational polarization, characterized by the interaction and spontaneous alignment of existing dipole moments. Their polarization is reversible by an applied field. Rather than rotating, the dipoles align and reverse align by cooperative ion displacements. This polarization mechanism relaxes in the infrared.

In ferroelectrics, a strong temperature dependence of dielectric constant is associated with crystallographic phase transitions and has to be suppressed or circumvented in order to make practical capacitors. The dielectric constant of materials of this class is also highly voltage sensitive, decreasing with increased dc field, but increasing with an increasingly strong ac signal.

Interfacial polarization occurs in multiphased materials when the phases have different conductivities. Grain boundary barrier-layer capacitors, for example, contain conductive grains separated by insulating grain boundary regions. Because of this heterogeneity an internal distortion of the imposed electric field is produced which from outside is evidenced as a large dielectric constant. The dimensional scale in interfacial polarization is much greater than atomic or molecular; the field-distorting charges can span a grain diameter or the dielectric thickness.

The response times of interfacial polarizations are determined by differences of dielectric constant and conductivity of the phases. When all phases are highly resistive, the polarization may occur only at very low frequencies; when one component has a very low resistance, as in the GBBL capacitor, the dielectric relaxation can be in the high radio-frequency range. The effect of voltage can be to increase, decrease, or not affect k', depending on the matrix material characteristics. Similarly, temperature increase can either raise or lower the permittivity. Figure 13 provides a representation of k' versus frequency for a hypothetical dielectric material that possesses all four types of polarization.

B. Energy Dissipation

The lower section of Figure 13 depicts the absorption of electrical energy for each type of dielectric polarization as a function of frequency. Each of the processes has an associated energy loss. The energy losses have peak values in the relaxation frequency ranges of their polarization mechanisms. But even at frequencies well below or well above the relaxation of any polarization mechanism, a finite energy loss or inefficiency is associated with any real energy storage mechanism.

In an ideal capacitor, application of a dc field is followed by the flow of a dielectric absorption current which falls to zero as the

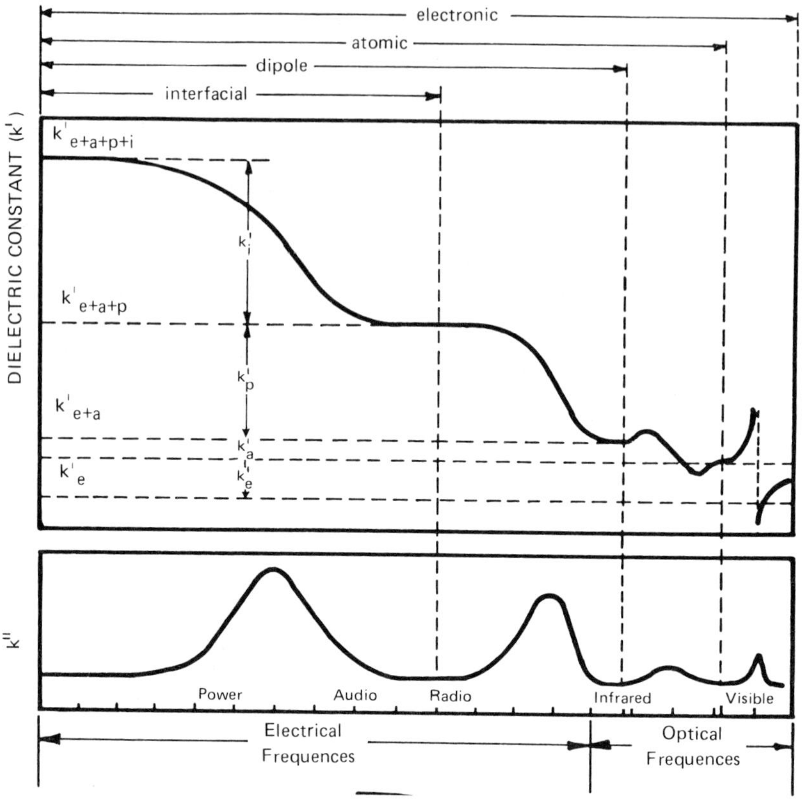

Figure 13 Dielectric polarization mechanisms and their frequency dependence. (From Ref. 94.)

energy storage function saturates. There is no energy loss. In an ac field in the same ideal capacitor, the absorption current leads the voltage by 90° and there are no dielectric losses.

In a real capacitor an increment of the dc current flow during absorption is not recoverable on discharge. Under ac, the phase angle between current and voltage is something less than 90°, the angular deviation from 90° being called the loss angle, δ. Tangent δ is a measure of efficiency of the capacitor and is termed the dissipation factor. Commonly tan $\delta \times 100$, called the "percent sipation," is used.

In addition to losses due to inefficiency of the polarization mechanisms, the finite conductivity of any real dielectric material adds an i^2R component to its loss mechanisms. Insulation resistance will be taken up later.

The shapes of the ideal dissipation factor and permittivity spectra were developed by Debye [95] based on a spherical dipolar molecule orienting and reorienting in a viscous medium. The loss factor goes through a maximum for every polarization mechanism due to the dissipation of field energy as frictional heat when the bound charges or dipoles or space charges move toward the polarized condition. The absorption peak is greatest for materials showing the greatest change in dielectric constant in passing through a dispersion region. Generally, the greater the number of polarization mechanisms in a given frequency region, the greater the dissipation factor of a dielectric.

C. Insulation Resistance

The energy gap for electronic conduction is so large in the compounds constituting the classical ceramic dielectrics—alumina, steatite, mica, and porcelain—that the most important conduction mechanism is ion migration. But generally even at the highest temperatures of usual capacitor use, ion conduction is not significant. In the case of glasses, however, or of a crystalline material with an impurity glassy phase, electrolytic conductivity by ions can be substantial. A commercial soda-lime-silica glass, for example, has conductivity of about 10^{-8} $(\Omega\text{-cm})^{-1}$ at 150°C. This type of glass is avoided in capacitors. Rather, devitrifying glasses are employed, or alkali-containing glasses whose alkali ions have been immobilized by associations with other network or modifier ions.

Conductivity behavior is of particular importance in the case of barium titanate, the prototypical material of most ceramic dielectrics. $BaTiO_3$ has a smaller band gap (3.0 eV) than do the classical materials mentioned above, but is intrinsically a good insulator. However, its crystal structure lends itself to several varieties of defects, any of which can lead to high electronic conductivity. Usually the carriers are n type, their presence associated with oxygen vacancies [96,97]. The oxygen vacancies, in turn, are due to dissociation at elevated processing temperatures or to aliovalent acceptor dopants present in the ceramic as raw material impurities or acquired during processing. For example, Al_2O_3, providing trivalent aluminum in substitution for Ti^{4+}, is a common $BaTiO_3$ raw material impurity and can also be acquired from attrition of mill jars or grinding media. Early in the technology of barium titanate-based capacitors, without the benefit of knowledge of its solid-state chemistry, it was recognized that addition of niobia is beneficial in increasing the resistance and prolonging the lifetime under dc load of a ceramic capacitor.

D. Field-Induced Degradation

More than by the short time value of its resistance on application of voltage, a capacitor is judged by the stability of the resistance value under sustained dc field and at elevated temperature. Figure 14 is a useful qualitative depiction of the varieties of response to electric field and temperature that are possible.

At low temperature and high field, the short horizontal line at the lower right of the diagram represents the weakly temperature dependent intrinsic breakdown strength. At higher temperatures, however, the breakdown voltage decreases rapidly with increasing temperature. An additional important aspect is that the breakdown voltage at a given temperature is time dependent (i.e., a mode of failure by degradation of the dielectric is recognized). In this mode the resistance of the capacitor decreases continuously with time until runaway i^2R heating causes thermal breakdown. The activation energy for conduction remains constant whatever the duration of the degradation process. This form of degradation has been a significant problem in the commercialization of high-dielectric-constant titanate ceramic dielectrics. As an example of the early magnitude of the problem, Gruver et al. [98] describe a 1951 titanate-based capacitor dielectric formulation with an initial k' value of 100. Under 1.1 kV/mm at 150°C, the resistivity of this material decreased by one order of magnitude in 500 hr; in contrast, a current commercial material [99] with a k' value of 1000–1500 is described as showing no measurable resistance change in 500 hr at 125°C and 10 kV/mm.

Aspects of the detailed mechanism of the degradation process are not clear. In general terms it can be described as follows. Oxygen vacancies are produced both by the quenching-in of dissociation occurring at high temperature during firing of the ceramic and by an electrolytic process in which vacancies are injected at the anode and diffuse under field toward the cathode. Ionic conduction has been reported to occur at as low as 100°C in $BaTiO_3$ [100]. The oxygen vacancies' (+2) charge is compensated by 3d electrons of Ti^{4+}, producing some content of Ti^{3+} ions. This introduces an additional, electron exchange conduction mechanism.

E. Dielectric Strength

The rated operating voltage of a capacitor is an engineering parameter that takes into account the intrinsic dielectric strength of the dielectric material as well as the contribution of ambient temperature to thermal breakdown. A design factor to ensure function over the specified lifetime is included.

The intrinsic dielectric strength is measured at low temperatures where current heating is not a factor. The mechanism of breakdown

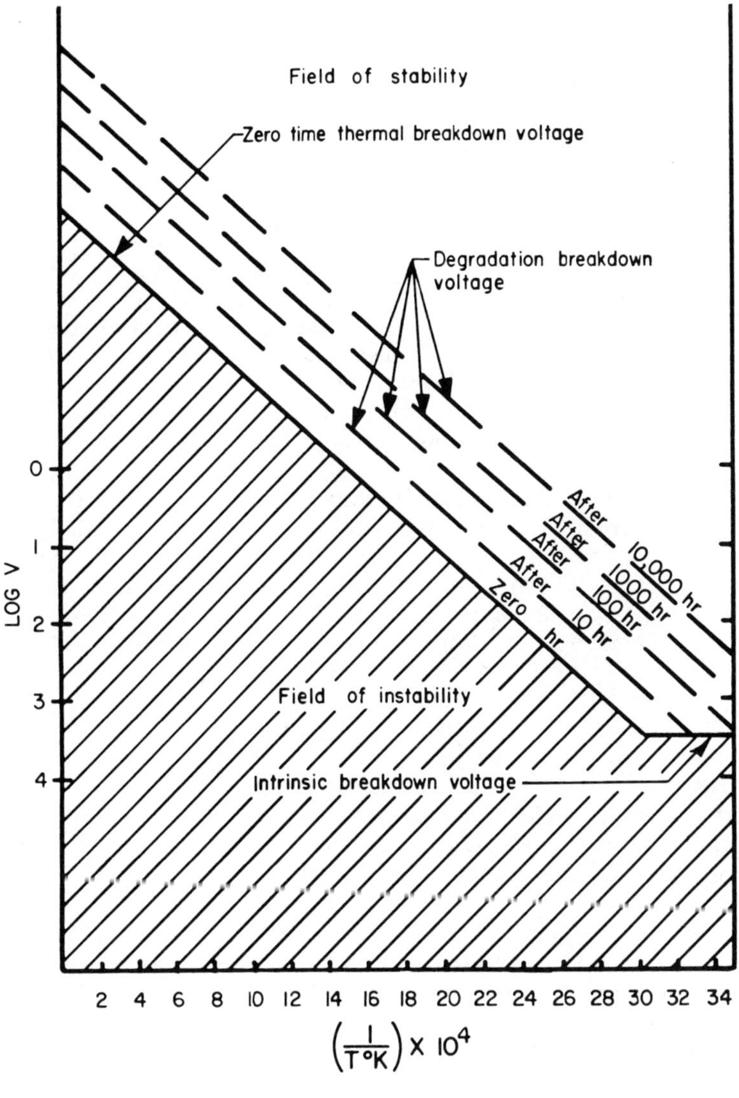

Figure 14 Ceramic dielectric degradation and breakdown under voltage. (From Ref. 98.)

is the emission of electrons under the applied field, which, in turn, free more electrons, ultimately producing an electron avalanche which punctures the test specimen. The field at which breakdown occurs is nearly temperature independent. Experimental determination of the breakdown field strength is complicated by edge effects or other inhomogeneity of the field experienced by the test specimen, by the geometry and materials of the electrodes used, the specimen thickness, the presence of pores or phase boundaries in the ceramic, and so on. The most reliable tests are performed with the specimen immersed in a dielectric fluid and using a specimen cross section especially contoured to give uniform field. More relevant to the conditions under which capacitors are used, dielectric breakdown tests commonly are carried out at room or elevated temperature. In these cases the field strength at which breakdown occurs is governed by leakage resistance and the rate of heat generation in the test specimen relative to its rate of heat dissipation. Breakdown values therefore can vary widely depending on the test conditions. Even with nominally identical conditions, microstructural variations in test specimens necessitate care that breakdown test results be statistically meaningful.

In summary, the maximum voltage-withstanding ability of a ceramic capacitor is governed not only by the intrinsic breakdown strength of its dielectric material but also by the capacitor's design and construction.

F. Aging and Piezoelectricity

Two related capacitor performance characteristics, piezoelectricity and aging, stem from the ferroelectric nature of $BaTiO_3$. Piezoelectricity, by definition, accompanies ferroelectricity; that is, in the crystal symmetry classes in which ferroelectricity can occur, so does piezoelectricity. In a ceramic with randomly oriented grains and domains of $BaTiO_3$, there is no net piezoelectric effect. Should the capacitor, however, be subjected to a dc field of the order of 1 kV/mm, such as is experienced in a voltage-withstanding test, domain alignments can impart a net crystallographic orientation to the ceramic. Then the piezoelectric effect, direct and inverse, is experienced. The direct piezoelectric effect can be objectionable in low-voltage circuits, generating spurious voltages when the capacitor experiences a shock or acceleration. Preventives of the condition are to avoid field strengths strong enough for poling, to anneal out any residual polarization, or to employ ceramic compositions that display minimal piezoelectric characteristics. Examples of such compositions are fine-grained, nearly cubic $BaTiO_3$ or a ceramic whose ferroelectric phases have Curie points below the operating temperature range.

The piezoelectric effect in dielectric materials is subject to aging, an effect discussed below in relation to permittivity.

G. Aging of Dielectric Characteristics

Capacitors of Electronic Industries Association's class 2 are subject to a decrease of permittivity and of dissipation factor with time. This "aging" is a ferroelectric phenomenon and begins following the last heating above the Curie temperature. This may occur when end terminations are affixed, polymer encapsulants cured, or when capacitor chips are soldered into a circuit.

As a manifestation of ferroelectricity, aging is most pronounced in the higher-permittivity dielectrics. A typical aging characteristic for an X7R-type capacitor is shown in Figure 15. A capacitor with the Z5U characteristic, having three to four times higher permittivity, would age three to four times faster than the X7R. The tan δ value also ages, decreasing several times faster than the capacitance. Solid-solution dielectrics with axial ratios close to unity age less.

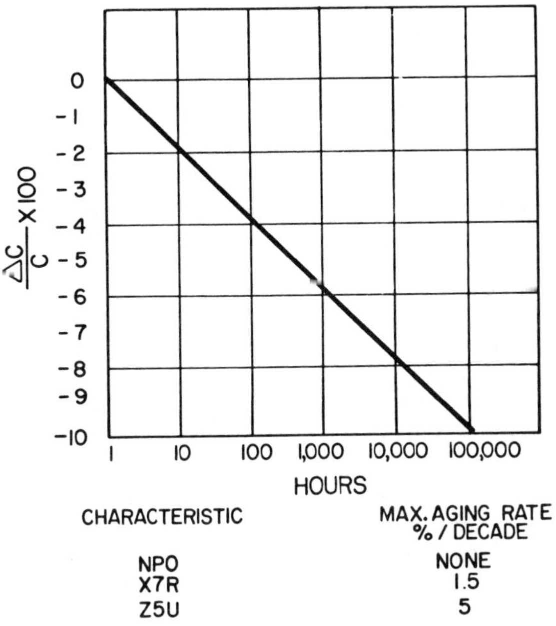

Figure 15 Aging of capacitance in ceramic capacitors. (From Ref. 101.)

Fine-grained dielectrics age less than coarse-grained material of the same composition.

The aging phenomenon has been the subject of study since the application of barium titanate in capacitors. This is understandable since the phenomenon effectively determines the tolerance within which a capacitance value can be guaranteed. The present perception of aging is primarily that of a relaxation of mechanical stresses. The stresses are generated by the strain of the cubic-to-tetragonal symmetry phase change; their relaxation takes place through the nucleation and growth of 90° domains [13]. The means of suppressing aging, therefore, are to inhibit domain nucleation by limiting the grains to submicron size and/or to reduce the c/a unit cell ratio and the associated intergrain stresses through the use of solid-solution compositions.

IX. PACKAGING OF CERAMIC CAPACITORS

Some degree of protection against the use environment is required for all ceramic capacitors. Without it, the dielectric characteristics of their hydrophilic surfaces would color those of the dielectric material itself. The protection can be applied to individual capacitors, to capacitor-containing subcircuits, or to entire circuits. In detail, the purposes of packaging are:

1. To exclude contaminants, particularly mobile ions, which can contribute to the surface conductivity or, if absorbed by a less than totally dense ceramic, to the bulk conductivity. In addition, where a coincidence of silver metallization, moisture, and electric field might lead to the generation of silver ions and their injection into and migration in and on the ceramic, coating or packaging excludes moisture and prevents this degradation. Proton injection has been implicated in the long-term degradation of dielectrics [98] and waterproof packaging forestalls that process as well.

2. To protect against mechanical mishandling, thermal shock, detachment of leads, and so on. Where the dielectric possesses some residual piezoelectricity, a shock-absorbent package insulates it from voltage-generating stresses.

3. To standardize. The package commonly takes forms that have become standard for other electronic components. This enables the use of common equipment and methods for placement on and attachment to circuit boards. Packages may take one or several of the following forms:

a. "Self-packaging," employed with multilayer capacitors in which the electrodes are totally internal except where their alternate sets of edges are exposed. These surfaces are metallized with a glass frit/powdered metal combination which serves as packaging as well as in a current collecting function. The two metallized faces usually are then solder coated, either as the final fabrication step for a leadless capacitor, or during attachment of soldered leads, or when a leadless capacitor is soldered to its circuit board. Ceramic capacitors of the self-packaged type may also be impregnated or coated with a silicone resin or other moisture repellent.

b. Axially leaded multilayer capacitors can be packaged in cylindrical sealed glass tubes of the type used for diodes and discrete transistors. Insertion of the capacitors into the tubes and bonding of the glass to the wire lead at each end of the device are done automatically.

c. Dip or conformal insulating coatings are used very commonly on single-layer (i.e., disk or wafer) leaded capacitors. Phenolic or epoxy thermoset coatings are most used in this low-cost packaging method. They are applied by immersing the part in a slurry of the monomer, draining, and heating to cross-link the coating material. An alternative approach is the fluidized-bed technique. The dry powdered coating material is fluidized with an airstream and allowed to impinge on the component, which has been preheated above the melting point of the coating material. The powder particles fuse together to coat the capacitor. Additional heating is sometimes needed to smooth the coat or cure the polymer.

d. When the capacitor is to be a leaded device with precisely defined exterior shape and dimensions, the packaging method may be one of the following:

1. *Potting in a premolded case.* Injection-molded cases of epoxy or silicone resin are common. After insertion of the body of the leaded capacitor, the case is filled with a liquid potting resin such as an epoxy. Commonly, an intermediate elastomer coating is applied to the part before potting. This takes up strains due to differences of thermal contraction of the ceramic and the potting material which might otherwise be great enough to detach end termination metallization or soldered-on leads. The shock-absorbing encapsulant also helps prevent generation of spurious voltages, as described previously.
2. *Molding in dies using high- or low-pressure techniques.* In liquid resin molding, the resin formulation is injected at 5–50 psi into a hot ($\leqslant$100°C) mold containing the capacitor. In compression or transfer molding, the temperature is at least 50°C higher and the parts being jacketed can experience 10,000 psi. Phenol formaldehyde, epoxy, or silicone resins are the most common. Capacitors can be packaged individually or in multiple units in industry-standard integrated circuit housings [single in-line (SIL) dual in-line (DIL)].

(a)

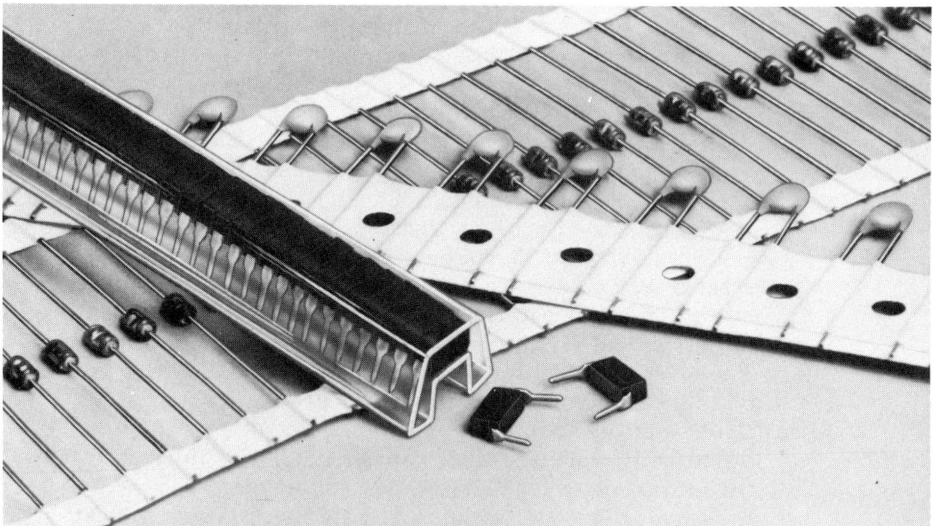

(b)

Figure 16 Ceramic capacitors packaged for automated assembly. (Courtesy of Centralab Inc.)

A. Capacitors for Automated Hybrid Circuit Assembly

The current trend in hybrid microelectronics technology is to mount discrete resistors, capacitors, and semiconductor devices, in the form of "chips," on both sides of substrates containing printed wiring patterns. The substrates may themselves be multilayer devices. Placement of the chips is automated and an adhesive holds them in place until they are soldered en masse. In line with this assembly concept, capacitor manufacturers offer components in forms ready for loading the automated assembly equipment. Once such form is a perforated paper tape with the dimensions of slit 8mm motion-picture film. The tape is three layered. Holes punched in a thick middle layer accommodate the chips being dispensed, with adhesive tape placed below and on top to contain the chips. The strip with its components is wound on standard film reels, a 7-in.-diameter reel containing about 4000 capacitor chips (Figure 16a). In older forms of packaging (Figure 16b) for automatic insertion, leaded capacitors held between strips of adhesive tape are prepared as "sticks" or wound into rolls. In the insertion machines the leads are cut free of tape, bent to shape, and the parts are "stuffed" into their boards and subsequently soldered in place. For yet another form of automatic insertion machinery, units packaged in SIL and DIL molded cases are supplied in preloaded stick magazines. One such stick can contain several hundred molded units.

APPENDIX: TYPICAL CERAMIC DIELECTRIC COMPOSITIONS

Levinson et al. [102] have presented lucidly the mix of science and empiricism that goes into the formulation of ceramic dielectrics to meet EIA or military capacitor specifications. In terms of volume of use, the most important categories of dielectrics are the Z5U, X7R/BX, and the NPO. Payne [103] has estimated the relative use of these in the United States as follows: NPO, 10%, BX/X7R, 40%, and Z5U, 50%. Most of these materials are based on $BaTiO_3$. The NPO-type is the class least likely to display any vestigial ferroelectric characteristics. Figure 17 illustrates the dependence of permittivity on temperature of representatives of the three classes of materials.

Z5U Dielectrics

For maximum capacitance the $BaTiO_3$ Curie peak has been shifted close to room temperature and broadened. Shifting is accomplished

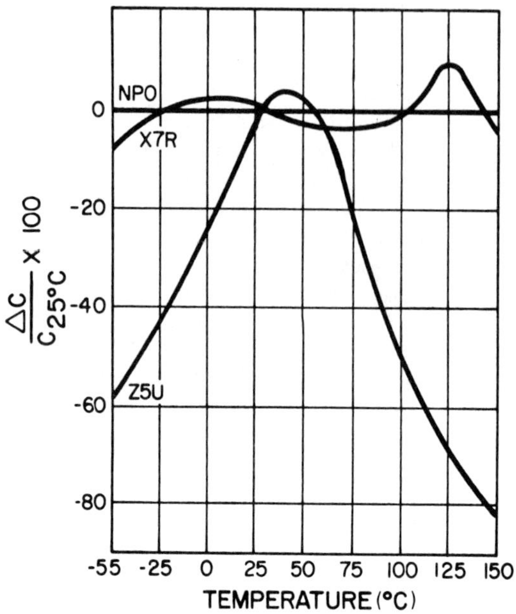

Figure 17 Temperature coefficient of capacitance versus temperature for NPO, X7R, and Z5U capacitors. (From Ref. 101.)

with solid-solution additives (e.g., $SrTiO_3$) and alkaline earth zircontates and stannates. Broadening is accomplished by superimposition of a number of noncoincident peaks and by depressor additives, for example, aliovalent materials, which tend not to form solid solutions but to concentrate in grain boundaries: Bismuth stannate, titanate, zirconate, or oxide, titanates of iron, nickel, copper or manganese, and magnesium zirconate and sodium niobate. Typical Z5U formulations are presented in Table 1 and X7R/BX formulations are presented in Table 2.

NPO Dielectrics

Originally based on TiO_2, these compositions now tend to include $BaTiO_3$ to counter the negative temperature coefficient of rutile and to increase the permittivity. To a degree, rare earth oxides are used for the same purpose. NPO dielectrics appear microstructurally to be, not solid solutions of $BaTiO_3$, but other titanate compounds [104]. As such they do not display ferroelectric characteristics.

The compositions of Tables 1 to 3 represent basic capacitor dielectric compositions, in most of which the characteristics of

Table 1 Z5U Dielectric Compositions

	Composition (wt %)			
	1	2	3	Comment
$BaTiO_3$	84–90	65–80	72–76	Base material
$CaZrO_3$	8–13	–	–	Shifter
$MgZrO_3$	0–3	–	–	Depressor
$SrTiO_3$	–	7–11	5–8	Shifter
$CaTiO_3$	–	7–11	4–6	Depressor
$BaZrO_3$	–	7–11	7–10	Shifter
$CaSnO_3$	–	–	2–4	Shifter
Other[a]	1–3	8–13	0–3	
k' (25°C, 1 kHz)	5700–7000	5500–6500	11,500–13,000	
tan δ	$\leqslant 0.03$	$\leqslant 0.03$	$\leqslant 0.03$	

[a]Among the miscellaneous additives are acceptor materials such as Nb_2O_5, Sb_2O_5, Ta_2O_5, or rare earth oxides, to compensate n-type depressors and impurities.
Source: Ref. 102.

Table 2 X7R/BX[a] Dielectric Compositions

	Composition (wt %)			
	1	2	3	Comment
$BaTiO_3$	90–97	85–92	86–94	Base material
$CaZrO_3$	2–5	4–8	—	Shifter
$BaCO_3$	0–5	—	—	Stoichiometry adjustment
$SrTiO_3$	—	3–6	—	Shifter
Bi_2O_3	—	—	5–10	Depressor, flux
Other	2–5	1–4	2–6	See note a in Table 1
k' (25°C, 1 kHz)	1600–2000	1800	1400–1500	
tan δ	<0.025	<0.025	<0.015	

[a] X7R is the EIA designation; BX is the U.S. military designation for capacitors with substantially the same characteristics. The principal difference is more rigorous BX requirements for life-test performance and voltage coefficient of capacitance.

Source: Ref. 102.

Table 3 NPO Dielectric Compositions

	Composition (wt %)		
	1	2	3
$BaTiO_3$	41–49	39–47	15–21
TiO_2	48–54	41–47	26–34
ZrO_2	–	8–13	–
$Nd(CO_3)_4$	–	–	59–45
Other	3–7	2–5	Small
k' (25°C, 1 kHz)	35	30	62
tan δ	<0.002	<0.002	<0.002

Source: Ref. 102.

ferroelectric $BaTiO_3$ have been shaped to the required properties. They are the most widely used types of compositions. The more recent approaches based on relaxator materials, crystallizing glasses, base metal–compatible dielectrics, or grain boundary polarization effects, which have been described in the text, as yet represent minority methods.

REFERENCES

1. *Encyclopedia Britannica, 10*:136 (1980).
2. H. Thurnauer, *Am. Ceram. Soc. Bull., 56*:219 (1977).
3. H. Thurnauer, *Am. Ceram. Soc. Bull., 56*:861 (1977).
4. J. C. Maxwell, *Electricity and Magnetism*, Vol. 1, Clarendon Press, Oxford, 1892.
5. S. P. Mitoff, *Adv. Mater. Res., 3*:305 (1968).
6. K. Lichtenecker, *Phys. Z., 10*:1005 (1909).
7. L. K. H. Ven Beek, *Prog. Dielectr.* 7:71 (1967).
8. B. M. Wul and F. L. Vereschagen, *Dok. Akad. Nauk SSSR, 48*:634 (1946).
9. K. Okazaki, *Am. Ceram. Soc. Bull., 61*:932 (1982).
10. A. R. Von Hippel, *Dielectrics and Waves*, Wiley, New York 1954, p. 203.
11. W. J. Merz, *Phys. Rev., 76*:1221 (1949).

12. E. N. Bunting, G. R. Shelton and A. S. Creamer, *J. Res. Natl. Bur. Std.*, *38*:337 (1947).
13. W. R. Buessem, L. E. Cross and A. K. Goswami, *J. Am. Ceram. Soc.*, *49*:33 (1966).
14. E. Wainer and C. Wentworth, *J. Am. Ceram. Soc.*, *35*:207 (1952).
15. W. W. Coffeen, *J. Am. Ceram. Soc.*, *37*:480 (1954).
16. K. W. Plessner and R. West, *Prog. Dielectr.*, *2*:167 (1960).
17. E. C. Subbarao, *Non-stoichiometric Compounds*, Academic Press, New York, 1954, p. 296. Cited on p. 90, end of line 8.
18. T. C. Rutt, *Am. Ceram. Soc. Bull.*, *47*:873 (1968).
19. G. A. Smolenskii and V. A. Isupov, *Dok. Akad. Nauk SSSR*, *97*:653 (1954).
20. S. Swartz, T. Shrout, and C. Otteson, *Am. Ceram. Soc. Bull.*, *61*:359 (1982).
21. S. J. Jang, L. E. Cross and K. Uchino, *J. Am. Ceram. Soc.*, *64*:209 (1981).
22. *Electronic Industries Association RS-198, American Standard Requirements for Ceramic Dielectric Capacitors, Classes 1 and 2*, American Standards Association, New York, 1958.
23. R. W. Berry, P. M. Hall and M. T. Harris, *Thin Film Technology*, Van Nostrand Reinhold, New York, 1968, p. 371.
24. J. V. Biggers, G. L. Marshall and D. W. Strickler, *Solid State Technol.* , *13*:63 (1970).
25. L. C. Hoffman, U.S. Patent 3,656,984, 1972.
26. L. C. Hoffman and T. Nakayama, U.S. Patent 3,666,505, 1972.
27. R. J. Bacher, U.S. Patent 4,071,881, 1978.
28. R. J. Bacher, U.S. Patent 4,089,038, 1978.
29. E. Barringer and H. K. Bowen, *Am. Ceram. Soc. Bull.*, *61*:336 (1982).
30. G. N. Howatt, R. G. Breckenridge and J. M. Brownlow, *J. Am. Ceram. Soc.*, *30*:237 (1947).
31. A. Rodriguez, U.S. Patent 3,004,197, 1961.
32. A. J. Deyrup, U.S. Patent 2,389,420, 1945.
33. K. H. Ballard, U.S. Patent 2,395,442, 1946.
34. A. J. Deyrup, U.S. Patent 2,413,549, 1946.
35. G. N. Howatt, U.S. Patent 2,486,410, 1949.
36. C. I. Bradford, B. L. Weller and S. A. McNeight, *Electronics*, *20*:106 (Dec. 1947).
37. P. W. Lee and B. L. Weller, U.S. Patent 2,779,975, 1957.
38. T. C. Rutt, U.S. Patent 3,679,950, 1972.
39. T. C. Rutt, U.S. Patent 4,071,880, 1978.
40. T. C. Rutt and J. A. Stynes, *IEEE Trans. Parts, Hybrids, Packag. PHP-9*, 144 (1973).
41. G. Goodman, *Adv. Geram.*, *1*:230 (1981).
42. L. C. Hoffman, U.S. Patent 3,390,981, 1968.
43. L. F. Miller, *Proc. Electron. Components Conf.*, 1968, p. 52.

44. R. E. Newnham, *Proc. Workshop Reliab. Multilayer Ceram. Capacitors*, U.S. Dept. of Commerce, 1982.
45. S. Padwal and H. J. deBruin, *J. Electrochem. Soc.*, *129*:1921 (1982).
46. J. E. Sheard, U.S. Patent 3,872,360, 1975.
47. J. Piper, U.S. Patent 3,612,963, 1971.
48. G. Maher, U.S. Patent 3,619,220, 1971.
49. G. Maher, U.S. Patent 3,682,766, 1972.
50. G. Maher, U.S. Patent 3,811,937, 1974.
51. I. Burn, *J. Mater. Sci.*, *17*:1398, 1982.
52. D. A. Payne and S. M. Park, U.S. Patent 4,158,219, 1979.
53. D. Payne and S. M. Park, U.S. Patent 4,218,723, 1980.
54. A. Herczog and S. D. Stookey, U.S. Patent 3,195,030, 1965.
55. A. Herczog and M. Layton, U.S. Patent 3,490,887, 1970.
56. G. A. Smolenskii and A. I. Agranovskaya, *Sov. Phys.—Solid State*, *1*:1429 (1960).
57. V. A. Bokov and I. E. Myl'nikova, *Sov. Phys.—Solid State*, *3*:613 (1961).
58. O. Iizawa, S. Fujiwara, H. Ueoka, et al., U.S. Patent 4,235,635, 1980.
59. H. Tanei, A. Ikegami, N. Taguchi, et al., U.S. Patent 4,308,571, 1981.
60. J. M. Herbert, U.S. Patent 2,750,657, 1956.
61. J. M. Herbert, *Trans. Br. Ceram. Soc.*, *62*:645 (1963).
62. C. E. Wicks and F. E. Block, *U.S. Bur. Mines Bull. 605*, 1963.
63. I. Burn and G. H. Maher, *J. Mater. Sci.*, *10*:633 (1975).
64. N. H. Chan and D. M. Smyth, *J. Electrochem. Soc.*, *123*:1584 (1976).
65. J. Daniels, *Philips Res. Rep.*, *31*:505 (1976).
66. N. G. Eror, I. Burn and G. Maher, U.S. Patent 3,920,781, 1975.
67. K. W. Wagner, *Arch. Elektrotech.*, *2*:371 (1914).
68. J. Volger, *Prog. Semiconduct.*, *4*:207 (1960).
69. A. Hornung, *Proc. Electron. Components Conf.*, 1968, p. 250.
70. R. R. Roup and C. E. Butler, U.S. Patent 2,520,376, 1950.
71. R. R. Roup and J. S. Kilby, U.S. Patent 2,841,508, 1958.
72. O. Saburi, U.S. Patent 3,299,332, 1967.
73. W. Noorlander, U.S. Patent 3,386,856, 1968.
74. S. Waku, U.S. Patent 3,473,958, 1969.
75. H. Brauer and R. Kuschke, U.S. Patent 3,569,802, 1971.
76. H. Brauer and R. Kuschke, Ger. patent 1,614,605, 1974.
77. S. Waku, A. Nishimura, T. Murakami, A. Yamaji, T. Edahiro and M. Uchidate, *Rev. Electr. Commun. Lab.*, *19*:665 (1971).
78. W. Heywang, *J. Mater. Sci.*, *6*:1214 (1971).
79. J. Daniels and K. H. Härdtl, *Philips Res. Rep.*, *31*:489 (1976).
80. J. Daniels, *Philips Res. Rep.*, *31*:505 (1976).
81. D. Hennings, *Philips Res. Rep.*, *31*:516 (1976).

82. J. Daniels and R. Wernicke, *Philips Res. Rep.*, *31*:544 (1976).
83. H. Ihrig and W. Puschert, *J. Appl. Phys.*, *48*:3081 (1977).
84. R. Wernicke, *Phys. Status Solidi*, *47*:139 (1978).
85. H. P. Frederikse, W. R. Thurber and W. R. Hosler, *Phys. Rev.*, *134*:A442 (1964).
86. H. P. Frederikse and W. R. Hosler, *Phys. Rev.*, *161*:822 (1967).
87. K. Ozawa, Japan Early Patent Disclosure 77-10596, 1977.
88. I. Burn and S. Neirman, *J. Mater. Sci.*, *17*:3510 (1982).
89. R. Wernicke, *Adv. Ceram.*, *1*:261 (1981).
90. H. Schmelz, *Powder Metall, Int.*, *7*:176 (1975).
91. S. M. Park and D. A. Payne, *Am. Ceram. Soc. Bull.*, *58*:732 (1979).
92. M. E. Lines and A. M. Glass, *Principles and Applications of Ferroelectrics and Related Materials*, Clarendon Press, Oxford, 1977.
93. B. Jaffe, W. R. Cook and H. Jaffee, *Piezoelectric Ceramics*, Academic Press, New York, 1971.
94. E. J. Murphy and S. O. Morgan, *Bell Sys. Tech. J.*, *16*:493 (1937).
95. P. Debye, *Polar Molecules*, Chemical Catalog Company, New York, 1929.
96. N. H. Chan and D. M. Smyth, *J. Electrochem. Soc.*, *123*:1584 (1976).
97. S. A. Long and R. N. Blumenthal, *J. Am. Ceram. Soc.*, *54*: 577 (1971).
98. R. M. Gruver, W. R. Buessem, C. W. Dickey, et al., *State of The Art Review on Ferroelectric Ceramic Materials*, U.S. Air Force Tech. Rep. AFML-TR-66-164, 1966, p. 112.
99. G. H. Maher, *Proc. Electron. Components Conf.*, 1977, p. 391.
100. D. D. Glower and R. C. Heckman, *J. Chem. Phys.*, *41*:877 1964).
101. Technical Bulletin, "The Capacitor," AVX Corp., Myrtle Beach, S.C., p. 7.
102. S. Levinson, H. U. Anderson, W. Payne, et al., *Multilayer Ceramic Capacitors—Notes for Professional Development Program*, AIChE, New York, 1980, pp. 3–66 - 3–76.
103. S. Levinson, H. U. Anderson, W. Payne, et al., *Multilayer Ceramic Capacitors—Notes for Professional Development Program*, AIChE, New York, 1980, p. 4–162.
104. R. L. Bolton, "Temperature Compensating Ceramic Capacitors in the System Baria–Rare Earth Oxide–Titania," Ph.D. thesis, University of Illinois, 1968.

3

Piezoelectric and Electrooptic Ceramics

GENE H. HAERTLING
Motorola, Inc., Albuquerque, New Mexico

I. INTRODUCTION

Ceramics have long been a part of human life, almost as long as man has existed, yet it is only within the last 50 years that this segment of the "world of materials" has begun to mature and achieve a sound, scientific basis. Why this is true is not easily understood, but one reason is that ceramics are highly complex, less than ideal materials which do not lend themselves well to theoretical investigation. For many years, theoreticians avoided the study of ceramic materials, their properties and their behavior, and understandably, preferred instead to treat the material phenomena associated with metals, single crystals, and the like. In our present highly technological age, it is not possible to ignore ceramics, however complex they are, since more and more applications are being found for these materials in devices affecting our everyday lives in such areas as electronic components, environmental sensors, gas ignitors, intrusion alarms, loudspeakers, ultrasonic cleaners, and medical diagnostic equipment. In fact, many of these applications have come about as a result of the special nature and properties of ceramics, which set them apart from other materials such as metals or plastics.

This special nature and properties of ceramics is treated in detail in following sections, but the relationship of ceramics to other materials can readily be illustrated as shown in Figure 1. In general, all solid materials can be grouped into three major categories: metals, organics (plastics), and inorganics. Similarly, the inorganic solids may be subdivided into the three categories of glasses, single crystals, and ceramics, depending on their microscopic and macroscopic structure. Glasses are distinguished from other inorganic materials by their amorphous nature or lack of long-range internal structure, whereas single crystals are characterized by a well-ordered internal structure on both a long-range (10^{-4}–1 cm) and a short-range (10^{-8}–10^{-4} cm) scale. Ceramics* differ from both of these materials in that they are composed of an aggregate of randomly oriented crystallites (small single crystals of less than 100 μm or 10^{-2} cm average diameter) intimately bonded together to form a solid. Although the crystallites do possess an ordered internal structure just as single crystals do, their random orientation produced during the high-temperature sintering or coalescing operation

*The general field of ceramics is sometimes regarded to cover a wide range of inorganic materials, including glasses and single crystals; however, we here are using the term "ceramics" in the stricter sense of an inorganic, polycrystalline solid.

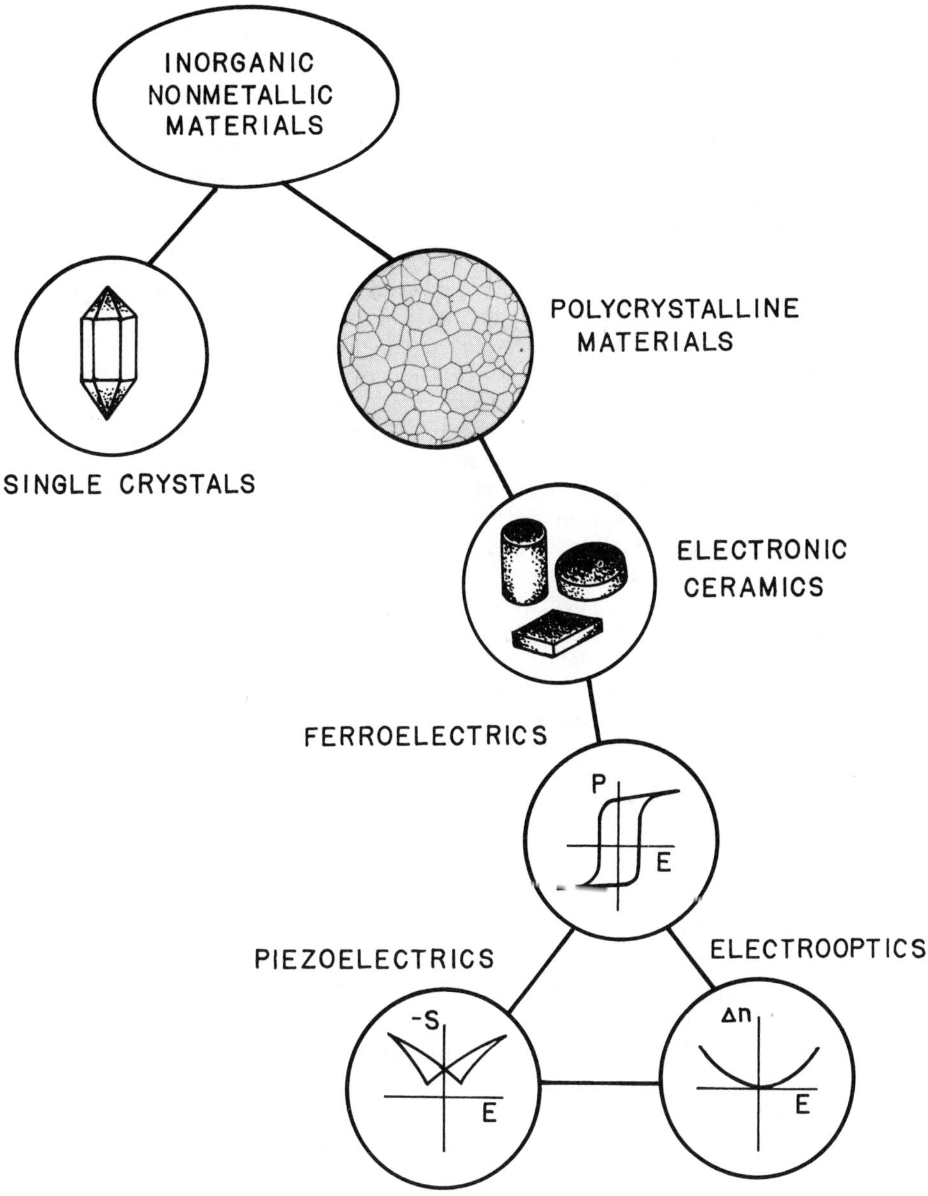

Figure 1 Relationship of ceramics and electronic subclasses to all inorganic, nonmetallic solids.

causes the ceramic to have an overall isotropic (equal in all directions) character in regard to its properties. This isotropic character can be modified, if desired, by special techniques employed during the sintering operation (crystallite orientation by mechanical, electrical, or thermal crystallization means) or after cooling to room temperature. Many of the unique properties of ceramics are thus a direct result of the properties of the small crystallites themselves, their orientation with respect to each other, or the grain (crystallite) boundaries between them.

Ceramics may encompass a wide range of polycrystalline materials; however, our interests here are centered on those types known as electronic ceramics. These materials are highly specialized, being prepared from specifically formulated compositions (usually not found in nature), processed under controlled conditions and fabricated into complex shapes with specific properties, particularly electrical properties. Among these are included insulators, ferrites, capacitors, substrates, varistors, PTCs (positive-temperature-coefficient materials), ferroelectrics, pyroelectrics, piezoelectrics, and electrooptic ceramics. We deal here specifically with piezoelectrics, ferroelectrics, and electrooptic materials. The inclusion of piezoelectrics under the category of ferroelectrics as given in Figure 1 is not rigorous since piezoelectricity is the more general phenomena; however, today's usage of the term ferroelectric tends to be all-inclusive since most ceramics exhibiting pyroelectric, piezoelectric, and electrooptic behavior are ferroelectric. The actual relationship between these effects is discussed in succeeding sections.

II. PIEZOELECTRIC CERAMICS

A. Symmetry Considerations

Piezoelectricity is a property possessed by a select group of materials. It was discovered in 1880 by Pierre and Jacques Curie during their systematic study of the effect of pressure on the generation of electrical charge by crystals such as quartz, zinc blende, tourmaline, and Rochelle salt; however, the term "piezoelectricity" (pressure electricity) was first suggested by W. Hankel in 1881. Cady [1] defines piezoelectricity as "electric polarization produced by mechanical strain in crystals belonging to certain classes, the polarization being proportional to the strain and changing sign with it." Two effects are manifest in piezoelectricity: the direct effect and the converse or inverse effect. The direct effect is identified with the phenomenon whereby electrical charge (polarization) is generated from a mechanical stress, whereas the converse effect is associated with the mechanical movement generated by the application of an electrical field. Both effects are useful and have found application in present-day devices.

An understanding of the concept of piezoelectricity must necessarily begin deep down inside the structure of the material; and for our purposes here, let's consider a single crystal or single crystallite in the case of a ceramic. This crystal has a definite chemical composition, and hence is made up of ions (atoms with positive or negative charge) which are constrained to occupy positions in a specific repeating relationship to each other, thus building up the structure or lattice of the crystal. The smallest repeating unit of the lattice is called the unit cell, and the specific symmetry possessed by the unit cell determines whether or not it is possible for piezoelectricity to exist in the crystal. Furthermore, the symmetry of a crystal's internal structure is reflected in a symmetry of its external properties.

The elements of symmetry that are utilized by crystallographers to define symmetry about a point in space, for example, the central point of a unit cell, are (1) a point (center) of symmetry, (2) axes of rotation, (3) mirror planes, and (4) combinations of these. Utilizing these symmetry elements, all crystals can be divided into 32 different classes or point groups, as shown in Figure 2. These 32 point groups are subdivisions of seven basic crystal systems, which are, in order of ascending symmetry, triclinic, monoclinic, orthorhombic, tetragonal, rhombohedral (trigonal), hexagonal, and cubic. Of the 32 point groups, 21 classes do not possess a center of symmetry (a necessary condition for piezoelectricity to exist) and 20 of these are piezoelectric. One class, although lacking a center of symmetry, is not piezoelectric because of other combined symmetry elements. It can be seen that a lack of a center of symmetry is all important for the presence of piezoelectricity when one considers that a homogeneous stress is centrosymmetric and cannot produce an unsymmetrical result (e.g., a vector quantity such as electric polarization) unless the material lacks a center of symmetry whereby a net movement of the positive and negative ions with respect to each other as a result of the stress produces an electric dipole (polarization). For piezoelectricity, this effect is linear and reversible; the sign of the charge produced is dependent on the direction of the stress (tensile or compressive).

When one now considers a ceramic consisting of a multitude of tiny, piezoelectric crystallites at random orientation, quite a different situation from that of a single crystal is obtained. This random orientation causes the overall ceramic to be inactive, with no piezoelectricity detectable until some means is found to polarize the ceramic as a whole entity. With piezoelectric crystallites this is possible only through techniques such as extrusion, hot forging, or directional recrystallization which will orient the crystallites during fabrication.

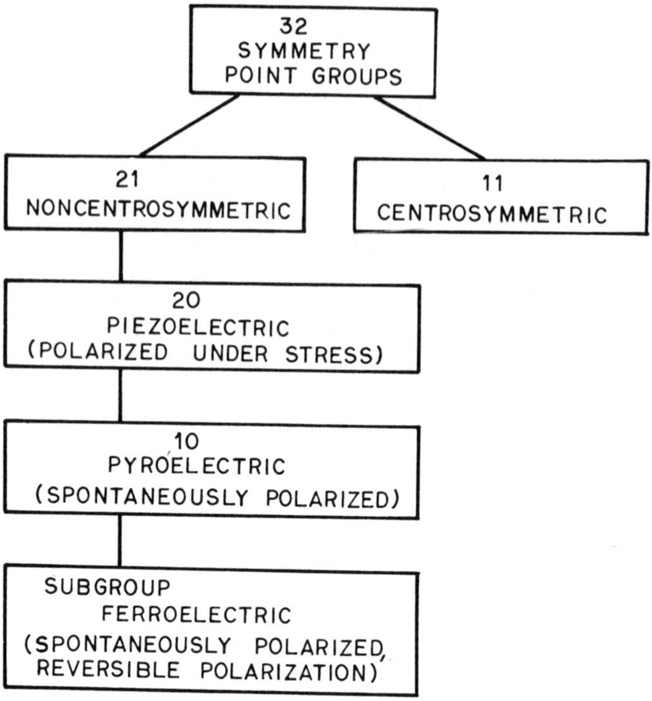

Figure 2 Interrelationship of piezoelectrics and subgroups on the basis of internal crystal symmetry.

A much easier method, called "poling," is commonly employed; however, it requires special compositions that are not only piezoelectric but also ferroelectric.

B. Basic Relationships

The equations of state that describe a piezoelectric ceramic in regard to its electric and elastic properties are, in their general form, as follows:

$$D = dT + e^T E \quad (1)$$

$$S = s^E T + dE \quad (2)$$

where D is the dielectric displacement (consider it equal to polarization), T the stress, E the electric field strength, S the strain, d a piezoelectric coefficient, s the material compliance (inverse of stiffness), and e the dielectric constant. Equation (1) describes the direct effect and Eq. (2) the converse effect. Both of these equations, in matrix form, actually describe a set of equations that relate these properties along different orientations of the material. Because of the detailed nature of the many equations involved, the reader is referred to several sources on the subject [2–5]. Suffice it to say that since we are dealing here with a piezoelectric solid, the equations relate a given property to the interacted mechanical and electrical states of the material.

A convenient method of uniquely specifying such directional properties is by the use of subscripts which define orientation directions as described in Figure 3. The 3 direction is usually taken as the poling (polar) axis and the 1 and 2 directions as orthogonal axes. Shear effects are indicated by the subscripts 4, 5, and 6. Coefficients that relate two variables (e.g., a d constant) are indicated with two subscripts, the first of which refers to the electrical direction (electric field or dielectric displacement) and the second to the mechanical direction (stress or strain). Also, when a given property is being measured while holding another quantity

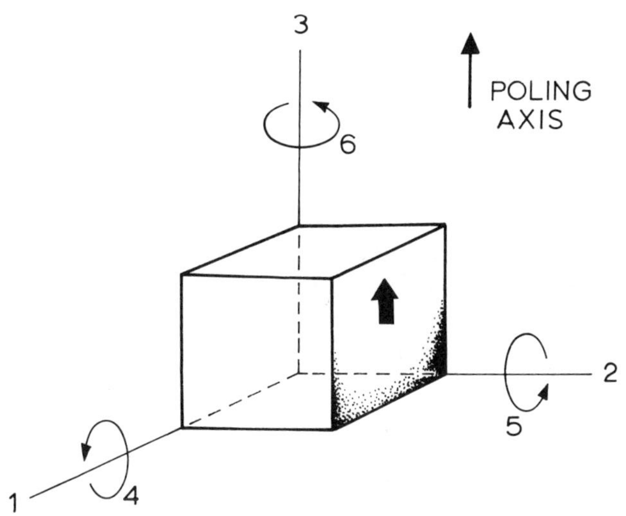

Figure 3 Notation of axes for poled ceramic element.

constant, a superscript symbol is used to indicate the quantity held constant. Typical examples of often used properties with appropriate subscripts or superscripts are given below:

d_{31}
- 1 → stress or strain direction
- 3 → displacement or electric field direction (electrodes are perpendicular to thc 3 axis

s^E = compliance at constant electric field (electrodes shorted)

s^D = compliance at constant displacement (open circuit)

ε^T = dielectric constant at constant stress (free)

ε^S = dielectric constant at constant strain (clamped)

d_{33} = charge or strain coefficient, longitudinal charge density/applied stress, C/N (strain developed/applied field, m/V)

d_{31} = charge or strain coefficient, lateral

s_{31} = mechanical compliance, lateral, m^2/N

g_{31} = voltage coefficient, lateral, V–m/N (open-circuit field/applied stress)

g_{15} = voltage coefficient, shear, V–m/N (open-circuit field perpendicular to 1 axis/applied shear stress around 2 axis)

K_3 = relative dielectric constant, longitudinal, no units ($K_3 = K_{33} = \varepsilon_{33}/\varepsilon_0$)

ε_0 = dielectric constant of free space, $\varepsilon_0 = 8.85 \times 10^{-12}$ F/m

k_{33} = electromechanical coupling factor, longitudinal, no units

k_{31} = electromechanical coupling factor, lateral, no units

k_t = electromechanical coupling factor, thickness, no units

k_p = electromechanical coupling factor, planar (radial), no units (thin disk only)

σ = Poisson's ratio of cross contraction, -0.3 approx., no units

ρ = material density, k_g/m^3

The piezoelectric charge coefficients (d_{33} and d_{31}) and the voltage coefficients (g_{33}, g_{31}, and g_{15}) describe the electromechanical parameters of the piezoelectric ceramic material. For both the direct and the converse effects, the d coefficient is the proportionality constant between displacement and stress, or strain and electric field, according to the relationships

$$d = \frac{D}{T} = \frac{S}{E} \tag{3}$$

where the d coefficient is numerically identical for both effects. High d coefficients are desirable for those materials which are utilized in motional or vibrational devices such as sonor and sounders. The g coefficient is related to the d coefficient via the dielectric constant according to

$$g = \frac{d}{K\varepsilon_0} = \frac{E}{T} = \frac{S}{D} \tag{4}$$

High g coefficients are desirable for materials intended to produce voltages in response to mechanical stress, such as in gas igniters and phonograph pickups.

The piezoelectric coupling factor (k_{33}, k_{31}, k_t, and k_p) is a convenient and direct measurement of the overall strength of the electromechanical effect (i.e., the ability of the ceramic to convert one form of energy to another). It may be defined as the square root of the ratio of energy available in electrical form to the total mechanical energy input (direct piezoelectric effect) or the square root of the ratio of energy available in mechanical form to total electrical energy input as in the converse effect. Since the conversion of electrical to mechanical energy (or vice versa) is always incomplete, k is always less than unity. For ceramics, k_p is a typical measurement in comparing materials, values ranging from 0.35 for barium titanate to 0.72 for lead zirconate titanate. It is most readily measured by determining the frequency difference between maximum impedance (parallel antiresonance, f_a) and minimum impedance (series resonance, f_r) of an electroded ceramic disk in an electrical resonant circuit. The most common coupling factors are calculated from the following equations:

$$k_{33} = \sqrt{\frac{\pi}{2} \frac{f_r}{f} \tan \left(\frac{\pi}{2} \frac{f_a - f_r}{f_a} \right)} \qquad \text{(cylindrical rods)} \qquad (5)$$

$$k_{31} = \sqrt{\frac{\frac{\pi}{2} \frac{f_a}{f_r} \tan \left(\frac{\pi}{2} \frac{f_a - f_r}{f_r} \right)}{1 + \frac{\pi}{2} \frac{f_a}{f_r} \tan \left(\frac{\pi}{2} \frac{f_a - f_r}{f_r} \right)}} \qquad \text{(rectangular plates)} \qquad (6)$$

$$k_p = \sqrt{(1 - k_p^2)\ \text{(Bessel function)}} \qquad \text{(disk only)} \qquad (7)$$

Because of its complexity, the solution to Eq. (7) is usually presented in graphical form as shown in Figure 4. Measured values of f_a and f_r are then used to determine k_p from the graph.

Other equations utilized in calculating piezoelectric material properties are as follows:

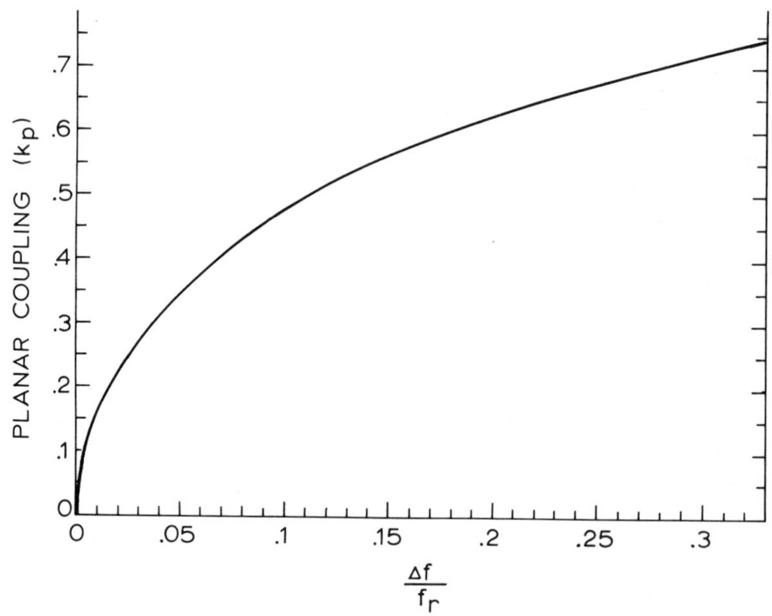

Figure 4 Planar coupling factor for poled thin disk.

$$s_{33}^{D} = \frac{1}{4\rho f_a^2 \ell^2} \qquad \text{(cylindrical rods)} \qquad (8)$$

$$s_{33}^{E} = \frac{s_{33}^{D}}{1 - k_{33}^2} \qquad \text{(cylindrical rods)} \qquad (9)$$

$$s_{11}^{E} = \frac{1}{4\rho f_r^2 \ell^2} \qquad \text{(rectangular plates)} \qquad (10)$$

$$s_{11}^{D} = s_{11}^{E}(1 - k_{31}^2) \qquad \text{(rectangular plates)} \qquad (11)$$

$$d_{33} = k_{33} \sqrt{\varepsilon_0 K_3 s_{33}^{E}} \qquad \text{(cylindrical rods)} \qquad (12)$$

$$d_{31} = k_{31} \sqrt{\varepsilon_0 K_3 s_{11}^{E}} \qquad \text{(rectangular plates)} \qquad (13)$$

$$g_{33} = \frac{d_{33}}{\varepsilon_0 K_3^T} \qquad \text{(cylindrical rods)} \qquad (14)$$

$$g_{31} = \frac{d_{31}}{\varepsilon_0 K_3^T} \qquad \text{(rectangular plates)} \qquad (15)$$

All of the properties above may be realized in a piezoelectric ceramic, which is, in reality, a poled ferroelectric ceramic material.

We have seen before that some convenient and reliable means must be employed to polarize the ceramic after fabrication in order to render it active. Such a means is called "poling" or polarizing the ceramic under the influence of an electric field. In this technique, electrodes are applied to the ceramic, and a sufficiently high electric field is applied to the material such that the dipoles within the individual crystallites are reoriented and aligned in the direction of the electric field. During this process, there is a small expansion of the material along the poling axis and a contraction in both directions perpendicular to it. An illustration of this process is shown in Figure 5. The strength of the electric poling field, often in combination with temperature, is an important factor in determining the extent of alignment and hence the resulting properties. In actuality,

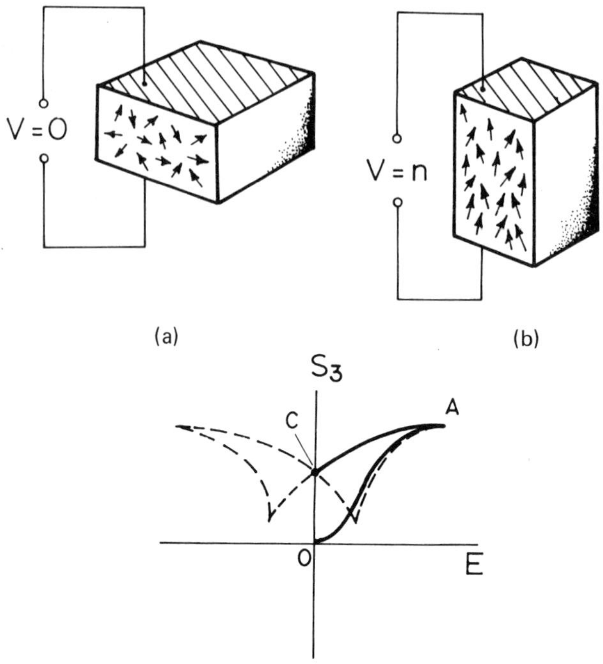

Figure 5 Dimensional changes of a ceramic as a result of poling: (a) before poling (virgin state); (b) after poling. The longitudinal strain as a result of poling and switching is illustrated in part (c), where the heavy black line traces the poling strain S_3 from a virgin state 0 to saturation state A and remanent state C.

alignment is never complete, but depending on the type of crystal structure involved, the thoroughness of poling can be quite high, ranging from 83% (tetragonal) to 91% (orthorhombic phase) of the single crystal [6]. Once poled, the ceramic acts much like a single crystal, with the whole body acting as a single entity. Since all ceramic bodies must be poled to render them useful as piezoelectric devices, we should note that they are ferroelectric as well as piezoelectric.

The symmetry elements of a poled ferroelectric (piezoelectric) ceramic are an axis of rotation of infinite order in the direction of poling (polar axis) and an infinite set of planes parallel to the polar axis as reflection planes [3]. In crystallographic notation this symmetry would be described as ∞mm and is equivalent to the dihexagonal polar crystal class, 6mm (class 26). For ferroelectric electrooptic ceramics, these same symmetry conditions hold, with the poling axis being collinear with the optic axis.

III. FERROELECTRIC CERAMICS

If we again refer to Figure 2, we see that there are 10 crystal classes, out of a possible 20, that are designated as pyroelectric. This group of crystals possesses the unusual characteristic of being permanently polarized within a given temperature range. Unlike the more general piezoelectric classes, which produce a polarization under stress, the pyroelectrics develop this polarization spontaneously and form permanent dipoles in the structure. This polarization also changes with temperature—hence the term *pyroelectricity*. Pyroelectric crystals such as tourmaline and wurtzite are often called polar materials, referring to the unique polar axis existing within the lattice. The length of the polar axis (dipole moment) varies with temperature, changing sign as the temperature is either elevated or lowered. Present-day pyroelectric devices utilize this effect in such applications as intrusion alarms, thermal imaging, and geographical mapping.

A subgroup of the spontaneously polarized pyroelectrics is a special category of materials known as ferroelectrics. Materials in this group are characterized as crystals that possess a spontaneous dipole, and this dipole is reversible by an electric field of some magnitude less than the dielectric breakdown of the material itself. Because of the empirical nature of determining the reversibility of the dipoles, as detected in a hysteresis loop measurement, one cannot predict the existence of ferroelectricity in a new material with much accuracy. We do see, however, that the basis for the existence of ferroelectricity rests primarily on structural (symmetry) considerations. The number of actual ferroelectrics today are known to number in the thousands when one includes the large number of ceramic solid-solution compositions; and it is no longer the great "accident of nature" that it was once thought to be. Ferroelectricity was first discovered in Rochelle salt by Valesek in 1920. Further significant developments in the history of ferroelectric materials came in the 1940s when ferroelectricity was discovered in single crystal and ceramic barium titanate and in the 1950s when the ferroelectric lead zirconate titanate ceramic solid-solution compositions were developed [5].

Ferroelectric ceramics have a number of properties which make them very useful in a variety of applications. These include (1) a high dielectric constant, (2) high piezoelectric constants, (3) relatively low dielectric loss, (4) high electrical resistivity, (5) moisture insensitivity, (6) high electromechanical coupling, (7) medium hardness, (8) fairly high pyroelectric coefficients; and in some special compositions, also (9) high optical transparency and (10) high electrooptic coefficients. Although these properties do not always combine to produce an optimum effect in any one application, one can see that the number of desirable properties provides the possibility for many new and unique applications. This is particularly true when one considers the interactive properties of (1) electromechanical (piezoelectric) behavior relating mechanical and electrical properties, and (2) electrooptic effects relating electrical and optical properties. In addition, these properties in ceramics can be continuously modified over specified ranges because many ceramic compositions form continuous solid solutions (solid solubility) of one or more compounds in each other. Less obvious, but perhaps equally important, is the fact that most ceramics can be fabricated into complex shapes and are more economical to produce.

The most outstanding feature of a ferroelectric ceramic is its hysteresis loop (i.e., a plot of polarization versus electric field). It describes the nonlinear polarization switching behavior as a function of field. Figure 6 illustrates a typical hysteresis loop together with the associated strain behavior. The initial application of an electric field to a virgin ferroelectric electroded disk produces very little effect until the electric field becomes sufficiently high to switch the dipoles in the crystallites. At this field, the polarization (measured as voltage on a large, known, parallel capacitor) changes sharply and reaches saturation at higher fields. Reducing the field to zero leaves the material with a net permanent polarization known as remanent polarization (P_R). As the field is reversed, polarization is first reduced to zero and then changes direction (sign) as the field produces saturation polarization in the opposite direction, thus tracing out a hysteresis loop. The field at which the polarization equals zero (changes sign) is known as the coercive field E_c. Poling consists of applying one-quarter of a hysteresis loop, leaving the material in a remanently polarized condition at P_R.

The accompanying strain involved in traversing the hysteresis loop is also given in Figure 6. This strain loop is sometimes referred to as a butterfly loop because of its appearance. In this figure, lateral strain ($-S_1$) is plotted as a function of electric field. Characteristics worth noting on this plot are (1) the permanent strain (lateral contraction, in this case) produced on first polarizing a virgin material, (2) the strain reversal that takes place during switching or domain (a homogeneous group of dipoles of like polarization) reorientation, and (3) the linearity of the curve over an electric field range of polarization saturation in one direction and just prior to domain switching in the opposite direction. Piezoelectric ceramics are utilized over the linear range of the curve, that is, within safe limits to assure that no backswitching occurs. If switching occurs, piezoelectric activity usually decreases drastically,

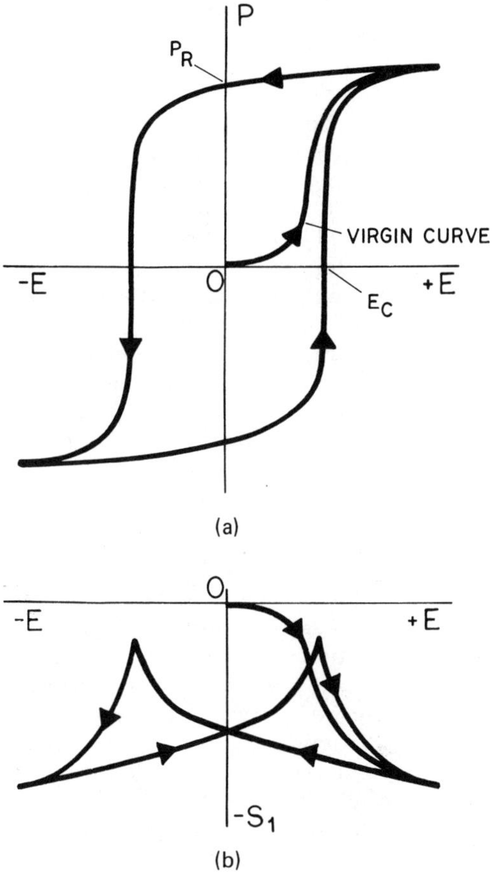

Figure 6 (a) Typical hysteresis loop for a high-coercive-field ceramic; (b) associated lateral strain (contraction).

making the device useless. In fact, for high-electric-drive applications such as sonar, special ceramic compositions are used which greatly resist depoling.

Although a large number of ferroelectric ceramics are used in piezoelectric applications today, it should be pointed out that devices which are strictly dependent on the ferroelectric (polarization switching) properties of ceramics have not found acceptance in the marketplace. This has been due primarily to (1) undesirable fatigue effects resulting from stress (strain)-induced microcracks developed during switching, and (2) the lack of a definite electrical switching field (E_c), which can cause loss of remanent (memory) polarization states in the material over a period of time when subjected to partial reversing fields. Undoubtedly, these detractors would need to be overcome before ferroelectric switching devices will be successful.

IV. ELECTROOPTIC CERAMICS

A. History

Ferroelectric materials in single-crystal form have long been known to possess high optical transparency and useful electrooptic properties, particularly when used in conjunction with polarized light. Such single crystals, however, have serious disadvantages either in their susceptibility to moisture, their lack of optical uniformity over large areas, their high cost, or their difficulty in being produced with proper orientation or chemical stoichiometry. Ferroelectric ceramics, on the other hand, are generally strongest in the areas mentioned here, but historically were known to be opaque or nearly opaque, thus making them impractical for optical applications. One just did not seriously think in terms of utilizing the optical properties of the ceramic because porosity, grain boundaries, and internal refraction at domain walls (boundaries) all combine to produce more than enough light scattering to render the optical properties ineffective for light modulation. More recent work has now shown that this need not be the case, and this area presently represents the newest and most challenging field to open up for applications of ferroelectric ceramics.

Valuable insight for improving the optical characteristics of traditional ferroelectric ceramics, particularly PZT,* was gained by the materials work carried on at Sandia Laboratories during the

*PZT, the acronym for lead zirconate titanate, is a registered trademark of Vernitron, Inc.

period 1961–1965. Studies on the hot pressing of Bi-doped PZT indicated that substantial improvements in optical translucency could be made by hot pressing the material to very near theoretical density [7,8]. As shown in Figure 7, other electrical properties, such as dielectric constant (K), planar coupling (k_p), remanent polarization (P_R), and dielectric breakdown (E_b) were also enhanced as a function of increasing density (decreasing porosity), but none quite as dramatically as optical transmission (T).

A history of the evolution of the electrooptic ceramics is outlined in Table 1. Most significant was the pioneering work of Land et al. [9–12] at Sandia Laboratories on the light-scattering and birefringent properties of PZT ceramics in the mid-1960s which laid the groundwork for further major developments in the field of transparent electrooptics ceramics in the 1970s. From the results of a series of investigations, they demonstrated that (1) electrically poled, thin polished plates of ferroelectric ceramics transmit visible light; (2) these materials exhibited either electrically variable light-scattering behavior or electrically variable optical birefringence (optical phase

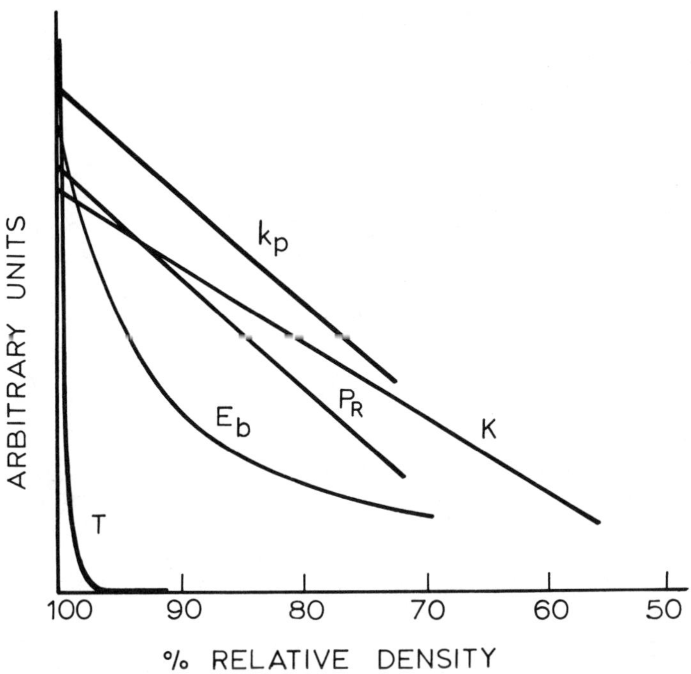

Figure 7 Some typical properties of ferroelectric ceramics as a function of density.

Table 1 History of the Development of Electrooptic Ceramics

Year	Development
1965	Hot-pressing techniques developed for doped PZT ceramics
1966	Optical scattering properties of FE materials measured, polarization dependence defined
1967	Optical birefringent properties of FE materials measured, polarization dependence defined
1968	Reduced scattering measured in La-doped PZT
1969	Development of transparent PLZT ceramics; phase diagram and properties defined
1970	Development of oxygen hot-press techniques
1970	Development of chemical process for E/O ceramics
1970	FERPIC strain-biased device for image storage
1971	E/O goggles for eye protection; transverse-mode device
1971	CERAMPIC device for image storage
1971	Longitudinal mode in quadratic PLZT reported
1972	PLZT ceramics sintered to optical quality
1972	FERICON device for image storage
1974	Photoinduced $\overline{\Delta n}$ changes in PLZT reported
1975	Vacuum/oxygen-sintered transparent PLZT ceramics
1976	Photoferroelectric image storage devices
1979	Ion implantation in image storage devices

retardation); (3) their light transmission characteristics depend on both the thoroughness (magnitude) and direction of poling; and (4) small local areas of the ceramic can be switched independently of adjacent areas. Furthermore, it was found that ferroelectric ceramics scatter light more efficiently at grain sizes larger than about 2 μm average diameter, whereas optical birefringent effects were most noticeable at small (less than 2 μm) grain sizes, where scattering does not mask the optical anisotropic behavior of the crystallites. In large-grain ferroelectrics, light is preferentially scattered along the polar direction.

Aided by early materials work at Clevite Research (now Vernitron, Inc.) and Bell Telephone Laboratories, continual improvements were made by workers at Sandia Labs in enhancing the translucency of PZT ceramics doped with Ba, Sn, and La. Of the dopants, La showed the most promise and was selected for further study. By late 1969, the lanthanum-modified PZT system (designated as PLZT) yielded the first transparent ferroelectric ceramic [13]. Significant improvements in the transparency and optical clarity of the PLZT materials were made within a year, through the development of (1) a hot-pressing technique utilizing a pure oxygen environment, and (2) a chemical coprecipitation process for the preparation of the powders. Both techniques were vital to the material quality improvement which was necessary for the successful application of these materials to electrooptic devices [14]. Since that time, various new methods have been developed for the processing [15,16] and fabrication [17–22] of optical ceramics. Considerable work was also done on exploring new compositions [23–29] and understanding the electrooptic phenomena in ceramics [30–37]; however, the PLZT materials still remain the standard of the industry for quality excellence.

The first known devices utilizing electrooptic ceramics were proposed as early as 1967 [9]. They involved both domain reorientation and the polarization dependence of light scattering and optical birefringence in these materials. A number of specific devices emerged over the ensuing years; each new type of device generally employing a newly discovered electrooptic effect [38–44].

B. Basic Relationships and Phenomena

The electromagnetic spectrum embraces a wide range of wavelengths, from the very short gamma rays to the long radio waves; however, the portion of the spectrum that the human eye can detect is quite small, ranging from 390 to 770 nm (10^{-9} m). This portion of the spectrum, called the visible range, consists of seven distinctive colors as follows:

	Range	Average
Violet	390–430	410
Indigo	430–460	445
Blue	460–500	480
Green	500–570	535
Yellow	570–590	580
Orange	590–650	620
Red	650–770	710

When combined, as in normal sunlight, these wavelengths will produce ordinary white light. Light of a given wavelength is referred to as monochromatic. Its frequency of vibration is related to wavelength according to the relationship

$$f = \frac{c}{\lambda} \tag{16}$$

where c is the velocity of light in a vacuum (3×10^{10} cm/sec) and λ is the particular wavelength. Thus, if one knows the wavelength of light, the frequency can easily be calculated. This frequency never changes regardless of the density of the medium through which the monochromatic light passes; however, the velocity of propagation and wavelength do change [45]. For a higher-density medium such as glass, the velocity is lower, and consequently, the wavelength is proportionally smaller.

The ratio of the velocity of light in a vacuum to that in any other medium is known as the index of refraction. Expressed mathematically, it is

$$n = \frac{c}{c_m} \tag{17}$$

where n is the index of refraction of the material, c the velocity of light in a vacuum, and c_m the velocity of light in the material. Since the velocity of light in dense media is less than that in vacuum, values of n are always greater than unity and range from 1.0003 (air) to approximately 3.0 for very dense solids. The index of refraction for PLZT ceramics is very high, $n = 2.5$.

The power of a substance to refract or "bend" light waves is sometimes described as refringence, and substances that have two different indices of refraction are said to possess the property of birefringence. Cubic (isotropic) crystals do not possess birefringence; however, crystals from all the other systems do. Those crystals having one unique axis (optic axis) along which the index of refraction is different from that along the other orthogonal axes are known as unaxial birefringent crystals. The magnitude of the difference between the two indices of refraction is referred to as the birefringence, or simply, Δn. It is defined as

$$\Delta n = n_e - n_o \tag{18}$$

where n_e is the index of refraction of a light wave vibrating in a plane parallel to the optic axis and traveling in a direction perpendicular to it, and n_o is the index of a wave vibrating parallel to either of the other axes and traveling in the direction of the optic axis. Since the value of n_e may be either higher or lower than n_o, birefringence may take on positive or negative values.

Ferroelectric PLZT ceramics possess crystalline phases which belong principally to the tetragonal and rhombohedral crystal classes. These materials are uniaxial and optically negative, with Δn values typically ranging from about 0.018 to near zero [13]. These values are low compared to some natural crystal birefringences, but it is the low distortion of the unit cell (∿1%), manifesting itself in low birefringence, which makes these ferroelectrics useful as electrooptic materials by allowing easy reorientation of domains. Since the electrooptic properties of the PLZT materials are intimately related to their ferroelectric properties, domain reorientation produces a change in the optical properties of the ceramic. Furthermore, the magnitude of the observed electrooptic effect is dependent on both the strength and direction of the electric field.

PLZT ceramics display optically uniaxial properties on a microscopic scale, and also on a macroscopic scale when polarized with an electric field. In ceramic materials where a statistical array of randomly oriented crystallites exist, the macroscopic or effective birefringence is designated by $\overline{\Delta n}$. On a macroscopic scale, $\overline{\Delta n}$ is equal to zero before electrical poling and has some finite value after poling, depending on the composition and the degree of polarization. The $\overline{\Delta n}$ value is a meaningful quantity in that it is related to the optical phase retardation in the material.

The phenomenon of optical phase retardation in an electrooptic ceramic is described in Figure 8. Linearly polarized monochromatic light, on entering the electrically energized ceramic, is resolved into two perpendicular components c_1 and c_2 whose vibration directions are defined by the crystallographic axes of the crystallites acting as one optical unit. Because of the different refractive indices, n_e and n_o, the propagation velocity of the two components will be different within the material and will result in a phase shift called retardation. The total retardation Γ is a function of both $\overline{\Delta n}$ and the path length t according to the relationship

$$\Gamma = \overline{\Delta n} t \tag{19}$$

where Γ is the total retardation and t is the thickness of the plate. When sufficient voltage is applied to the ceramic, half-wave retardation is achieved for one wave component relative to the other, resulting in a net rotation of the vibration direction of the linearly polarized light by 90°. This rotation effect will either allow the light (still linearly polarized in the case of halfwave retardation) to be transmitted by the second polarizer in the crossed condition or

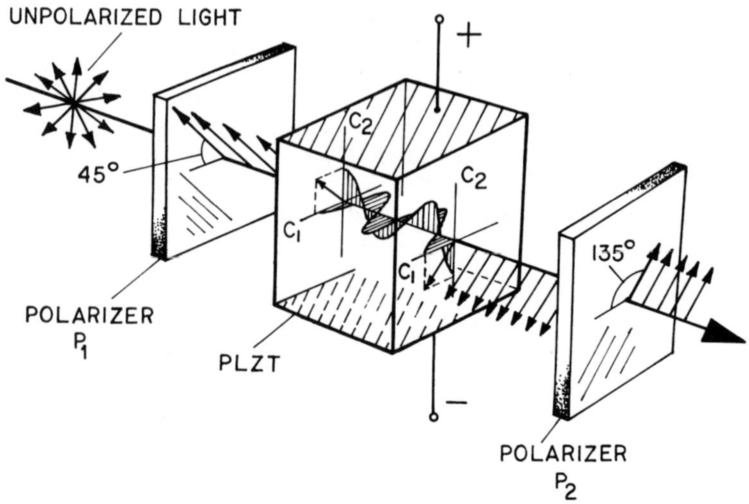

Figure 8 Optical phase retardation in an activated electrooptic ceramic. The open state at half-wave voltage is shown.

be blocked if the second polarizer is in the parallel position. Electrically switching the electrooptic ceramic from a state of zero retardation to half-wave retardation will create an ON/OFF light shutter. It should be noted in Figure 8 that the vibration direction of the polarizers are normally oriented at an angle of 45° to the poling axis of the ceramic in order to achieve maximum effect; if they are aligned with the poling axis, no effect will be observed.

The retardation effect is illustrated in the vector diagrams of Figure 9. As shown in the figure, if no rotation or $n\lambda$ rotation ($n = 1, 2, \ldots$) has taken place on exit of the wave from the ceramic, there is an effective cancellation of the two orthogonal waves in the direction of the crossed polarizer P_2; however, if $n\lambda/2$ rotation ($n = 1, 3, \ldots$) occurs, reinforcement of the two waves in the direction of the crossed polarizer P_2 results in a fully open state [46]. Conditions of $n\lambda/4$ wave retardation produces circular polarization, while $n\lambda/8$ or less yields the most general condition of elliptical polarization. Both of these conditions permit partial light passage through the shutter, whereas half-wave retardation ($n\lambda/2$) achieves a fully open state.

The general relationship that describes the transmitted light intensity of the configuration of Figure 8 is

$$I_T = I_i \sin^2 \frac{\Gamma\pi}{\lambda} = I_i \sin^2 \frac{\overline{\Delta n} t\pi}{\lambda} \qquad (20)$$

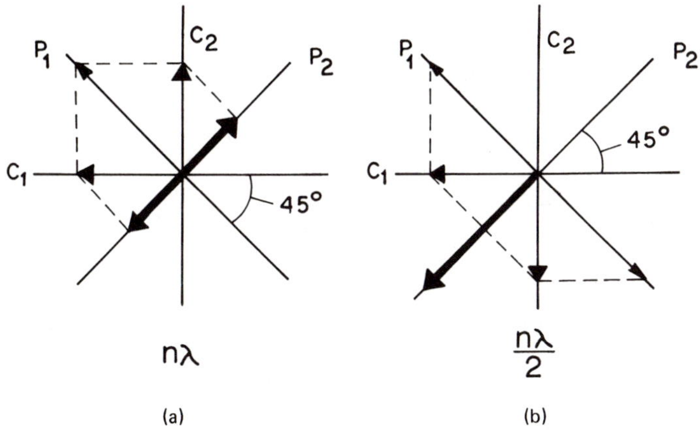

Figure 9 Vector diagrams illustrating (a) cancellation (closed state) and (b) reinforcement (open state) of light waves traveling through the electrooptic ceramic; P_1 and P_2 are crossed polarizers, and C_1 and C_2 are polarization directions (C_2 is the polarization axis) in the ceramic.

where I_T is the transmitted light intensity, I_i the incident light intensity, and λ the light wavelength. When $\Gamma = 0$ or integral multiples of λ, I_T is at a minimum; I_T is a maximum when $\Gamma = n\ \lambda/2$, where n is an odd integer. In electrooptic ceramics, the total retardation is varied by electrically changing the birefringence (Δn) of the material. It should be noted that the sine-squared function above becomes a cosine-squared function when parallel polarizers are used.

As stated previously, the conditions above apply for monochromatic light only. One should now imagine the same situation occurring simultaneously for all the wavelengths making up the visible spectrum when white light is used as the illumination source. The net effect is one of generating a repeating order of colors consisting of yellow, red, blue, and green. A Kerr color chart showing these interference colors is given by Bloss [45]. Such a chart in graph form is shown in Figure 10. It is a graph of plate thickness versus color order for different constant values of birefringence. Although commonly used by mineralologists for the identification of natural minerals, we may use the same chart equally well for the electrooptic materials, recognizing that in our case the birefringence values are not usually constant but can be varied with electric field. The particular electrooptic material employed here for our discussion is a PLZT composition 9/65/35 (La/Zr/Ti ratio) which is isotroptic in its unactivated state and becomes ferroelectric and electrooptic under the influence of an electric field.

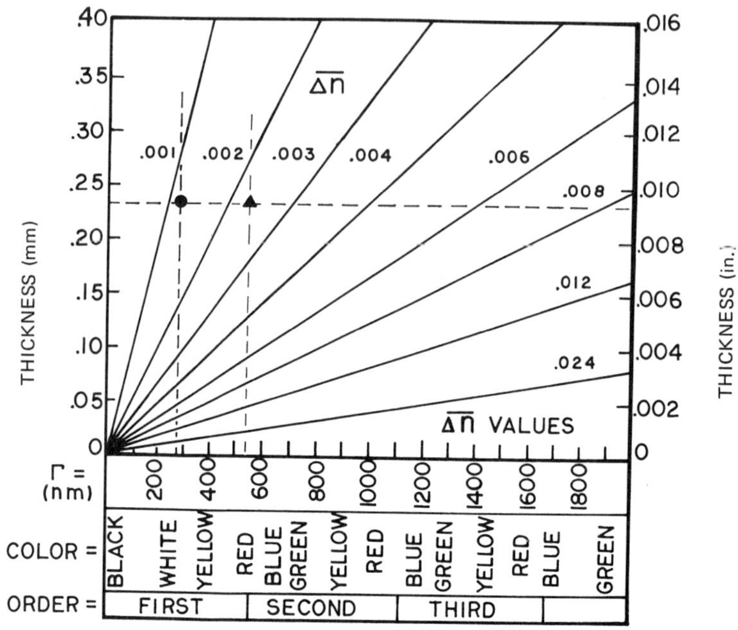

Figure 10 Kerr optical interference color chart.

The generation of color in these materials, on increasing voltage, usually follows a sequence:

1. *Black*: as a result of no applied voltage to the ceramic, hence no retardation, only the crossed polarizers blocking the light.
2. *Gray*: partial passage of light at retardation of less than a quarter-wave.
3. *White* (clear): maximum passage of light at half-wave retardation for green light, which is the midpoint (average) of the visible spectrum; all wavelengths are passed to varying degrees because all the wavelengths making up white light are retarded less than a full-wave blocking condition.
4. *Yellow*: the first full-wave closed condition is achieved for the shortest wavelength (i.e., blue); all other wavelengths are transmitted to varying degrees; if blue is subtracted from the spectrum, a yellow color results.
5. *Red* (magenta): produced from a partial closed condition of blue and a full closed condition for green, leaving a predominately red (some blue) color transmitted.

6. *Blue* (cyan): produced from a transmitting blue and partially transmitting green with a closed condition for red (this is the beginning of the second-order colors).
7. *Green* (yellowish): produced from a blocking of the first-order red and second-order blue; transmitting predominately green with some yellow.
8. *White*: a blending of overlapping colors in the second order, usually producing a near-white color.

These same colors may be produced by introducing an optical bias (retardation) plate such as a quarter-wave mica plate or full-wave red gypsum plate used in ordinary polarizing microscopes. By doing so, of course, the black OFF condition is no longer achievable.

The electrooptic ceramic materials, after application of an electric field, behave in three distinct manners: (1) as memory materials with remanent electrooptic states, (2) as linear electrooptics (Pockel cells) possessing a linear relationship between $\overline{\Delta n}$ and E, and (3) as Kerr-type electrooptics having a quadratic relationship between $\overline{\Delta n}$ and E. A convenient method of characterizing the latter two materials is by their respective electrooptic coefficients, r_c and R, respectively [13]. These may be derived from the following relationships:

$$r_c = \frac{-2\Gamma}{n^3 tE} = \frac{-2\overline{\Delta n}}{n^3 E} \tag{21}$$

$$R = \frac{-2\Gamma}{n^3 tE^2} = \frac{-2\overline{\Delta n}}{n^3 E^2} \tag{22}$$

where Γ is the total retardation, $\overline{\Delta n}$ the birefringence, n the index of refraction, t the plate thickness, and E the electric field.

The linear effect may be thought of as the optical analog of the piezoelectric effect in square-loop ferroelectrics. A plot of Δn versus E for these materials results in a butterfly loop very similar to that for the piezoelectrics, indicating a linear relationship between $\overline{\Delta n}$ and strain [47]. Once poled, the ferroelectric domains are oriented and not switched again; however, subsequent application of an electric field (within prescribed limits) will produce a birefringent electrooptic effect. Since there is no domain switching, hysteresis effects are very minimal in these materials.

The quadratic, slim-loop ferroelectrics are quite similar to solid-state Kerr cells. All the birefringence is created by the electric field (or the associated strain), and birefringence is present in the material only as long as an electric field is applied. This electrooptic effect is quite large and thought to be a result of the electric field enforcing an energetically close, ferroelectric phase [32,35]. The similarity of the $\overline{\Delta n}$ versus E behavior to that of $-S_1$ versus E leads one to believe that there is a close relationship of this effect with the electrostrictive (field-induced strain, where strain is proportional to even powers of electric field) behavior [47]. The recent results of Kirkby [48] show that this is indeed true, with the strain-optic effect accounting for approximately 65% of the total quadratic electrooptic effect. The stress sensitivity of these materials can easily be demonstrated by slightly flexing a polished plate under polarized light. Stress-induced optical birefringence is immediately noticeable. In fact, stress sensitivity is so acute that special measures must be taken to eliminate stress (strain) during fabrication and use.

V. MATERIALS

A. Compositional Systems: General

Ferroelectric ceramics have historically been formulated from a number of compositions and solid solutions, including barium titanate, lead zirconate titanate (PZT), lead niobate, bismuth titanate, sodium potassium niobate, and lead titanate. Foremost of these has been barium titanate, which dates from the early 1940s and is still in use today. Its primary applications are centered in the capacitor industry; however, it has largely been supplanted by the PZTs for transducer (piezoelectric) applications. This is because the PZT compositions (1) possess higher electromechanical coupling coefficients than barium titanate; (2) have a higher Curie point, which permits higher-temperature operation or higher-temperature processing during fabrication of devices; (3) can easily be poled; (4) possess a wide range of dielectric constants; (5) are relatively easy to sinter; and (6) form solid solutions with many chemical compositions, thus allowing a wide range of achievable properties.

It is not within the scope of this chapter to discuss the many different compositional systems that are in use today for transducer applications. Listed below, however, are the major ones, which center around the PZT system:

$PbZrO_3$-$PbTiO_3$
$Pb(Zr,Ti)O_3$ + additives of Nb, Sb, Bi, La, Fe, Ta, Cr, Co or Mn
$(Pb,Sr)(Zr,Ti)O_3$

$(Pb,Ba)(Zr,Ti)O_3$
$Pb(Zr,Ti,Sn)O_3$
$Pb(Mg,Nb)O_3$-$PbZrO_3$-$PbTiO_3$
$Pb(Co,Nb)O_3$-$PbZrO_3$-$PbTiO_3$
$Pb(Mn,Sb)O_3$-$PbZrO_3$-$PbTiO_3$
$PbTiO_3$ + additives of Mn, La, Nd, and In
$(Pb,La)(Zr,Ti)O_3$

Lead zirconate titanate ceramics are almost always used with a dopant, modifier, or other composition in solid solution. This is commonly done to improve on the properties of the basic PZT ceramic material for specific applications.

Examples of this practice include:

1. *Donor* additives such as Nb^{5+} replacing Zr^{4+} or La^{3+} replacing Pb^{2+} in order to counteract the natural p-type conductivity of the PZT and thus raise the electrical resistivity. They are usually compensated by A-site vacancies. These additives enhance domain reorientation, and materials produced with these additives are characterized by square hysteresis loops, low coercivity, high remanent polarization, high dielectric constant, maximum coupling factors, high dielectric loss, high compliance, and reduced aging [49]. Typical applications are in the area of high sensitivity, such as hydrophones, phonograph pickups, sounders, and loudspeakers.
2. *Acceptor* additives such as Fe^{3+} replacing Zr^{4+} are compensated by oxygen vacancies and usually have only limited solubility in the lattice. Domain reorientation is limited, and hence ceramics with acceptor additives are characterized by poorly developed hysteresis loops, lower dielectric constant, low dielectric loss, low compliance, and high aging rates. Typical applications are in high-power devices such as sonar and ultrasonic tranducers.
3. *Isovalent* additives such as Ba^{2+} or Sr^{2+} replacing Pb^{2+} or Sn^{4+} replacing Zr^{4+}, in which the substituting ion is of the same valency and approximately same size as the replaced ion. Solid-solution ranges with these additives are usually quite high and may result in lower Curie points and increased constants. Hysteresis loops may be poorly developed without additional additives. Other properties include lower dielectric loss, low compliance, and higher aging rates. These ceramics are used in high-drive applications.

It should be mentioned here that it is common practice in formulating compositions to include more than one type of additive to a given composition in order to achieve a given set of properties. In fact, in following the literature for this area, one can easily be overwhelmed by the large number of additives that can and are being made to the basic PZT system. The PZT system was first investigated by Shirane et al. [50] Japan, and the first piezoelectric compositions from this sytem were developed by Jaffe [5,51] in the United States.

B. PLZT Compositional System

One compositional system that embraces both the piezoelectric and the electrooptic materials is the lanthanum-modified PZT (PLZT) system. Therefore, it will be selected here as a typical system which is useful for both applications.

The solid-solution system that forms the basis of the PLZT materials is a series of compositions resulting from the complete miscibility of lead zirconate and lead titanate in each other and modified by the solubility of substantial amounts of lanthanum oxide in the crystalline lattice. A general formula for all compositions in the PLZT system is

$$Pb_{1-x}La_x(Zr_yTi_{1-y})_{1-x/4}O_3$$

where lanthanum ions replace lead ions in the A site of the perovskite ABO_3 ionic structure shown in Figure 11. Since La^{3+} (added as

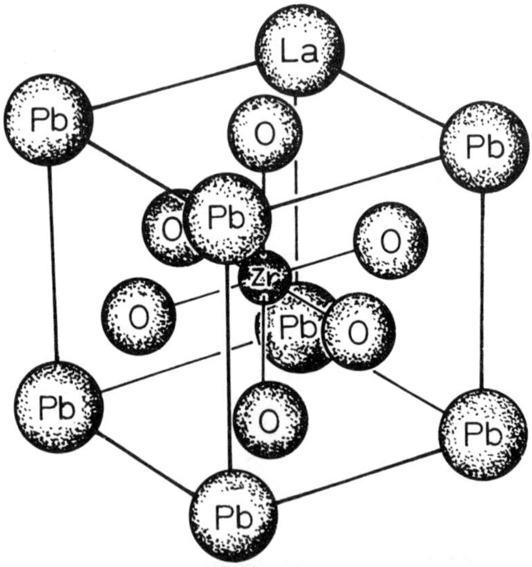

Figure 11 Perovskite ABO_3 crystal unit cell.

La_2O_3) substitutes for Pb^{2+}, electrical neutrality is maintained by the creation of lattice site vacancies. Although the formula above presumes that all the vacancies are in the B sites, the actual location of these vacancies in either the A^{2+} sites or B^{4+} sites of the unit cell has not yet been completely resolved despite numerous studies on the subject. It is most probably that both A- and B-site vacancies exist [52]. If both A- and B-site vacancies are present in the lattice, it is to be expected that the formulation above would provide excess Pb^{2+} ions which are expelled from the lattice (as PbO vapor) during the densification process at elevated temperatures. This behavior does, indeed, occur; and in fact, this excess PbO contributes to achieving full density by forming a liquid phase at the grain boundaries and by inhibiting grain growth during the initial stages of densification. Both of these effects are beneficial to the attainment of theoretically dense material by eliminating residual porosity before it becomes entrapped within the grains.

Ceramics usually formulated from the conventional A-site formula,

$$Pb_{1-3x/2}La_x(Zr_yTi_{1-y})O_3$$

must be sintered with excess PbO atmosphere in order to achieve nearly full densification. They characteristically pick up PbO from the atmosphere, indicating the additional likelihood that a combined A-site and B-site vacancy structure is actually present in these materials.

The PLZT phase diagram is given in Figure 12. As can be seen from the diagram, the effect of adding lanthanum to the PZT system is first, one of maintaining extensive solid solution throughout the PZT system, and second, one of decreasing the stability of the ferroelectric phases (reducing the Curie points) in favor of the nonferroelectric cubic and antiferroelectric (nonpolar phase possessing adjacent, oppositely aligned polar unit cells) phases. At a 65/35 ratio of $PbZrO_3$ to $PbTiO_3$, a concentration of 9.0% La is sufficient to reduce the stable region of the FE rhombohedral phase to below room temperature. Thus a material of this composition (designated as 9/65/35 or simply, 9065) is nonferroelectric and cubic in its virgin state. The crosshatched area existing along the FE-PE (cubic) phase boundary denotes a region of diffuse, metastable ferroelectric phases that can be electrically induced with a sufficiently high electric field. Materials within this region exhibit the quadratic electrooptic behavior.

The solubility of La in the PZT structure is a function of composition and is related directly to the amount of lead titanate present. The compositional dependence of the solubility limit is indicated by the dashed line adjacent to the mixed-phase region (double

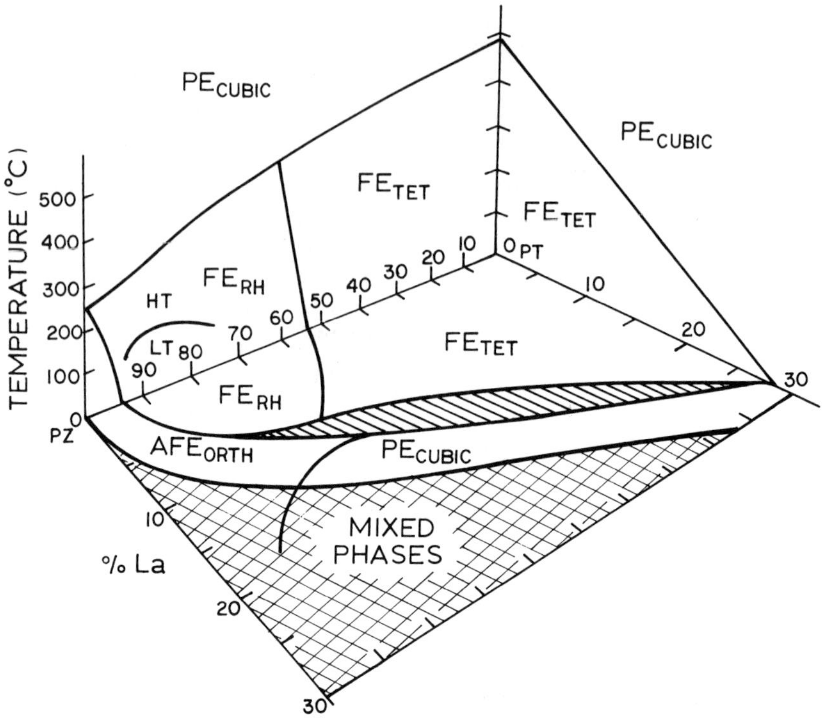

Figure 12 Room-temperature PZT and PLZT phase diagrams.

crosshatched area) in Figure 12. For the two end-member compositions, lead zirconate and lead titanate, these limits are 4 and 32 at % La, respectively. The solubility limits for intermediate compositions are proportional to their Zr/Ti ratios. Compositions in the mixed-phase region are easily identified by their opaque appearance, in contrast to the transparency characteristic of the single-phase, cubic materials.

Modification of the PZT system by the addition of lanthanum oxide has a marked beneficial effect on several of the basic properties of the material, such as increased squareness of the hysteresis loop, decreased coercive field, increased dielectric constant, maximum electromechanical coupling coefficients, increased mechanical compliance, and enhanced optical transparency. The latter of these properties, optical transparency, was discovered only in recent years but came about as the result of an in-depth study of various additives to the PZT system. Results from this work indicate that La^{3+}, as a chemical modifier, is unique among the off-valent additives in producing transparency. The reason for this behavior is still not fully understood; however, it is known that lanthanum is, to a large extent, effective because of its high solubility in the PZT oxygen octahedral structure, thus producing an extensive series of single-phase solid-solution compositions. The mechanism is believed to be one of lowering the distortion of the unit cell, thereby reducing the optical anistropy of the crystalline lattice and at the same time promoting uniform grain growth and densification of a single-phase, pore-free structure.

Compositions within the PLZT system for piezoelectric applications are usually confined to less than 5% La, whereas the electrooptic ceramics possess approximately 6% La or more. The majority of compositions, in either case, are located at or near the morphotropic (temperature-independent) phase boundary which separates the FE rhombohedral and FE tetragonal phases. This boundary is especially important because most of the properties of interest are optimized along it (e.g., dielectric constant, remanent polarization, piezoelectric coefficients, and planar coupling are all increased to their maximum values). Compositions located on the tetragonal side of the phase boundary are usually "hard" (high-coercive-field) materials and those on the rhombohedral side are "soft" (low-coercivity) materials. Specific applications dictate whether hard or soft materials are used. Applications for compositions along the RE rhombohedral—AFE orthorhombic phase boundary are very few in number and are limited to special devices utilizing a pressure-enforced FE-to-AFE phase transition [53]. Pyroelectric applications exist for a number of compositions throughout the system which possess reduced Curie points, including some that utilize the reversible pyroelectric effects between the rhombohedral LT (low temperature) and HT phases [54, 55]. Compositions rich in lead titanate are especially useful for SAW (surface acoustic wave) devices [56]. Recent work has also indicated that compositions in the AFE region of the phase diagram may find application as high-stability capacitors under high electric fields [57].

Electrooptic compositions in the PLZT phase diagram are generally divided into three application areas: quadratic, memory, and linear. As mentioned previously, the quadratic materials are compositionally located along the phase boundary separating the ferroelectric and paraelectric phases, principally in the crosshatched area. Memory compositions having stable, electrically switchable optical states are largely located in the FE rhombohedral phase region, and the linear materials possess nonswitching, linear electrooptic effects are confined to the area encompassing the tetragonal phase.

C. Processing and Fabrication

Ferroelectric ceramics are traditionally made from powders formulated from individual oxides; however, the newer electrooptic materials utilize a chemical coprecipitation technique. The processing method that one selects to prepare the powder depends, to a large extent, on cost; but even more important is the end application (i.e., whether piezoelectric or electrooptic). Understandably, the electrooptic ceramics require a higher purity, more homogeneous, and high-reactivity powder than do the piezoelectrics because inhomogeneities can be detected optically much more easily than electrically. As a result, different powder process techniques have evolved in the two cases. Piezoelectric ceramics are still prepared from the most economical, mixed-oxide (MO) process, whereas the optical PLZT materials utilize a specially developed chemical precipitation (CP) process involving liquid precursors. Although not yet achieved, the trends in this area are toward the development of one unified process that will meet the objectives of both types of materials. There is, in fact, a commonality in these objectives, since the more recent piezoelectric devices demand higher-quality material (essentially zero porosity) and the electrooptic devices require a more economical process.

The present general processing steps involved in the powder preparation and fabrication of both piezoelectric and electrooptic ceramics are given in Figure 13. As can be seen, a majority of the steps involved in the two processes are quite similar. Essential differences lie in the areas of initial oxide mixing versus coprecipitation and atmosphere sintering versus hot pressing. Calcining (a high-temperature chemical reaction of ingredients to form a final composition) is a critical step in both processes.

The conventional MO process consists of (1) weighing the oxides to the appropriate accuracy, (2) blending in a liquid medium, (3) drying to completeness, (4) calcining the powder at approximately 900°C for 1 hr, (5) milling in a liquid medium, (6) drying with binder addition, (7) cold pressing a specific shape, (8) high-temperature firing (sintering) at approximately 1275°C for several hours, (9) shaping, (10) electroding, and (11) final poling.

In the CP process, the high-purity liquid organometallics (tetrabutyl zirconate and tetrabutyl titanate) are first premixed with lead oxide in a high-speed blender and then precipitated by adding the lanthanum acetate, water-soluble solution while blending. The resulting slurry, consisting of a finely divided suspension of mixed oxides and hydroxides in an alcohol-water solution, is dried to completion at approximately 80°C. Subsequent steps in the process involve (1) calcining the powder at 500°C for 16 hr, (2) wet milling the powder for several hours to promote additional homogeneity, (3) drying without binder, (4) cold pressing the powder, (5) hot pressing in oxygen, (6) shaping and slicing, (7) polishing, and (8) electroding.

A typical hot-pressing setup is shown in Figure 14. The arrangement shown is typical of many different methods of heat and applying pressure. Experience has shown that a simple, uniaxial,

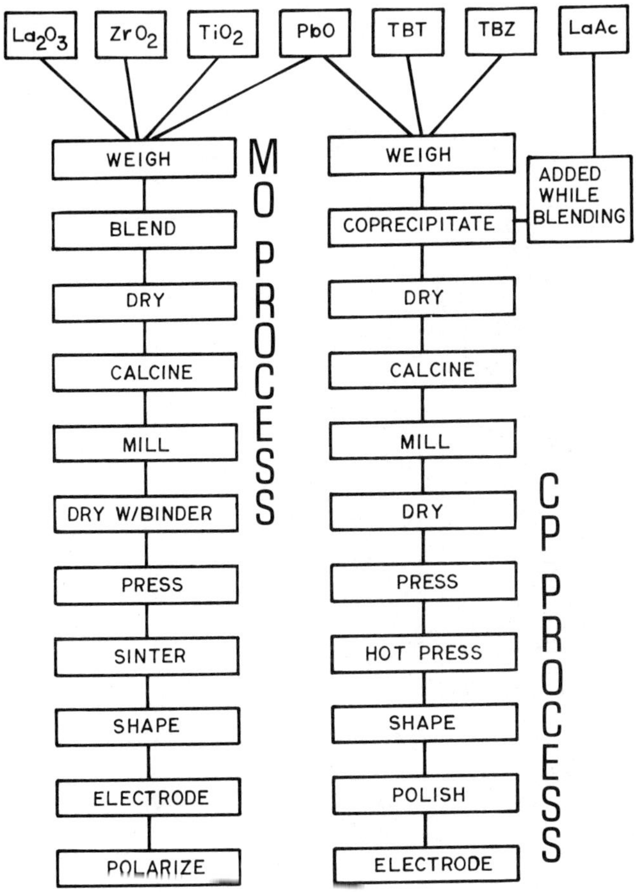

Figure 13 Flowsheets of the mixed oxide (MO) and chemical coprecipitation (CP) processes.

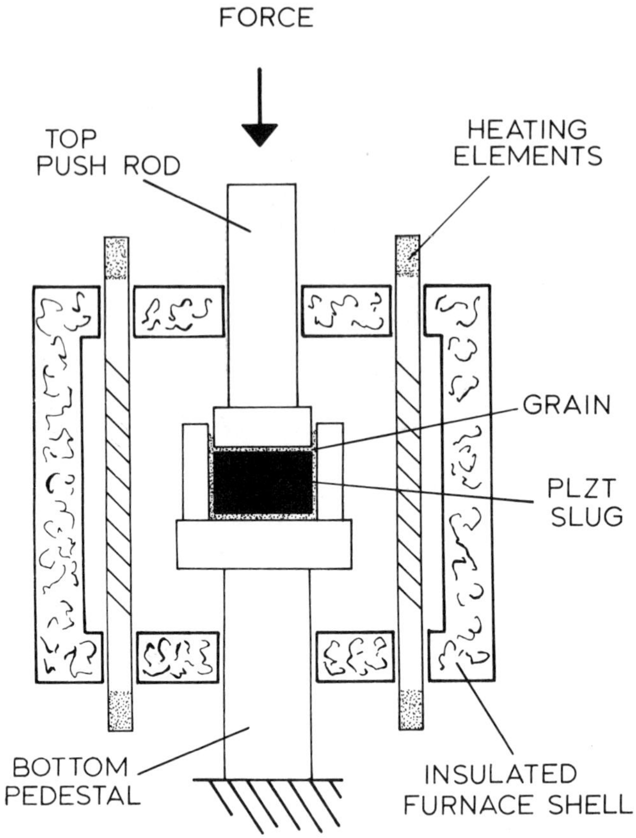

Figure 14 Typical hot-pressing setup for fabricating PLZT electrooptic ceramics.

single-ended hot-pressing arrangement is both reliable and economical in producing consistent, optical-quality material. The preformed powder is hot pressed in a silicon carbide mold and surrounded by refractory grain (magnesia or zirconia) to prevent high-temperature reactions with surrounding materials. During the heating portion of the hot-press cycle, the slug is evacuated and back-filled with oxygen. Typical hot-pressing conditions are 1250°C for 18 hr at 2000 psi. After hot pressing, the slug is extracted from the mold and polished for optical evaluation. This method of vacuum/oxygen hot pressing is now being used successfully for fabricating optical-quality PLZT slugs up to 6 in. in diameter [58]. An alternative method of hot pressing in flowing oxygen rather than vacuum/oxygen is also known to produce optical quality, but it is generally limited to slug sizes less than 2 in. in diameter. Typical examples of hot-pressed ceramics are given in Figure 15.

The two most critical steps in the fabrication of these ceramic materials are powder preparation and firing, and of these, the first is predominant. Although one can readily determine many measurable characteristics of the powders, such as particle size, surface area, composition, crystalline structure, and morphology, powder reactivity (the tendency of particles to interdiffuse and coalesce at some elevated temperature) is perhaps the most important, but is still the least understood. In fact, a primary objective of the various chemical preparation techniques is to increase powder reactivity, thereby achieving full density (zero porosity) and maximum chemical homogeneity.

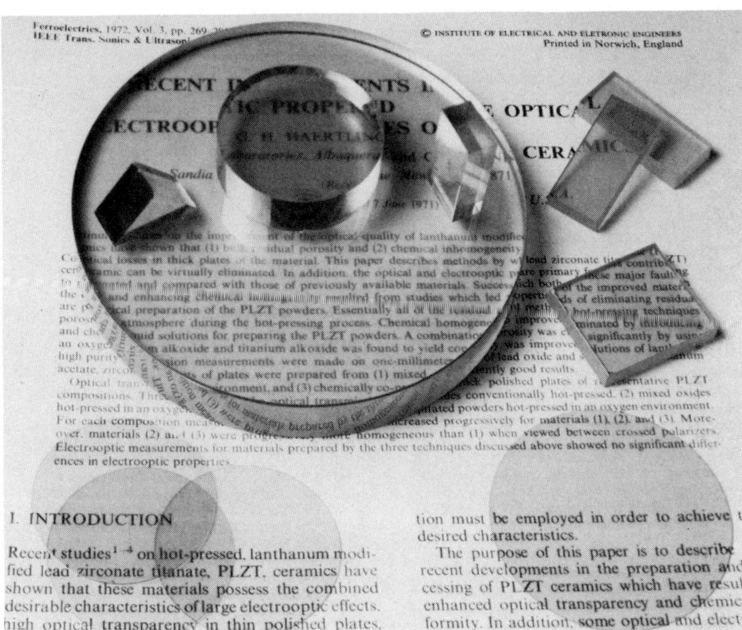

Figure 15 Typical examples of hot-pressed electrooptic PLZT ceramics, indicating size and high transparency.

VI. PROPERTIES

A. Microstructure

Compositions in the PLZT system, whether piezoelectric or electrooptic, are characterized by a highly uniform microstructure consisting of randomly oriented grains (crystallites) intimately bonded together. An example of such a microstructure is given in Figure 16. The sample was polished and thermally etched at 1150°C for 1 hr. As is typical for most hot-pressed microstructures, little or no entrapped porosity is evident. This is due to the influence of the external pressure during hot pressing, which aids in pore removal while the material is in a thermochemically active state at elevated temperatures. In actuality, some amount of porosity exists in all the materials, whether hot pressed or atmosphere sintered; however, it is of prime importance to the electrooptic ceramics because of their optical transparency. It can also be seen from the microstructure that the average grain size for this sample is approximately 8 μm and is the result of the conditions (temperature, time) selected during fabrication. Piezoelectric ceramics usually possess grain sizes in the range

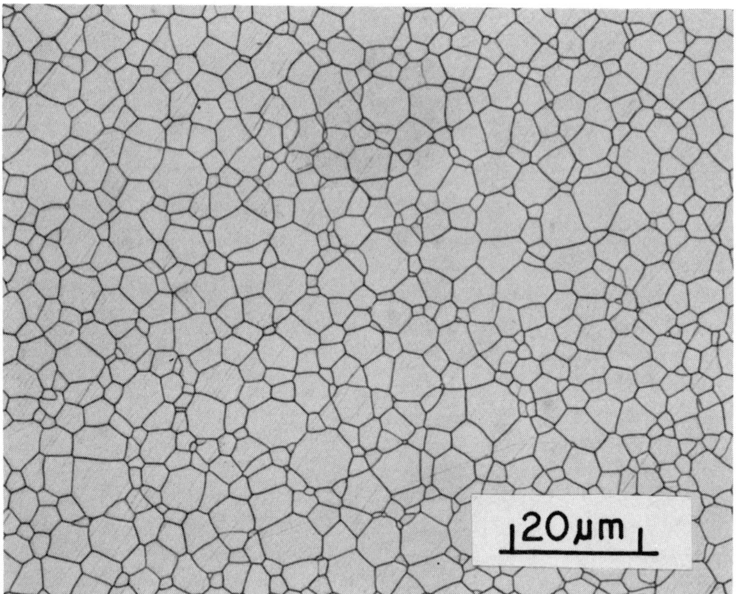

Figure 16 Typical microstructure of a hot-pressed electrooptic PLZT ceramic.

2–6 μm whereas the electrooptics cover a wider range, from 2 to 10 μm, depending on the specific electrooptic effect being optimized. A uniform grain size is a highly desirable feature from the standpoint of performance.

Another noteworthy feature of the microstructure of Figure 16 is the complete lack of a second phase, which often precipitates along the grain boundaries or at triple points between grains. Porosity may be considered to be a second phase (air), but here it is not referred to as such. Second phases usually consist of other, unwanted compounds which are present in the material and have developed either during calcining or sintering via precipitation or were inadvertently added to the batch during formulation or processing. Such phases are undesirable in the electrooptic ceramics, and even in minor amounts can produce opacity. In the case of piezoelectrics, a somewhat different set of circumstances exist. For piezoelectrics that are utilized as thin disks with thicknesses ranging from 0.002 to 0.005 in. (e.g., loudspeakers), second phases (including porosity) must be prevented as much as possible; on the other hand, for larger shapes, second phases can be tolerated if they are adequately controlled. Recent work by Newnham and coworkers at Penn State on the deliberate introduction of second phases for enhancement of piezoelectric properties has shown that in some cases, significant improvements can be made [59,60]. Connectivity is the key word here; that is, the volume of the second phase or phases and their pattern of dispersion throughout the material are very instrumental in achieving the desired properties.

Domain (electrical polarization) patterns can also be revealed in the microstructure of ferroelectric ceramics when chemical etching techniques are used. Examples of these patterns are shown in Figure 17. The domains show up as a two-level structure because one end of the electric dipole chemically etches faster than the opposite end. Distinctive features here are (1) the absence of etched grain boundaries; (2) domain structure crossing over grain boundaries, indicating little disorder at the boundary; (3) domain orientation being influenced by the direction of polishing (before etching), pointing out the ferroelastic (domain reorientation due to mechanical stress) behavior of these materials; and (4) the size of the domains, being in the order of 0.5–3 μm. It should be noted that the thermally etched samples in Figure 17 (b) reveal grain boundaries but no domains, whereas the chemical etched samples of Figure 17 (a) exhibit domain patterns and few or no grain boundaries. The reason is that at elevated temperatures where thermal etching (partial melting) takes place there are no domains since the material is in the nonpolar, cubic phase, whereas at room temperature where chemical etching occurs, domains have spontaneously formed when the material is brought below its Curie point.

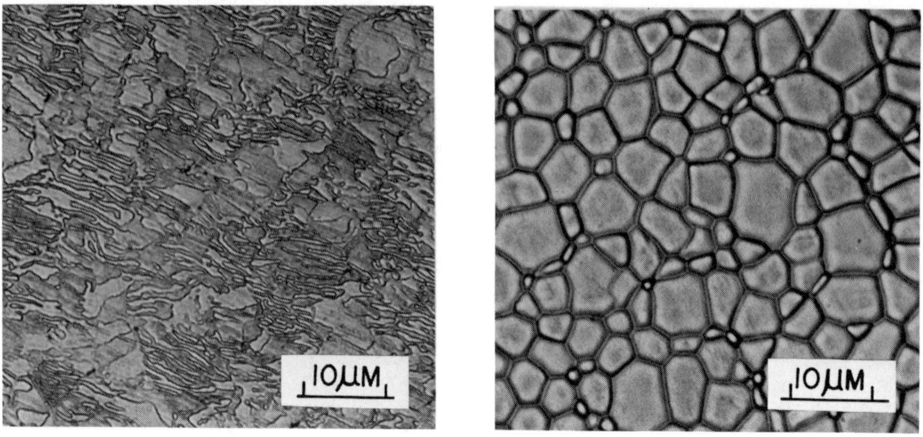

(a) (b)

Figure 17 Microstructure of PLZT electrooptic ceramics: (a) domain patterns in 7/65/35 revealed by chemical etching; (b) grain boundaries revealed by thermal etching.

B. Mechanical Properties

Ferroelectric ceramics, particularly some of the piezoelectrics, are actually structural elements that must withstand reasonably high stress levels in their use condition. Consequently, some knowledge of their mechanical strength is necessary in an optimum design. Although it is always better from a strength standpoint to utilize ferroelectrics (as is true of most ceramics) in compression rather than tension, this is not always possible, and tensile strength becomes a limiting factor. An early investigation by Dungan [61] on the pure tensile strength of several hot-pressed ferroelectric compositions in the PZT system has shown that tensile strengths are in the range 9500–14,000 psi as compositions vary from 52/48 (Zr/Ti) at the morphotropic phase boundary to 95/5 near lead zirconate. Additional studies on the fracture processes in ferroelectrics have pointed out that (1) flexure strengths can range from 10,000 to 24,000 psi for the PZTs; (2) internal stresses significantly affect strength; and (3) ferroelastic twinning, phase transformations, and microcracking are toughening mechanisms preceding failure [62].

The mechanical properties of the PLZT materials have not been investigated as thoroughly as the PZT; however, studies on strain-biased devices using these materials have shown that they have the ability to tolerate unusually high mechanical strains, approximately

3000 ppm (0.3%), before fracture occurs. A strain of this magnitude is, perhaps, an order of magnitude higher than most conventional ceramics. This is believed to be due to the ferroelastic properties of these ferroelectrics, which accommodate the strain and relieve the stress by domain reorientation. Such behavior should also result in a change in the polarization state of a poled ferroelectric and this was, indeed, shown by Maldonado and Meitzler to be the case for PLZT 7/65/35 [63].

C. Thermal Expansion

Thermal expansion is a universal property of all materials and consists of a volumetric expansion (or contraction) of the material as a function of changing temperature. The coefficient of linear expansion is the ratio of the change in length per degree Celsius to the length at 0°C. For solids, this coefficient is approximately one-third of the value of the coefficient of volumetric expansion.

Thermal expansion coefficients can very considerably in compositions throughout the PZT system, depending on the symmetry of the phases present. Figure 18 illustrates this for various selected Zr/Ti ratio compositions containing 2 at % niobia. The anomalous behavior below the Curie point is usually a result of subtle changes in the unit cell (e.g., in 95/5 the change is between the LT and HT rhombohedral phases; however, above the Curie point the linear expansions are regular and vary in a uniform manner from a value of 6.7×10^{-6}°C^{-1} at the morphotropic phase boundary (53/47 ratio) to 8.8×10^{-6}°C^{-1} for 95/5. In all cases, the Curie point is marked by a volume decrease in the vicinity of the Curie point.

Similar behavior can be seen for the PLZT compositions in Figure 19. In general, the inflection at the Curie point becomes less distinct as the concentration of La increases. For the 65/35 ratio series, this amounts to a complete loss of a recognizable Curie point at a composition of 9/65/35 since the Curie point has been lowered to below room temperature and has become extremely diffuse. The coefficients of linear expansion for all compositions above the Curie point are approximately 5.4×10^{-6}°C^{-1} [64]. This relatively low value of thermal expansion is desirable for increasing the resistance of the material to thermal shock.

D. Electrical Properties

Since nearly all the useful properties of ferroelectric ceramics (whether they are dielectric, mechanical, or optical properties) are related in some manner to their response to electrical stimuli, the

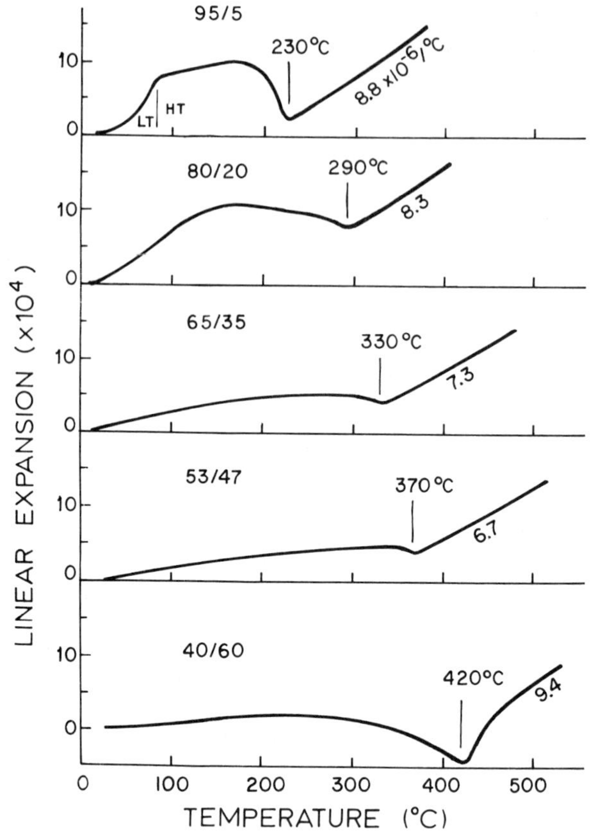

Figure 18 Thermal expansion characteristics of several niobia-doped PZT compositions with varying Zr/Ti ratio.

electrical behavior of these materials is important to their successful application in piezoelectric or electrooptic devices. Ferroelectrics, in general, are characterized by (1) higher dielectric constants (K = 500–4000) than ordinary insulating substances (K = 5–100), making them useful as capacitor and energy storage materials; (2) relatively low dielectric loss (0.1–5%); (3) high specific electrical resistivity ($\sim 10^{11}-10^{13}$ Ω-cm); (4) moderate dielectric breakdown strengths of 250–300 V/mil (100–120 V/cm) in thin disk; and (5) nonlinear electrical behavior (hysteresis loop) which results in an electrically variable dielectric constant (dielectric constant is not really a constant here, since it is equal to the slope of the P versus E hysteresis loop curve) or in electrically variable memory states. In addition, ferroelectrics possess mechanical and optical

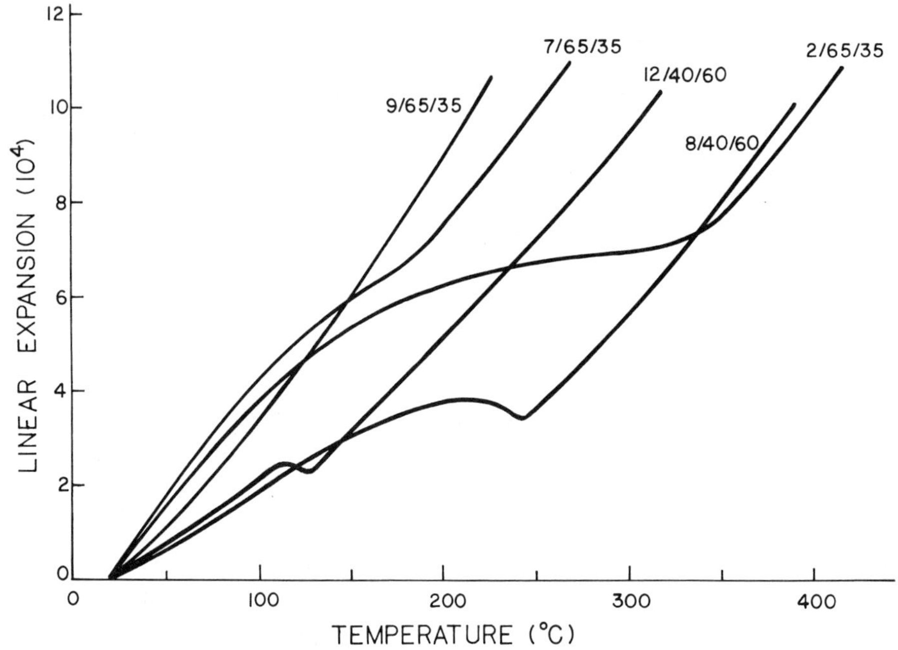

Figure 19 Thermal expansion properties of several PLZT electrooptic compositions with varying La and Zr/Ti ratios.

effects which are interactive with their electrical behavior in producing useful electromechanical and electrooptic properties.

It should be kept in mind that not all of these properties are optimized and realized in a given type of ferroelectric structure or chemical composition, and hence a variety of ceramic materials are manufactured and are available from several different companies throughout the world. As a result, no attempt will be made here to enumerate the specific properties of the various ceramics produced by these manufacturers. In enumerating the electric properties, distinction is usually made between large-signal (>5 V/mil) and small-signal (<1 V/mil) measurements (e.g., dielectric constant is normally a small-signal property, whereas the hysteresis loop is a large-signal property).

1. Dielectric Properties

Small-signal (1 kHz) dielectric constant values as a function of temperature for several selected compositions in the PLZT system are given in Figure 20. For the series with 65/35 Zr/Ti ratios, a consistent and gradual reduction in the Curie point (indicated by the peak in the dielectric curve) occurs up to a La concentration of 8 at %. Higher La contents reduce the height of the dielectric constant peak, making it more diffuse without significantly changing the temperature of the peak. As a rule, compositions near $PbZrO_3$ and $PbTiO_3$ are characterized by low dielectric constants (K = 150–300) and loss tangents of approximately 1% or less. Maxima in dielectric constants occur along the FE-cubic (K = 5700) and the FE tetragonal–FE rhombohedral (K = 4100) phase boundaries. Loss tangents in these regions also reach maximum values of 6% and 4%, respectively.

2. Hysteresis Loops

The hysteresis loop is the single most important measurement that can be made on a ferroelectric ceramic during the process of characterizing its electrical behavior. Indeed, the internal polarization

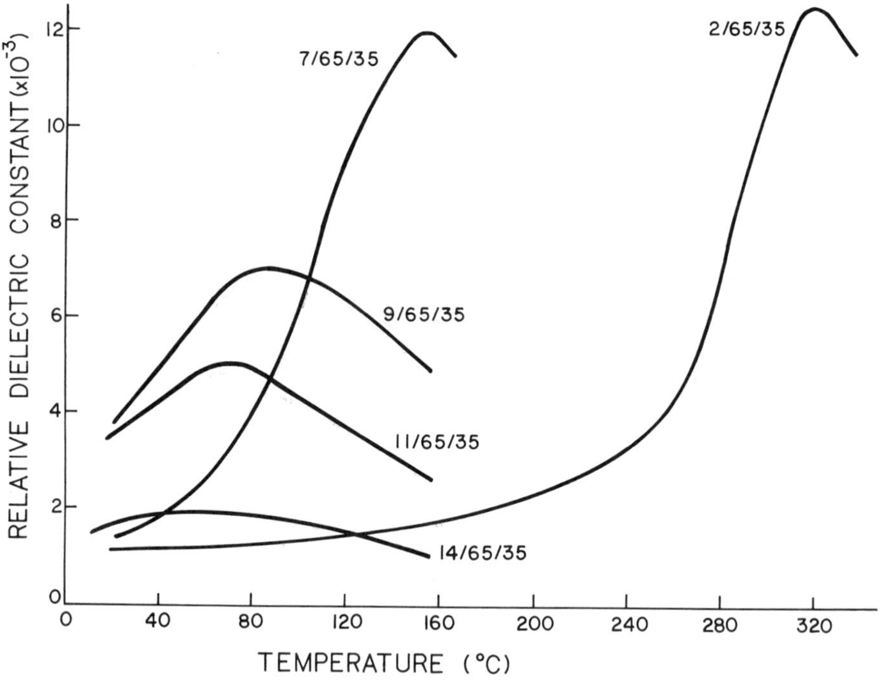

Figure 20 Temperature dependence of relative dielectric constant for several selected PLZT compositions with a 65/35 Zr/Ti ratio.

of the material must be switchable at some field less than electrical breakdown for the material to even qualify as ferroelectric. As mentioned previously, piezoelectrics and some electrooptics are routinely subjected to a partial hysteresis loop when they are permanently polarized for device applications.

Hysteresis loops come in all sizes and shapes, and like a fingerprint, identify the material in a special way. Therefore, one should become familiar with such a measurement. Although early workers in the field of ferroelectrics utilized a dynamic (60 Hz) measurement with a Sawyer-Tower circuit and an oscilloscope readout, more recent work has been done with essentially a dc ($\sim$1/10 Hz) measurement using an X-Y plotter readout [5,65].

Hysteresis loops are usually run on virgin (thermally depoled) electroded disks by varying the electric field ($E = V/t$) across the disk while monitoring the charge collected on a large (approximately 1000 times larger in capacitance than the sample), low-loss capacitor in series with the sample. This charge (actually measured as voltage, which relates to charge by the familiar $Q = CV$) on the low-loss capacitor is proportional to the charge on the sample, and thus one obtains a continuous plot of polarization as a function of electric field. In a strict sense, dielectric displacement rather than polarization is being measured; however, in the case of high-dielectric-constant materials such as ferroelectrics, the two quantities are very nearly equal and the term "polarization" is usually preferred.

A typical hysteresis loop for a soft (easily switchable) ferroelectric ceramic is given in Figure 21. Since a virgin ceramic is macroscopically isotropic (nonoriented crystallites) to begin with, no preferred orientation effects should be evident in the loop; and it should be symmetric about a point of origin that is selected for the measurement (this, of course, will not be true for previously poled samples). Starting at point O, the electric field is gradually increased, initially producing very little change in polarization. At some higher field, domain switching occurs fairly abruptly, and a significant increase in effective polarization takes place within the material. This internal polarization (dipole moment) is balanced by externally bound charge which is delivered from the electrical circuit. For square-loop materials, this polarization switching may occur over a relatively narrow field range of 1–5 V/mil; however, a more typical range is 10–20 V/mil.

As the electric field is further increased to point A at approximately $3E_c$, the hysteresis loop saturates at a polarization level which is dependent on composition as well as density and grain size [66]. Reducing the electric field to zero leaves the material in a polarized condition at point C with a remanent polarization value of P_R. Increasing the field to the opposite direction will first reduce the polarization to zero at its E_c, the coercive field of the material, and then saturates in the reverse direction at point D. Again, the

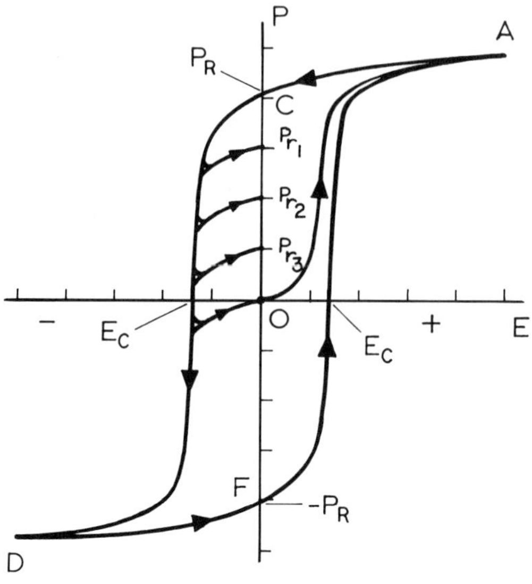

Figure 21 Typical hysteresis loop for a soft (low E_c) ferroelectric ceramic illustrating various characteristics of the loop.

process of reducing the field to zero traces a path to point F which is equal to $-P_R$. The hysteresis loop is completed by, once again, changing the field direction, increasing it to saturation in the original direction, and finally reducing it to zero at P_R. Points P_{r1}, P_{r2}, and P_{r3} are intermediate polarization states obtained by partial back-switching. Back-switching to a state of zero polarization (point O) leaves the material in a condition referred to as "electrical zero." Although electrically there is no net polarization in this state, it is not equivalent to the starting (virgin or thermally depoled) condition.

Several different types of hysteresis loops are exhibited by compositions in the PLZT system, as shown in Figure 22. These include (1) an antiferroelectric (AFE orthorhombic phase) loop consisting of a linear portion and a field-enforced FE region, (2) a square (FE rhombohedral phase) loop with a low coercive field, (3) a square (FE tetragonal) loop with a high coercive field, (4) a slim loop exhibiting field-enforced ferroelectric (SFE) behavior, and (5) a linear loop characteristic of normal non-FE dielectrics. Specific applications exist for materials possessing loops of each of these types. Table 2 tabulates some typical properties for several selected PLZT compositions.

E. Electromechanical Properties

The electromechanical properties of a piezoelectric (ferroelectric) material are of basic concern to the designer who is planning to utilize these materials in a new or existing application. Since the resulting effects produced by the interaction of the material's electrical and mechanical properties are specific and unique, an understanding of these effects is essential to a successfully designed component. In general, one should be familiar with and understand the concepts surrounding (1) the basic types of physical motion involved in material deformation, (2) motor-generator actions, (3) resonant and nonresonant operation, and (4) the equivalent circuit concept of the piezoelectric resonator. Each of these is discussed briefly in this section together with the specific electromechanical properties of the PLZT system.

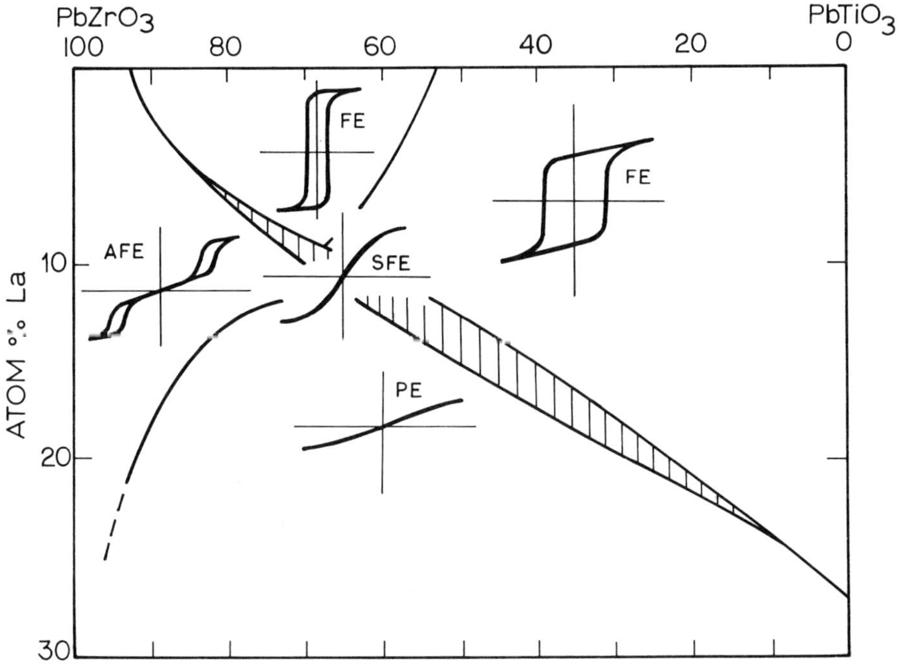

Figure 22 Different types of hysteresis loops exhibited by compositions in the PLZT system.

Table 2 Selected Properties of PLZT Compositions

Composition La/Zr/Ti	Density (g/cm^3)	T_c (°C)	P_R ($\mu C/cm^2$)	E_c [kV/cm (V/mil)]	K	Tan δ (%)
2/65/35	7.98	320	40.0	13.7 (34.7)	650	2.5
6/65/35	7.86	240	34.8	8.7 (22.0)	1210	1.6
7/65/35	7.84	150	34.0	5.3 (13.5)	1850	1.8
8/65/35	7.82	65	30.0	3.6 (9.0)	3400	3.0
9/65/35	7.80	5	0	0	5700	6.0
9.5/65/35	7.79	−10	0	0	5500	5.5
10/65/35	7.78	−25	0	0	5100	5.4
12/65/35	7.74	—	0	0	2200	4.6
14/65/35	7.45	—	0	0	1450	2.3

2/55/45	7.97	340	42.7	13.4 (34.0)	1830	2.4
3/55/45	7.95	312	40.0	13.0 (33.0)	2000	2.2
4/55/45	7.92	285	38.2	12.6 (31.5)	2290	2.2
5/55/45	7.90	250	37.7	9.5 (24.0)	3270	2.2
8/40/60	7.84	240	28.0	17.7 (45.0)	980	1.2
12/40/60	7.74	140	25.0	12.5 (31.8)	1300	1.3
15/40/60	7.65	40	18.0	7.5 (19.0)	5270	1.0
8/90/10	7.83	—	0	0	303	0.4
8/70/30	7.83	30	26.0	2.4 (6.2)	4050	4.7
8/62/38	7.82	105	32.8	4.1 (10.5)	3550	2.7
8/10/90	7.84	355	29.0	37.5 (90.0)	355	1.0

1. Piezoelectric Deformations

The basic deformations of a piezoelectric ceramic are best illustrated by showing a change in physical dimensions as a result of the application of a varying electric field on an electroded and poled plate or disk. For ceramics, it is essential to know that original poling direction and the direction of the applied field (indicated by the electrodes) in order to predict the motion of the material. As we have seen previously, a permanent expansion occurs in the direction of the electric field (polar direction) on initially poling the virgin or thermally depoled ceramic material; and now when considering a dimensional change as a result of the influence of the electric field, it is very important to keep in mind the sign (+ or −) of the applied field in relation to the sign of the original poling field. If the sign of the applied field is the same as that of the original field, expansion will occur in that direction and contraction will occur in the other orthogonal direction; and when the sign of the applied field is negative with respect to the original field, opposite deformations will take place.

Typical examples of the deformations occurring in plates or disks are given in Figure 23. The deformations are highly exaggerated for illustrative purposes since the actual strains are usually very small; [i.e., on the order of a few micrometers ($1\ \mu m = 1 \times 10^{-6}$ m or 40×10^{-6} in.)]. In Figure 23 (a) thickness expansion or contraction in a plate (or length expansion in a long rod) occurs when the electric field is alternately positive or negative, whereas opposite strains (contraction or expansion) take place simultaneously in the lateral dimensions of the plate. Since contraction is considered to be a negative quantity, the lateral electromechanical coefficient that describes this action (d_{31}) is negative and the longitudinal d_{33} coefficient is positive. Similar action occurs in the case of a thin, hollow cylinder or tube (i.e., a continuously curved plate) with electrodes applied to the interior and exterior surfaces. Under the influence of an alternating field, the tube increases and decreases in diameter (hoop mode) as well as in length.

In the case of the radial mode shown in Figure 23 (b), all lateral dimensions of the thin disk are equivalent, and thus the disk expands and contracts in the thickness direction while concurrently producing radial (diametral) dimensional changes.

Shear deformations are obtained by initially polarizing the material with temporary electrodes in one direction, removing the electrodes, and then applying permanent electrodes in an orthogonal direction. Subsequent application of an electric field produces a shearing action, as illustrated for the thickness shear mode in Figure 23 (c).

The bender action shown in Figure 23 (d) is a special case of the thickness/length extensional mode shown in part a. When two

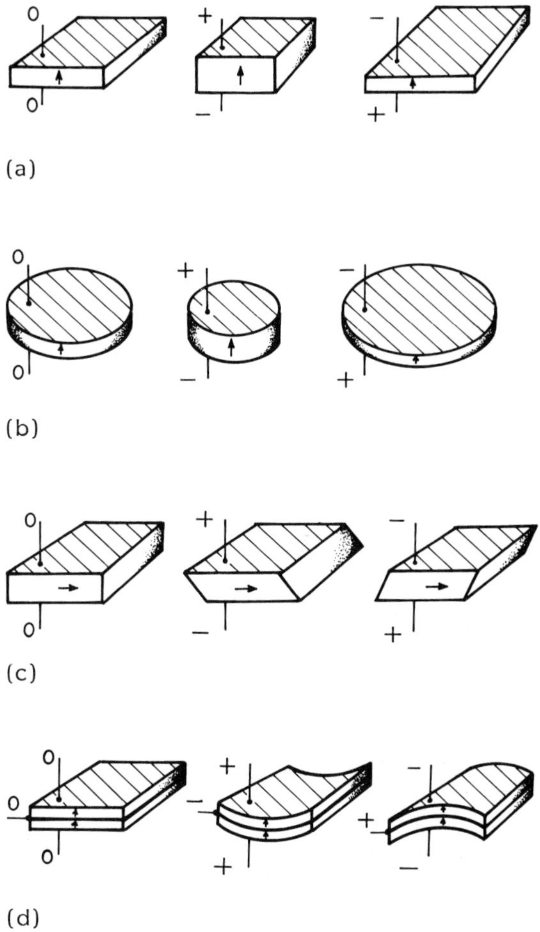

Figure 23 Typical mechanical deformation of poled, piezoelectric plates when subjected to an electric field: (a) thickness and length; (b) radial; (c) thickness shear; (d) bender.

plates are cemented together and activated as shown, one plate will contract (transverse contraction) and the other will simultaneously expand (transverse expansion), producing a bending action much like the familiar bimetallic strip in a thermostat or relay. A thin metal shim or vane is usually cemented between the two plates to provide structural integrity and stress/strain uniformity. The advantage of this mode over the other modes is the larger mechanical displacements (multiplication factors range from 100 to 1000) achievable through the lever action; however, the higher compliance and lower mechanical impedance limits the amount of force it can generate.

Benders may be formed from disks as well as rectangular plates. These circular benders are desirable in certain applications, such as loudspeakers, because they have the highest energy conversion efficiency and possess the largest dielectric capacitance for a given resonant frequency.

The electrical connections to the bender may be arranged in two different ways, series or parallel, because of the conductive center vane which is sandwiched between the two piezoelectric plates. This center vane allows the plates to be poled in the same direction (parallel type) or in opposite directions (series type) and also permits flexibility of the applied electric signal during operation. The bender shown in Figure 23 (d) is of the parallel type, being initially poled with the positive and negative terminals across the top and bottom surfaces, and subsequently operated with one terminal connected to both major surfaces and the other terminal to the center vane. From an energy standpoint, there is no difference between a series-type bender and a parallel type; however, a choice between the two types is usually made on the basis of voltage and mechanical displacement considerations. The parallel operation scheme provides the largest output displacement for a given input drive voltage (e.g., loudspeakers), whereas the series operation provides the largest output voltage for a given input displacement (e.g., phonograph pickup).

Benders may consist of two thin piezoelectric plates cemented together as mentioned previously; or in the simplest case, only one piezoelectric plate, which is bonded to a metal backing plate. Of these two cases, the double piezo plate configuration produces the largest bending action for a given voltage and yields the highest performance; however, it is usually higher in cost. Benders are traditionally called "bimorphs" (two piezo plates) or "unimorphs" (one piezo plate) as a result of the early work carried out in this area at the Clevite Corporation. This nomenclature is now a registered trademark of Vernitron, Inc.

2. Motor-Generator Actions

A piezoelectric ceramic can operate either as a motor or a generator depending on its configuration and application. As a motor, the material composition and configuration are usually optimized for producing maximum displacements with a minimum input voltage (converse piezoelectric effect); as a generator, the same considerations are made for producing maximum output voltage with a minimum of input stress or strain (direct effect). Common examples of motors are ultrasonic cleaners, sonar, sounders, loudspeakers, pumps, and positioners. Examples of generators include gas ignitors, phonograph pickups, microphones, contact fuses, sensors, and accelerometers. In some instances, both types are utilized in a single device, such as filters, delay lines, and piezoelectric transformers.

3. Resonant-Nonresonant Operation

In general, any material of a given size and shape has a natural resonance or frequency at which it will vibrate freely. When put into motion with a mechanical force at its resonant frequency, the body will vibrate at a higher amplitude than at other frequencies and will continue to vibrate until damping forces cause a cessation of movement. This is, of course, true also for piezoelectric solids, but due to the coupling between mechanical and electrical properties, these materials can be electrically stimulated to produce mechanical vibrations at many frequencies either below, above, or at resonance. Most piezoelectric devices are operated below resonance, but some notable exceptions are quartz oscillators for frequency control, ceramic IF resonators, and piezoelectric transformers for high-voltage generation. In addition, some must operate efficiently over wide range of frequencies (e.g., tweeters operating over the range 3–30 kHz). In considering resonant or nonresonant operation, it should be noted that many of the piezoelectric equations are limited to the lower frequencies and do not apply at or near resonance.

4. Equivalent Circuits

The concept of utilizing equivalent electrical circuit techniques for the solution to mechanical problems has long been in use and has been applied with considerable success to piezoelectric elements [67, 68]. The application of this concept involves an analogy between electrical and mechanical quantities:

Electrical quantity	Mechanical quantity
Charge	Displacement
Voltage	Force
Current	Velocity
Capacitance	Compliance
Inductance	Mass
Impedance	Impedance

It is this analogy that is used to set up an electrical representation of a piezoelectric element as shown in Figure 24. The bottom

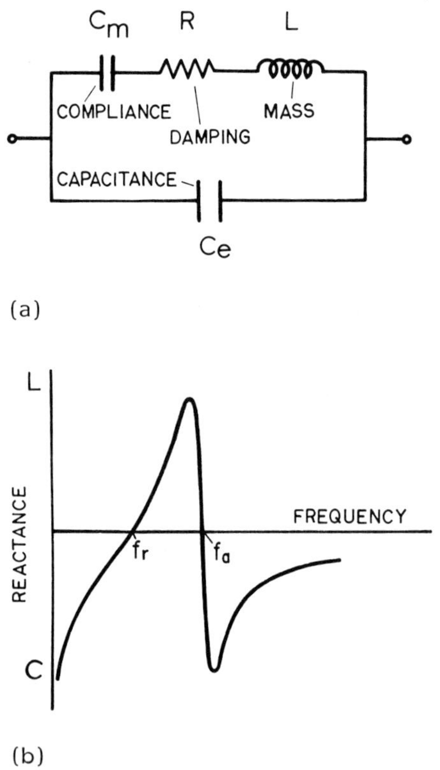

Figure 24 Equivalent circuit of a piezoelectric ceramic near a fundamental resonance: (a) the top branch of the circuit represents the mechanical portion of the ceramic and the bottom branch is the electrical portion; (b) the electrical reactance of a piezoelectric ceramic as a function of frequency.

branch of the circuit represents the dielectric (capacitance) portion of the circuit, and the top branch represents the mechanical portion of the resonant element transformed into the appropriate electrical quantities via the electromechanical coupling factor. The electrical behavior of this piezoelectric resonator is given in Figure 24. As can be seen, below resonance the element behaves as a capacitor; at series resonance it possesses pure ohmic resistance; between series resonance and parallel antiresonance, the element has inductive reactance; and at high frequencies, it again behaves capacitively. The frequency spread between series (f_r) and parallel (f_a) resonance is a measure of the electromechanical coupling of the material (i.e., the larger the spread, the higher the coupling).

5. Piezoelectric Properties

Selected compositions within the PLZT system possess some of the highest electromechanical coupling coefficients attainable in ceramic materials. Some typical values of k_p, d_{33}, and g_{33} for the PLZTs are given in Table 3. For comparison purposes, ceramic barium titanate properties are also listed.

Maximum reported values of k_p and d_{33} are 0.72 and 710×10^{-12} 10^{-12} C/N, respectively. These are found in compositions located within the phase boundary region separating the ferroelectric rhombohedral and tetragonal phases. Over the years, there has been considerable speculation concerning the reasons for the fortuitous maximum in coupling at this phase boundary [69,72,73]. These may be summarized as being due to (1) the existence of a mixture of phases at the boundary, (2) a concurrent maximum in dielectric constant at the boundary, (3) maximum orientational polarization attainable by an electric field in the boundary region, and (4) a maximum in mechanical compliance, permitting maximum domain reorientation without physically cracking. All these factors may contribute to this desirable and important effect; however, more recent work reported by Okazaki [74] and Multani et al. [75] seems to discount the existence of two phases at the boundary when the materials are carefully prepared by chemical means.

F. Optical Properties

The optical transparency of the PLZT ceramics is highly dependent on the concentration of lanthanum oxide in the material. Both low and high concentrations are not conducive to producing high optical transparency, but in each case it is for a different reason. At low levels of La, the higher birefringence and the existence of ferroelectric domain walls combine to produce significant light scattering,

Table 3 Electromechanical Properties of PLZT Compositions

Composition La/Zr/Ti	k_p[a]	d_{33}[b] ($\times 10^{12}$ m/V)	g_{33}[c] ($\times 10^{3}$ V-m/N)
2/65/35	0.450	150	23.0
7/65/35	0.620	400	21.6
8/65/35	0.650	682	20.0
4/55/45	0.670	490	21.6
5/55/45	0.705	578	18.2
6/55/45	0.620	417	16.6
7/55/45	0.565	405	15.1
8/40/60	0.343	—	—
12/40/60	0.470	235	12.3
7/60/40	0.720	710	22.2
$BaTiO_3$[d]	0.360	190	12.6

[a]From Ref. 13.
[b]From Ref. 70.
[c]Calculated from Eq. (4).
[d]From Ref. 71.

whereas at the higher levels of La, second phases precipitated in the grain boundaries lead to the high opacity. The specific concentrations of La that yield high optical transparency are dependent on the Zr/Ti ratio of the compositions. For a 65/35-ratio material, the high transparency range extends from about 8 at % La to approximately 16 at % La.

Thin (10–15 mil thick) polished plates of PLZT (9/65/35 La/Zr/Ti) characteristically transmit about 67% of the incident light and reflect 33%. When broadband antireflection coatings are applied to the major surfaces, transmission is increased to greater than 98%. The surface reflection losses are a function of the index of refraction of the PLZT according to the Fresnel formula,

$$R = \frac{(n - 1)^2}{(n + 1)^2} \tag{23}$$

where R is the reflection coefficient in air from a single surface at normal incidence of light and n is the index of refraction ($n \cong 2.5$ for PLZT). Since the index of refraction varies as a function of wavelength [76], surface reflection losses may also be expected to vary with the wavelength of light.

Optical absorption in the PLZT materials is wavelength dependent, becoming extremely high in the violet (short wavelength) end of the spectrum near 0.37 μm (370 nm). In the infrared portion of the spectrum, transmittance remains high out to approximately 6.5 μm and gradually decreases in a regular manner to 12 μm, where the material is fully absorbing.

The optical transmission characteristics of three PLZT compositions are given in Figure 25. The compositions were selected because they represent materials of distinctly different electrooptic behavior. Composition 9/65/65 is substantially more transparent than either of compositions 7/65/35 or 12/40/60. This is most noticeable in the blue end of the spectrum, where absorption and light scattering predominate.

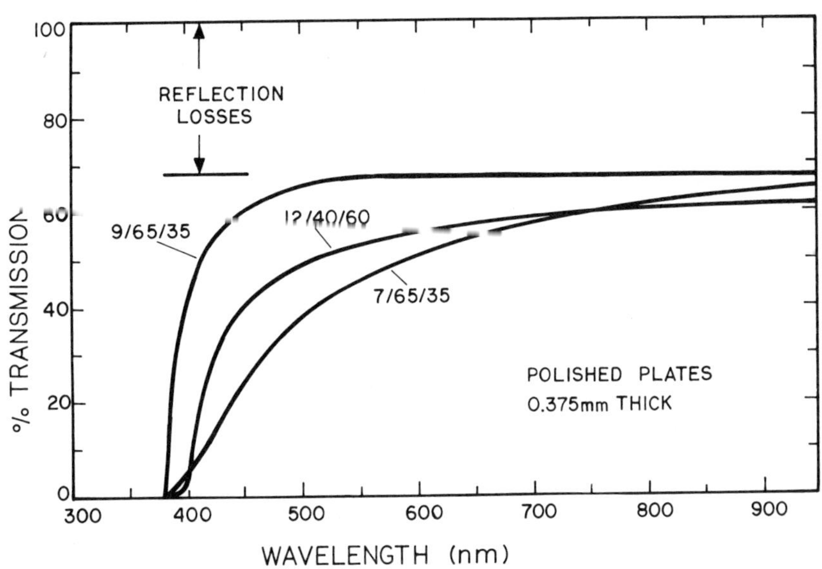

Figure 25 Optical transmission characteristics of selected PLZT compositions.

Both compositions 7/65/35 and 12/40/60 are ferroelectric and hence possess domain walls that scatter light. The tetragonal phase composition 12/40/60 is more transparent than the rhombohedral 7/65/35 composition, particularly in the electrically poled state. All samples shown in Figure 25 were measured in the virgin state.

G. Electrooptic Properties

1. Electrooptic Effects

Several distinct electrooptic effects have been found to be operative in the PLZT materials. The three most common types are illustrated in Figure 26: (a) quadratic birefringence, (b) linear birefringence, and (c) memory scattering. Also included in Figure 26 is a typical setup required for generating each effect together with the observed behavior shown in terms of light-intensity output as a function of electric field.

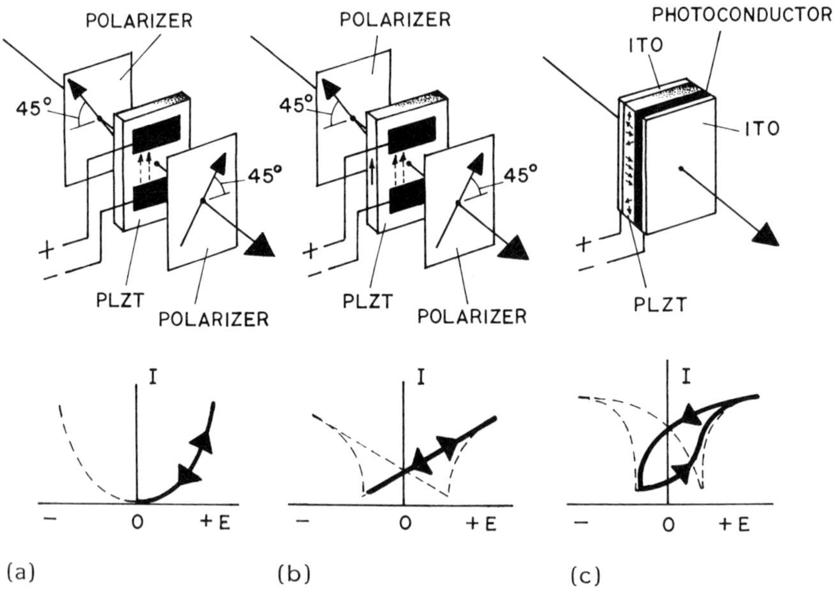

Figure 26 Operational configurations and typical light output response of (a) quadratic, (b) linear, and (c) memory PLZT materials. The heavy accented portions of the response curves represent the usable ranges.

The first and most widely applied of all the electrooptic responses is the quadratic (Kerr) effect. It is generally displayed by those materials that are essentially cubic phase but are located close to the ferroelectric rhombohedral or tetragonal phases. These materials, by virtue of their natural cubic symmetry, do not possess permanent polarization and are not optically birefringent in their quiescent state. As such, they contribute no optical retardation to an incoming polarized light beam; however, when an electric field is applied to the material, an electric polarization (and consequently, birefringence) is induced in the material and retardation is observed between crossed polarizers (ON state). On removing the electric field, the material relaxes again to its cubic state and is in the OFF state. Relaxation times to the OFF state vary with composition but generally range from one to 100 μsec. Turn-on times are on the same magnitude, and ON/OFF ratios as high as 5000 to 1 have been measured with high-efficiency polarizers.

The second type of behavior existing in the PLZTs is the linear (Pockels) effect, which is generally found in high-coercive-field tetragonal materials such as 12/40/60. This truly linear nonhysteretic effect has been found to be intrinsic to the material and not due to domain reorientation processes which occur in the quadratic and memory materials. The linear materials possess permanent remanent polarization; however, in this case the material is switched to its saturation remanence, and it remains in that state. Optical information is extracted from the ceramic by the action of an electric field that causes linear changes in the birefringence, but in no case is there polarization reversal in the material. The experimental setup for observing this effect, as seen in Figure 26, is identical to that for the quadratic response, except that the PLZT plate is prepoled to saturation remanence before using.

A third type of electrooptic behavior which is employed almost exclusively for image storage devices is that of electrically controlled light scattering in memory material. The experimental arrangement involved in observing this effect is also shown in Figure 26. No polarizers are employed since it is predominantly due to light scattering from domains within the material. The orientation of these domains are electrically alterable; and because light is preferentially scattered along the polar direction of the domains, the light transmitted by the PLZT plate is also electrically controllable. In addition, local areas can be polarized to different levels, leading to an ability for storing images with a gray-scale capability and a resolution of at least 30 line pairs per millimeter. Once a given local area is switched to a specific polarization state, it is permanently locked in until it is electrically switched to a new state or the material is heated above its Curie point (thermally depoled), which erases all the polarization states. The means by which local areas are switched independently of each other is provided by the photoconductor layer sandwiched between one of the transparent ITO (indium-tin oxide) electrodes and the PLZT. When light impinges on the photoconductor layer, it reduced is resistivity by several orders of magnitude, electrons from the voltage source are transferred from the ITO electrode to the PLZT, and the local polarization is switched to a new state. Erasure of the total image is performed by flooding the plate with light while the voltage is applied in the positive saturation direction. The maximum contrast ratio may be as high as 100 to 1.

Additional electrooptic effects in PLZT include memory birefringence, depolarization scattering, surface diffraction, and photo-induced changes in refractive index. Although all these effects are more specialized in character than those previously described, they possess reasonable potential for utilization in future applications. Table 4 summarizes all the known electrooptic phenomena occurring in the PLZT materials.

2. Electrooptic Coefficients

A convenient and physically significant comparison of the electrooptic effects in PLZT ceramics with other electrooptic crystals may be made by examining their respective quadratic and linear electrooptic coefficients, R and r_c. These coefficients are listed in Table 5 for various PLZT compositions as well as for several other selected electrooptic crystals. In general, it may be noted that the values for the PLZT ceramics are exceptionally high and exceed those of most other crystalline materials. Indeed, it is primarily these high electrooptic coefficients that make the PLZT materials especially attractive in electrooptic applications.

The temperature dependence of the R coefficient is given in Figure 27. Since R can be seen to vary by a factor of 3.5 from −40 to +80°C, the operating voltage can be expected to change by a factor of $\sqrt{3.5}$ or 1.9. In some specific applications, a change this large would require temperature compensation; however, for most less critical devices, compensation would not be necessary if an intermediate operating voltage were selected.

H. Aging

Any poled ceramic is theoretically in a constant state of depoling as a function of time. This characteristic, referred to as aging, fortunately does not vary linearly with time. It has been found empirically that most ceramic properties that are dependent on the poling process, such as coupling, dielectric constant, and resonant frequency, exhibit a fairly constant change during each decade of time. For example, planar coupling may decrease by 0.2% after the first

Table 4 Electrooptic Effects in PLZT Ceramics

E/O effect	Electrical address	PLZT type	Polarized light	Devices
Birefringence	Transverse	Nonmemory, memory	Yes	Shutters, modulators. color filters, goggles, displays, memories
Depolarization scattering	Longitudinal	Nonmemory	Yes	Shutters, modulators, displays
Birefringent fringing	Longitudinal	Nonmemory	Yes	Displays
Birefringence	Longitudinal (strain bias)	Memory	Yes	FERPIC, image storage, displays, shutters
Scattering	Longitudinal	Memory, nonmemory	No	CERAMPIC, image storage, modulators, displays
Diffraction (surface)	Longitudinal	Memory	No	FERICON, image storage
Photoferroelectric (space charge)	Transverse, fringing	Nonmemory, memory	Yes	Holographic recording, image storage
Photoelastic (stress-induced $\overline{\Delta n}$)	Transverse	Nonmemory	No	Optical waveguides, modulators

Table 5 Electrooptic Coefficients R and r_c for Selected Ceramics

Composition La/Zr/Ti	R ($\times 10^{16}$ m^2/V^2)	r_c ($\times 10^{10}$ m/V)
8.5/65/35	38.60	–
9/65/35	3.80	–
9.5/65/35	1.50	–
10/65/35	0.80	–
11/65/35	0.32	–
12/65/35	0.16	–
8/40/60	–	1.0
12/40/60	–	1.2
KTN (65/35)	0.17	–
$LiNbO_3$	–	0.17
KDP	–	0.52
SBN	–	2.10
$LiTaO_3$	–	0.22

Source: Refs. 13 and 77.

day of poling, an additional 0.2% after the next 10 days, and another 0.2% after the 100 days. Extending this up to 10,000 days, or 27 years, results in a total decrease in coupling of only 1%. Typical ranges of aging per time decade for planar coupling, relative dielectric constant, and resonant frequency are −0.2 to −2.3%, −0.6 to −5.8%, and +0.08 to +1.5%, respectively. The aging of optical memory states in the transparent ceramics have not yet been measured, but they are anticipated to be in a range similar to that for planar coupling.

Since the process of aging begins the moment the ceramic is polarized, it is very difficult to specify exactly a given property which is dependent on its polarization condition. One can, however, predict a given property change with some accuracy if the time of poling is known and the aging characteristics for that composition are given.

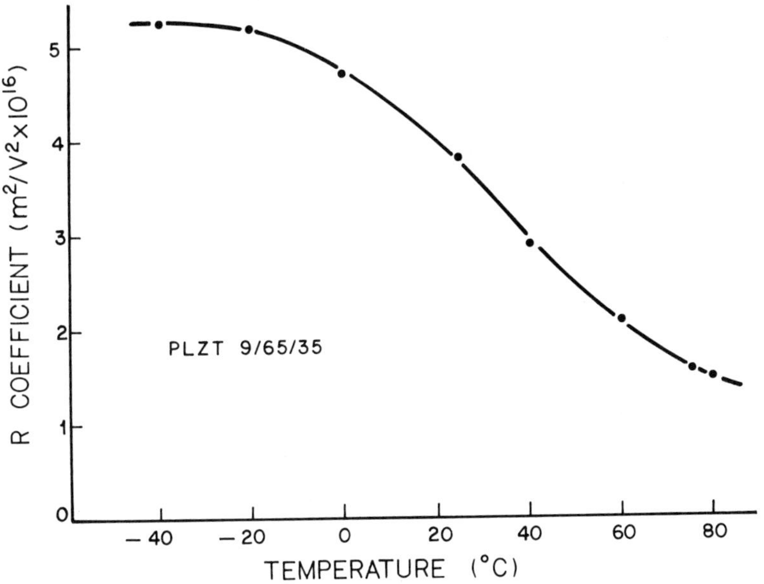

Figure 27 Temperature dependence of the transverse quadratic electrooptic R coefficient for PLZT 9/65/35. (From data given in Ref. 44.)

It is also important to know that aging can be interrupted or decelerated by repoling the ceramic, in which case aging begins all over again.

The process of aging may be accelerated by exposing the ceramic to (1) a strong electric depoling field, (2) temperatures that approach the Curie point, (3) high mechanical stresses, and (4) combinations of these factors. Therefore, care must be exercised in selecting a given ceramic for a specific use condition. Ceramic compositions are often formulated to be more resistant to a given depoling environment. A common rule of thumb on temperature environments (the most common depolarization mechanism) is that the ceramic should be subjected to temperatures no higher than one-half of the Curie point.

Aging in ferroelectric ceramics is a complicated process and may occur as a result of several different mechanisms. Jonker [78] summarizes these as (1) a gradual equilibration of the domains with minimization of elastic and dielectric free energy, (2) a segregation of impurities and vacancies on domain walls and crystallite boundaries, and (3) an ordering of impurities and vacancies inside the ferroelectric domains with respect to the polar axis. Okazaki [79], on the other hand, attributes aging primarily to space charge effects in which charge compensation gradually occurs inside every domain because of the effect of the depolarizing field which results from the ordered spontaneous polarization.

VII. APPLICATIONS

A. General

Like most materials, the successful application of piezoelectric, ferroelectric, and electrooptic ceramics is highly dependent on the relative ease with which these materials can be adapted to useful and reliable devices. This is, to a great extent, the reason that piezoelectric materials have been so successful over the years in finding increasing numbers of applications. Their simplicity, small size, low cost, and high reliability are very attractive features to the design engineer. Examples of piezoelectric applications include the following:

High-voltage generators: gas ignitors, spark pumps, impact fuses, flash bulb actuators
Ultrasonic generators: ultrasonic cleaners, sonar, atomizers, ultrasonic weldding, intrusion alarms, pest control devices, flaw detectors
Sounders: loudspeakers, tweeters, tone generators, head sets, buzzers, alarms
Sensors: phonograph pickups, accelerometers, hydrophones, auto diagnostic devices, flaw detectors
Resonators: IF filters, surface wave filters, delay lines, piezo transformers
Miscellaneous: relays, pumps, motors, fans, positioners, printers, touch controls

On the other hand, ferroelectrics (designated as materials undergoing electrical switching through a hysteresis loop) have been notably unsuccessful in securing any practical device applications. This is largely due to reliability problems stemming from fatigue (microcracking) and internal biasing (space charge) behavior, which leads to a variable and unpredictable switching threshold. Ferroelectric memories, energy conservation devices, and security locks are three applications that have been under study for a number of years.

The most recent area of applications to emerge for ceramics is electrooptics. Here, too, the key to success depends on simplicity, cost, and reliability. Two successful applications that have developed so far are in the category of optical shutters: (1) a wide-area eye-protective device for the military, and (2) a high-speed optical-strip-chart data recorder. Applications involving optical modulators, color filters and displays are presently being developed as an outgrowth of the optical shutter technology. Continuing work on image storage displays also promises to open up a new area of optically scanned displays for information storage and retrieval.

Another category of applications for ferroelectric-type materials is that of high-dielectric-constant capacitors. Although not the subject of this chapter, capacitors are extremely important and constitute a significant portion of the electronic ceramic business as a whole. Most ceramic capacitors are, in reality, high-dielectric-constant ferroelectric compositions which have their ferroelectric (hysteresis loop) properties suppressed with suitable chemical dopants while retaining a sufficiently high dielectric constant over a broad temperature range. Barium titanate was historically the first composition used for high-dielectric-constant capacitors, and it (or its variants) is still the industry standard today. In tune with shrinking electronic circuits, capacitor sizes are constantly being reduced through new technologies developed for multilayer and barrier layer capacitors.

Special induced-ferroelectric (relaxor) materials are currently being investigated for such applications as optical interferometers, phase-correction mirrors, and other multiposition optical components. These materials have the ability to generate high mechanical strains as a function of the applied electric field. This effect is called electrostriction and is so designated because of its similarity in behavior to the electrictive effect common to all dielectric materials where ferroelectricity and piezoelectricity are not involved. Electrostriction is differentiated from piezoelectricity in that strain for the latter is an odd function (linear) of the electric field and strain for electrostriction is an even function (quadratic) of the field (in this sense, the quadratic electrooptic PLZT materials can also be considered optical electrostrictive materials). Cross et al. [80] have shown that certain relaxor ferroelectrics offer advantages over normal piezoelectric transducers, such as (1) large strains comparable to the best piezoelectric ceramics, (2) excellent positional reproducibility, (3) no poling required, and (4) very low thermal expansion coefficients.

The multilayer technology used in the ceramic capacitor industry is one of the driving forces for the development of new electrostrictive devices. Although electric fields across alternate layers of the structure are in opposite direction, the total displacement is additive since the strain in electrostrictive materials is the same for both positive and negative fields. This simplifies the construction and operation of the devices while achieving displacements many times larger for the same voltage.

Piezoelectric composites represent the latest in technology trends for literally squeezing the last bit of high performance from a piezoelectric transducer. When one deliberately introduces a second phase in a material, such as in a composite, connectivity of the phases is a critical parameter. There are 10 connectivity patterns possible in a two-phase solid, ranging from 0-0 unconnected three-dimensional checkerboard pattern to a 3-3 interpenetrating pattern in which both phases are three-dimensionally self-connected. Some of these connectivity patterns are particularly well suited for decoupling longitudinal and transverse piezoelectric effects such that materials with significantly enhanced (up to a factor of 100 or more) piezoelectric properties are possible. Newham et al. [59] have demonstrated considerable ingenuity in designing, fabricating, and evaluating the many types of diphasic structures.

B. Specific Applications

1. Piezoelectric Positioners and Relays

Thin, electroded piezoelectric plates can be used as mechanical positioners and relays when the required movements are very small (i.e., in the range of a few hundred micrometers or less). A design engineer using such a device should be familiar with (1) the motional modes available from the plates, (2) the calculations required to predict the magnitude of the resultant motions, and (3) the means by which such small deflections are amplified into larger translations without an increase in operating voltage. Good sources for this information as well as construction details are given in References 81 to 83. Included below are some sample calculations.

Consider a poled and electroded rectangular piezoelectric plate as shown in Figure 28. This plate is to be operated as a micropositioner in a dc mode at a voltage of 100 V. When voltage is applied, the plate will expand or contract in the thickness direction depending on the polarity of the voltage with respect to the poled direction of the ceramic. Simultaneously, the length of the plate will contract and expand in similar fashion. The magnitude of these movements may be calculated from Eq. (3) remembering that the d coefficient is the ratio of resulting strain to applied electric field. For the thickness mode (d_{33} is appropriate) case, this may be expressed as

$$d_{33} = \frac{S}{E} = \frac{\Delta t}{t/(V/t)} = \frac{\Delta t}{V} \tag{24}$$

or

$$\Delta t = d_{33}V \tag{25}$$

indicating that the change in thickness of the plate is dependent on the material d coefficient and the voltage and is independent of the plate thickness. Substituting values into Eq. (25) and taking a d_{33}

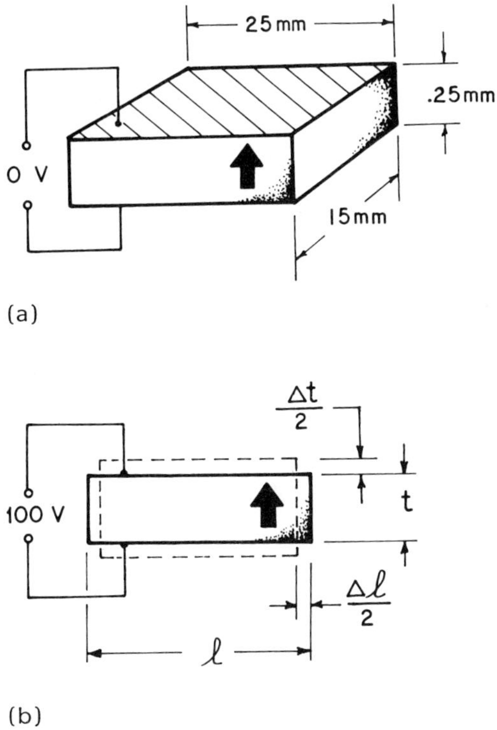

Figure 28 Poled, piezoelectric ceramic plate: (a) before applying voltage; (b) after applying 100 V.

coefficient of 578 × 10^{-12} m/V from Table 3 for composition 5/55/45 yields

$$\Delta t = 578 \times 10^{-12} \text{ m/V} \times 100 \text{ V} = 0.0578 \times 10^{-6} \text{ m} = 0.0578 \text{ µm}$$

or 2.3 µin.

This is, indeed, an infinitesimally small movement. This value can be increased, but in a practical sense, only by raising the voltage. There is, of course, a limit on how high one can raise the voltage because either depoling (back switching of the hysteresis loop) or dielectric breakdown will eventually occur.

In the length-extensional case, the transverse coefficient is given as

$$d_{31} = \frac{S}{E} = \frac{\Delta \ell}{\ell(V/t)} = \frac{\Delta \ell t}{\ell V} \tag{26}$$

or

$$\Delta \ell = d_{31} = \frac{\ell}{t} V \tag{27}$$

and in this case, the movement is multiplied by the dimensional factor ℓ/t. Again substituting values into Eq. (27) and assuming the appropriate d_{31} coefficient of -268×10^{-12} m/V (the negative sign indicates contraction), we obtain

$$\Delta \ell = -268 \times 10^{-12} \text{ m/V} \times \frac{25 \times 10^{-3} \text{ m}}{0.25 \times 10^{-3} \text{ m}} = 100 \text{ V}$$

$$= -2.7 \times 10^{-6} \text{ m}$$

$$= -2.7 \ \mu\text{m or } 108 \ \mu\text{in.}$$

This is an approximate 50-fold increase in movement for the same voltage, despite the fact that the d_{31} coefficient is only approximately one-half the value of d_{33}.

Now consider a rectangular plate of the same external dimensions but being composed of two oppositely poled plates, as in the bimorph shown in Figure 29. The deflection at the end of the cantilever mounted bimorph is given by Germano [81] as

$$h = \frac{3}{2} d_{31} \times \frac{\ell^2}{t^2} \times V \tag{28}$$

and substituting values into Eq. (28) yields

$$h = 1.5 \times -268 \times 10^{-12} \text{ m/V} \times \frac{(25 \times 10^{-3} \text{ m})^2}{(0.5 \times 10^{-3} \text{ m})^2} \times 100 \text{ V}$$

$$= 100.5 \times 10^{-6} \text{ m}$$

$$= 100.5 \ \mu\text{m or } 4016 \ \mu\text{in.}$$

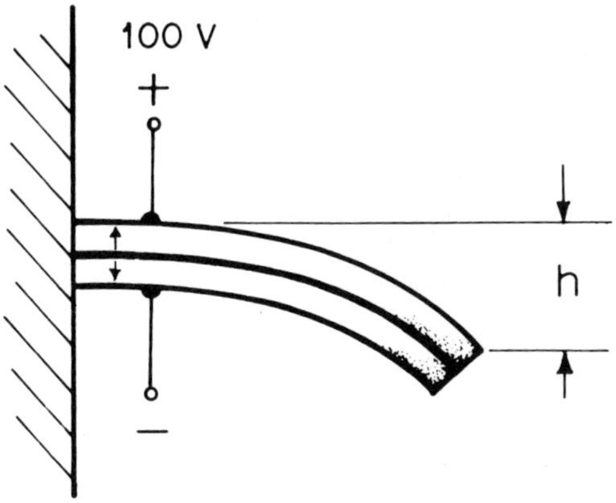

Figure 29 Series-type, cantilever piezoelectric bender illustrating output bending action during voltage application; reversing the voltage polarity causes the bender to deflect in the opposite direction (bending action is highly exaggerated for simplicity).

This notably larger value represents an additional multiplication factor of 37 over the length-extensional mode and a factor of 1740 over the thickness mode. It points out the reason why benders are used whenever possible since they provide the highest deflection for the least amount of voltage. Nevertheless, benders are not appropriate in some applications because they are mechanically compliant and are not able to generate significant forces.

2. Loudspeakers, Tweeters, and Sounders

One of the more recent successful applications for piezoelectric ceramics has been in the area of piezoelectric loudspeakers, particularly tweeters. Compared with the conventional electrodynamic tweeter, the piezo tweeter is lighter, shallower in design, and more power efficient, more reliable, has better transient response, and is cost competitive. These desirable features make them ideally suited for high-power audio applications, hi-fi and stereo audio systems, portable audio devices, and sound systems for the automobile. Figure 30 illustrates several different types of piezo speakers as well as construction details.

The piezo tweeter development is a notable achievement because it represents the successful application of several technologies to reach a common objective. These involved establishing proper operating principles and design criteria, development of high-quality ceramics, and introducing unique processing techniques for manufacturing.

(a)

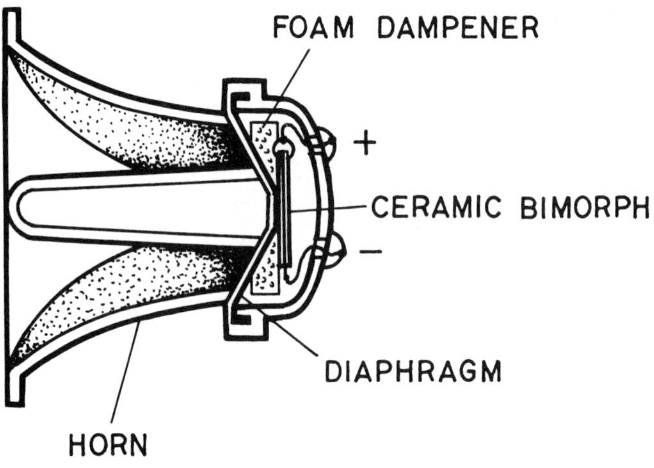

(b)

Figure 30 Examples of piezo tweeters manufactured by Motorola, Inc.: (a) models (top, left to right), 3.5-in. superhorn, satellite tweeter, 3.75-in. exponential horn and (bottom) 2 × 5-in. wide dispersion horn; (b) cross-sectional view of superhorn shown in part (a).

Work was begun by Schafft in the mid-1960s and subsequently reported by him and coworkers in later years [84–86].

The basic principle of operation involves a circular ceramic bender (bimorph) which drives a speaker cone in an unsupported "momentum drive" mode. As an alternating voltage is applied to the disk bender, it will dish in or out, depending on the polarity and the magnitude of the applied signal. The cone, which is lighter than the driver, will thus be moved in a reciprocating in-and-out motion by the momentum of the bender. A circular bender was chosen because (1) it is a high-compliance transducer which is best suited for acoustic devices that radiate into air; (2) it has the highest dielectric capacity, which is required for delivering power from a low voltage source; (3) the circular bender has the highest electromechanical coupling factor of all bender configurations; and (4) it has the highest flexural resonant frequency for a given dimension or volume. A typical frequency response curve for a 3.5-in. piezo superhorn tweeter using a 0.89-in. (22.6-mm)-diameter bender bimorph is given in Figure 31. From only a 2.8 V input, this speaker will generate over 100 dB sound pressure on a microphone placed 0.5 m away. The frequency response is fairly flat from 4 to 30 kHz.

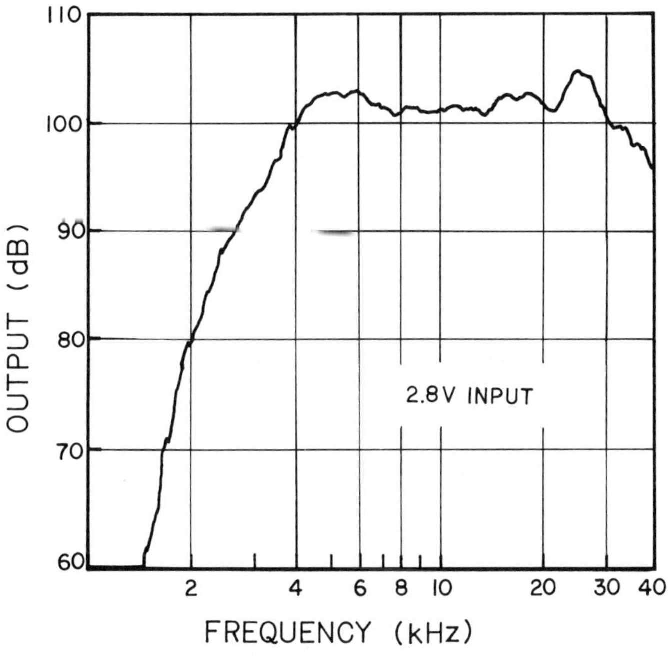

Figure 31 Typical frequency response of a piezo superhorn tweeter.

The ceramic drivers (benders) used in the piezo tweeters require the highest-quality material available on the market today. Since some of the individual ceramic disks may be as thin as 3 mils (0.075 mm), porosity must be at a bare minimum and grain size should be tightly controlled. Specifications [85] established for the ceramic included a voltage-sensitivity merit factor:

$$\text{Voltage sensitivity} = K \times k_p^2 \qquad (29)$$

and an energy output merit factor:

$$\text{Energy output} = E_D^2 \times K \times k_p^2 \qquad (30)$$

where E_D is the magnitude of the electric field (60 Hz applied signal) at which k_p has dropped to 90% of its initial value of 0.63 or greater. Both of these empirical figures of merit should be maximized for the highest tweeter performance. Tweeters today are manufactured in millions of units per year by a number of different vendors.

3. Gas Ignitors

A poled piezoelectric ceramic disk is a neat, compact, portable power supply if one only had the means to extract the electrical power from it. As we have learned, this should be possible by following the example of the Curie brothers and discovering the direct piezoelectric effect all over again! Piezoelectric impact gas ignitors utilize this effect in various applications involving gas appliances, such as stoves, heaters, patio grills, and cigarette lighters. It is a simple and effective device which is already popular in Europe and Japan and is becoming increasingly so in the United States. A simple example of a gas (spark) ignitor is given in Figure 32 together with a typical construction view.

The ignitor works by applying a force to a properly poled and electroded ceramic whose electrodes are connected to terminals in a preset air gap similar to an automobile spark plug. As the pressure (pressure = force per unit area) builds up on the ceramic from the mechanical input, the voltage across the ceramic increases until the air gap between the terminals breaks down, producing a spark or series of sparks. On releasing the force, voltages of like magnitude but of opposite sign will be produced, yielding additional sparks. If the force is applied too slowly, no sparking action occurs because of electrical leakage across the electrodes of the ceramic. This is

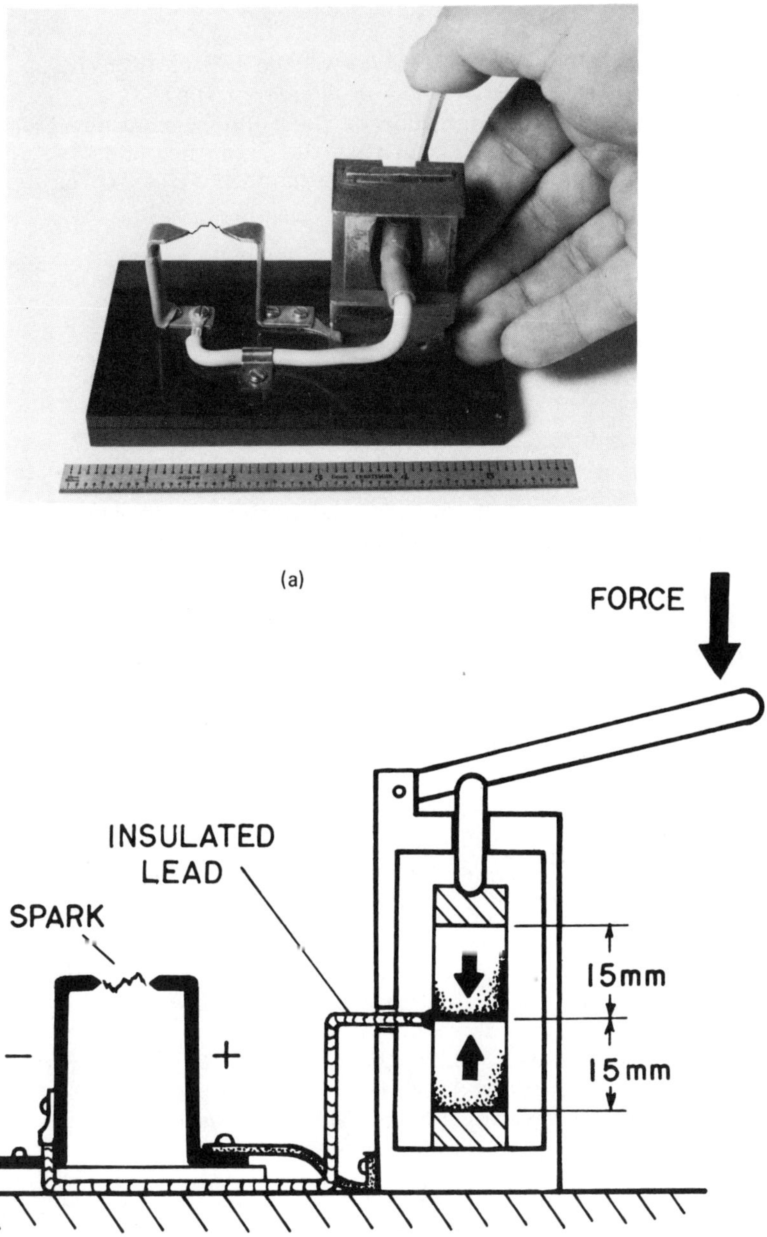

(b)

Figure 32 (a) Laboratory model, squeeze-type spark ignitor; (b) construction view of the device.

the reason why many ignitors are of the impact (momentary force) variety rather than of the slower squeeze type.

To calculate the magnitude of the voltage generated from a given device, we should first recall that the g coefficient is the ratio of the open-circuit electric field to the applied stress (pressure) according to

$$g = \frac{E}{T} = \frac{V}{t/T} \tag{31}$$

Rearranging Eq. (31), we obtain

$$V = g \times t \times T \tag{32}$$

indicating that the voltage produced is directly proportional to the material g coefficient, the thickness t of the ceramic, and the magnitude of the applied stress. Assuming an applied stress of 8000 psi (5516×10^4 N/m^2) and obtaining a g_{33} value of 23×10^{-3} V-m/N from Table 3 for composition 2/65/35, we substitute values into Eq. (32):

$$\begin{aligned} V &= 23 \times 10^{-3}\ \text{V-m/N} \times 15 \times 10^{-3}\ \text{m} \times 5516 \times 10^{4}\ \text{N/m}^2 \\ &= 19{,}000\ \text{V} \end{aligned}$$

This voltage of 19 kV is easily high enough to produce a spark across a 6-mm (1/4-in.) gap and possesses sufficient energy to ignite most gases.

4. PLZT Shutters, Modulators, and Color Filters

Thin polished plates of PLZT, when used in conjunction with polarized light, make excellent wide-aperture electronic shutters. Their advantages over competing technologies such as mechanical shutters and LC (liquid crystal) light gates include (1) faster response time, in the low-microsecond range; (2) less vibration; (3) lighter weight; (4) thinner profile; and (5) wider operating temperature range, from −40 to +80°C. Their main disadvantages are (1) low-ON-state transmission of about 15%, primarily a result of the polarizers; and (2) high operating voltages required to reach a full ON state. Despite the low ON-state transmission characteristics, contrast ratios in excess of 5000 to 1 are regularly achieved because of the deep OFF state (0.002%) possible with high-efficiency polarizers. It has also been found that for a large number of applications, the ON-state transmission of 15% is quite acceptable. In the case of eye-protective devices, 15% transmission is adequate because of the logarithmic response of the eye to light intensity. A typical ON-state (half-wave voltage) transmission curve of a PLZT shutter as a function of wavelength is given in Figure 33 together with spectral response curves for the Polaroid HN-32 polarizers. As can be seen, the spectral response of the shutter device correlates with that of the polarizers. The PLZT response is flat (Figure 25) over the optical range. Also, the 15% transmission figure is an average value over the total optical spectrum, which can be expected to increase to approximately 20% with antireflection coatings on the PLZT and polarizers [77].

The polarizer-PLZT configuration usually employed in a shutter device is shown in Figure 34. The interdigital electrodes serve to decrease the required operating voltage by reducing the gap between the positive and negative finger electrodes. These electrodes are produced either by vacuum deposition on the surfaces or by machining and plating small grooves (1.5 mil deep) into the surfaces. Since the width of the fingers is small (1.5 mil or 0.0375 mm) and the gap fairly large (40 mils or 1 mm), the electrodes themselves block only

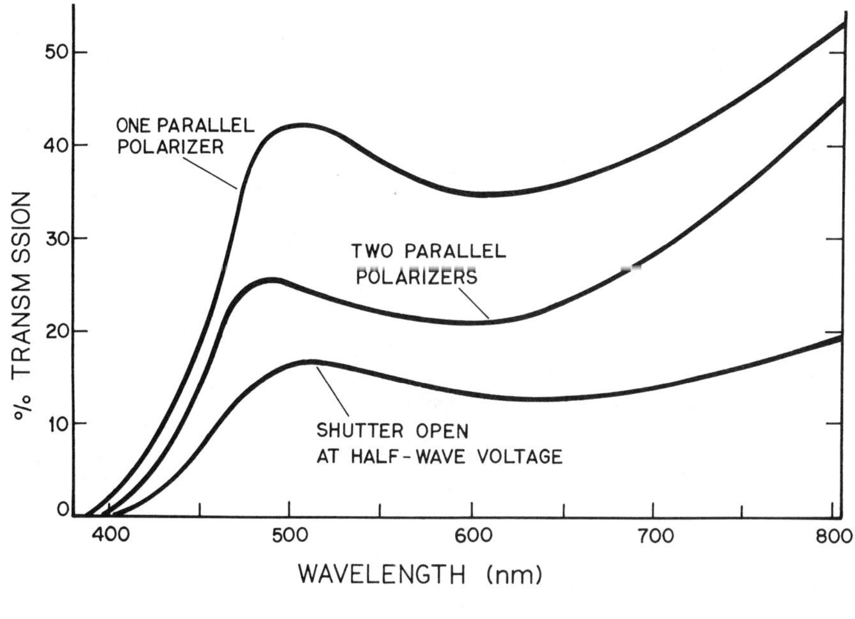

Figure 33 Spectral response of an activated PLZT 9/65/35 electrooptic shutter (25 mm diameter) using crossed HN-32 polarizers.

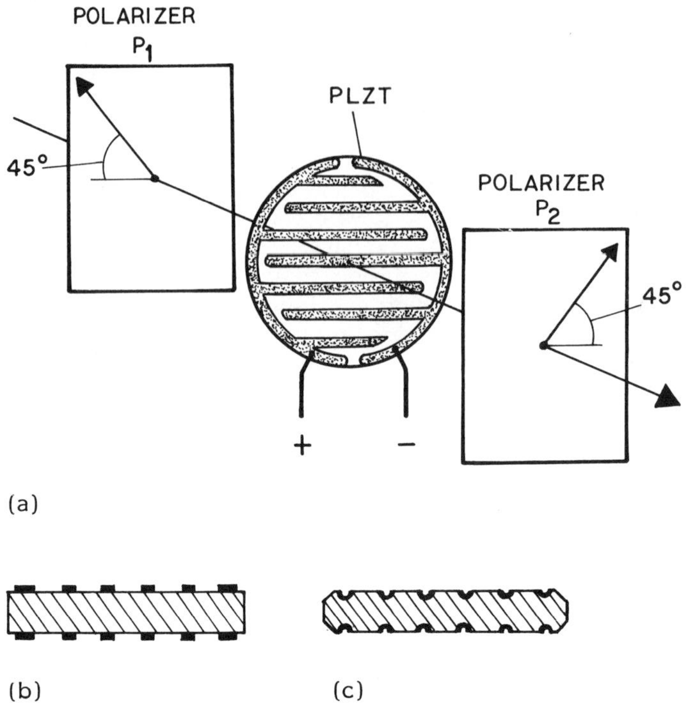

Figure 34 Typical device configuration: (a) an electrooptic shutter using crossed polarizers; (b) vacuum deposited, surface electrodes; (c) chemically plated, slotted electrodes.

a small percentage of the incident light. These electrodes are thin enough so that they can be optically defocused during viewing without any noticeable effect. To keep the electric fields between the electrodes in a more-or-less favorable direction (i.e., 90° to the direction of light transmission), it is desirable to maintain an electrode gap-to-plate thickness ratio of greater than 1. Reducing this ratio to substantially less than 1 will result in undesirable higher operating voltages. With a 1-mm (40-mil) gap on a 0.375-mm (15-mil)-thick plate as in a normal configuration, the gap-to-thickness ratio is a favorable 2.7.

The slotted electrodes which are partially embedded into the surface of the PLZT by approximately 0.0375 mm (1.5 mils) are usually preferred because they produce a more uniform transverse electric field and hence a more uniform transverse electrooptic effect. This leads to a beneficial reduction in half-wave voltage (ON state) of approximately 10% as well as minimizing of residual space charge effects near the electrodes [87]. Since these electrodes do not fully penetrate the thickness of the PLZT wafer, there is the question of just how far the electric field penetrates the wafer. The depth of penetration can be estimated from a measurement of the capacitance of the wafer. Knowing that the total finger length of the electrodes on a 50-mm-diameter wafer with 1-mm gaps is 314 cm, the capacitance of the double-sided electroded wafer is measured at 18.3×10^{-9} F and the dielectric constant of 9/65/35 is 5700 from Table 2; we can calculate the field depth as follows:

$$C = K \frac{A}{g} \times 8.85 \times 10^{-14} \tag{33}$$

where A is the area in cm^2, g the gap length in cm, C the capacitance in farads, K the relative dielectric constant, and the remaining factor is the permittivity of free space in F/cm. Identifying the area as the length of the finger electrodes times the depth of penetration d and rearranging Eq. (33), we have

$$\begin{aligned} d &= \frac{C \times g}{K \times \ell} \times \frac{1}{8.85 \times 10^{-14}} \\ &= \frac{18.3 \times 10^{-9}\ \mathrm{F}(0.1\ \mathrm{cm})}{5700 \times 314\ \mathrm{cm}} \times \frac{1}{8.85 \times 10^{-14}\ \mathrm{F/cm}} \\ &= 0.0115\ \mathrm{cm\ or\ 4.5\ mils} \end{aligned} \tag{34}$$

This depth of penetration is, of course, only an estimate; but it does point out that for a 0.375-mm (15-mil)-thick wafer with 0.0375-mm-deep slots, the electric field lines do not extend fully through the thickness of the wafer. For a 0.5-mm (20-mil) electrode gap, this depth of penetration is reduced slightly to 0.1 mm (4-mils), due largely to the reduced gap to thickness ratio of 1.33.

Using this value of field penetration, which effectively defines the true optical thickness in Eq. (19), and remembering that two surfaces are being activated in series (t = 2d), the operating voltage of a PLZT shutter or a color filter may now be calculated. This is first illustrated graphically for a shutter by referring to the Kerr retardation chart in Figure 10. Since thickness t is equal to twice the depth of penetration, or 0.23 mm, we follow the dashed horizontal line representing thickness at 0.23 mm until it intersects with the appropriate dashed vertical line representing the half-wave retardation (Γ) for a full ON (clear white) condition. This would be at a value of

Γ = 275 nm, which is one-half of the wavelength of green light (550 nm). A circular dot indicates this intersection at an estimated $\overline{\Delta n}$ value of 0.0012. Taking this value, substituting it into Eq. (22), and rearranging, we have

$$E = \left(\frac{-2 \times \overline{\Delta n}}{n^3 \times R} \right)^{\frac{1}{2}} \tag{35}$$

$$= \left[\frac{-2 \times 0.0012}{(2.5)^3 \times 3.8 \times 10^{-16} \ m^2/V^2} \right]^{\frac{1}{2}}$$

$$= 6.63 \text{ kV/cm or } 16.2 \text{ V/mil}$$

and since

$$V = E \times t$$

$$= 6.36 \text{ kV/cm} \times 0.1\text{-cm gap}$$

$$= 636 \text{ V}$$

This value of 636 V agrees fairly well with the value of 620 V obtained from the graph in Figure 2 of Reference 44. In practice, the agreement obtained between calculated and measured voltages for these devices depends heavily on the material uniformity (R coefficient) and the actual device configuration (gap-to-thickness ratio and electrode type). Therefore, in some instances, the disparity can be quite large. It should also be remembered that the value above is for a double-sided shutter; if a single-sided shutter is used, the voltages will be approximately doubled. As a result of the significantly higher voltages, one can readily see why single-sided shutters are rarely used. A method of designing shutters for minimum voltage is described in Reference 88.

When operating the shutter as an optical modulator, one is especially interested in the shape of the transfer function relating optical transmission and operating voltage. This is illustrated in Figure 35 up through the first maximum (half-wave voltage) and minimum for white light. The quadratic nature of the response (S-shaped) curve is readily evident up to the first maximum. At higher voltages, maxima and minima occur according to the combined effect for all the visible wavelengths in white light. Some of the individual wavelengths (red, blue, and green) are also shown in Figure 35. The half-wave voltage for this shutter/modulator is approximately 1000 V because it is made from a more temperature stable composition (9.5/65/35), which possess a lower R coefficient (1.5×10^{-16} m^2/V^2).

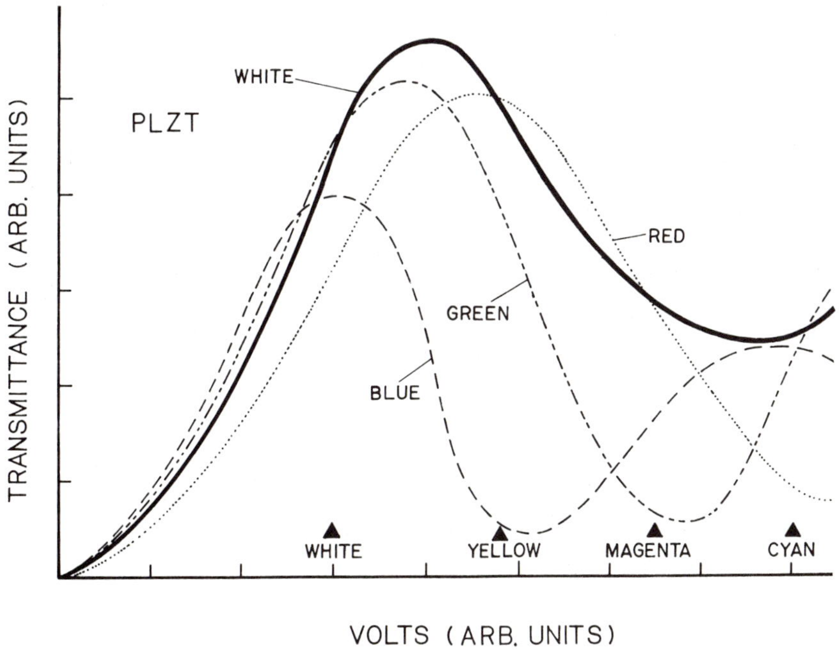

Figure 35 Optical transmittance curves as a function of operating voltage for a typical shutter when viewed with various colored light sources. The colors with the triangular markers are those observed at the various voltages when a white light source is used.

Operating voltages for color filters may also be calculated from Figure 10 because color generation is simply a further phase retardation of the same shutter configuration. Assuming the same effective thickness as in the preceding example, the intersection of the horizontal dashed line with the vertical dashed line representing a red color is indicated by a triangular marker. This $\overline{\Delta n}$ value is estimated to be 0.0024. Substituting this value into Eq. (35), we have

$$E = \left[\frac{-2 \times 0.0024}{(2.5)^3 \times 3.8 \times 10^{-16}\ \mathrm{m^2/V^2}}\right]^{\frac{1}{2}}$$

$$= 9.0\ \mathrm{kV/cm\ or\ 22.9\ V/mil}$$

and

$$V = 9.0\ \mathrm{kV/cm} \times 0.1\text{-cm gap}$$

$$= 900\ \mathrm{V}$$

Thus a first-order red can be produced at a voltage approximately 50% higher than that required to turn the shutter to a clear white ON state. Voltages for the other colors of yellow, red, and blue can be calculated similarly. In practice, since the higher voltages are usually undesirable, a static retardation plate may be inserted in series with the PLZT plate to reduce this voltage by providing an optical bias to the system. Such plates are commercially available and usually consist of a stretched clear plastic film embedded in a clear plastic backing plate. They are manufactured to provide a given amount of phase retardation such as quarter-wave of green light, half-wave, and full-wave retardation. If a full-wave blue plate is utilized as a color bias (blue is then the OFF condition color), less voltage will now be required to generate other colors because the $\Delta\Gamma$ now required to achieve the other colors is less. These voltages may also be calculated from Figure 10 and Eq. (35). The actual colors produced from such a color filter are very broadband. Typical spectral response curves are given in Figure 36.

An important point to note from Figure 10, in relation to the colors generated, is that the colors indicated in the figure are those

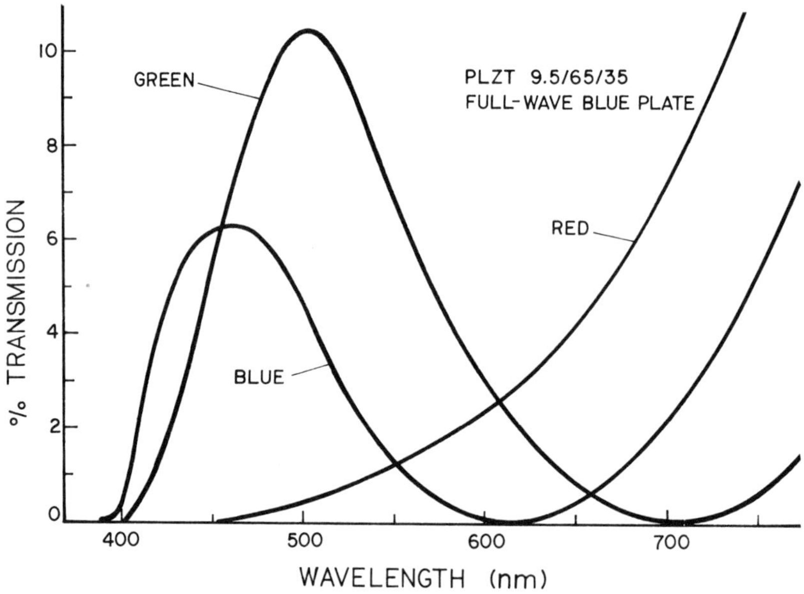

Figure 36 Spectral response curves of a typical PLZT shutter/color filter with a full-wave blue retarder plate inserted adjacent to the PLZT; a blue color is generated with no voltage applied and red and green with correspondingly higher voltages.

actually obtained under crossed polarizers. When we speak of a full-wave red retardation plate, for instance, this refers to the color obtained, not to the amount of retardation for the color red. In actuality, a red plate possesses a full-wave retardation for green (560 nm); and as we learned previously, a full-wave retardation condition under crossed polarizers is a blocking condition for that wavelength. Therefore, subtracting a green color from the spectrum results in a magenta or red color.

In some instances it is preferred to use the second- or third-order colors instead of the first order. This is accomplished simply by inserting more than one full-wave plate in the optical path between the polarizers. The orientation of these plates must, of course, be maintained in their proper relationship to the polarizers in order to achieve the desired colors. It is important to remember that retardation of wave plates are additive or subtractive depending on orientation with respect to each other. In addition, the retardation of the PLZT shutter, in a color-biased system, may also either add to or subtract from the color bias retardation. Therefore, a design engineer for such a system must be cognizant of the polarizers (type and orientation), PLZT (composition, orientation, and voltage), and retardation plates (number, type, and orientation).

5. Eye-Protective Devices

a. Thermal/Flashblindness Goggles. Since 1975, Sandia National Laboratories has been involved in the design and development of thermal/flash protective devices for aircraft personnel of the U.S. Air Force using PLZT technology. The final product, officially designated as the EEU-2/P Flashblindness Flyers Goggles, is shown in Figure 37. It possesses several advantages over its predecessor, a liquid photochromic system, among which are (1) smaller size, (2) less weight, (3) solid state, (4) faster response, and (5) higher portability. It has been in production for the last three years and is the first PLZT device to reach this stage.

The flashblindness goggles are basically transverse-mode shutters of the configuration shown in Figure 26a. The shutters are operated in the fully open or energized state until a light hazard is detected by means of sensors mounted behind the viewing lenses. When this occurs, the energizing voltage is rapidly discharged, causing the goggles to revert to the closed state. When the threat is removed, the PLZT is reenergized and the open state is restored. The open and closed states of the device have typically 20% and 0.0015% (4.8 optical density) transmissions, respectively. Closure time is less than 150 μsec. An additional feature of the goggles is an electronic feedback loop which keeps a constant and controlled amount of light passing through the lenses in the open state when the goggles are not triggered [44].

Figure 37 PLZT thermal/flash protective goggle developed by Sandia National Laboratories for the U.S. Air Force.

b. Electronic Welding Helmet. Personnel eye safety devices, such as the electronic welding helmet, inspection goggles, and safety flip-down glasses, are some devices that are under development at the present time. All these devices operate in a manner very similar to that of the flashblindness goggles. They are transverse-mode shutter devices assembled in a configuration as described in Figure 26a. The light sensors and power supply are mounted externally to the PLZT shutters, which act as the viable-density filter plate. When an arc is struck or some other similarly intense light source is activated, the sensors detect this change in light level and remove the voltage from the shutters, causing them to darken instantaneously to a shade previously set by the operator. When the arc is interrupted, the shutters quickly and automatically reactivate to their full ON condition. Since this automatic action eliminates the necessity of raising and lowering the helmet, the mask can be worn in the down position at all times, thus increasing productivity and preventing accidental eye burns from neighboring welding operations. A wide range of filter plates, from shade 4 (5.2% transmittance) to shade 13 (0.0012%), is possible.

6. Optical Data Recorder

The first commercial application of a PLZT device was an optical data display recorder manufactured by Bell and Howell. This device is shown in Figure 38. The CEC HR-2000 Datagraph works on the principle of a digitally controlled electrooptic shutter using polarized light and a PLZT ceramic plate as the electrooptic material [89]. By selectively passing or blocking light through a linear array of hundreds of tiny light gates or shutters, each of which is controlled by driver electronics, the input data signals are accurately reproduced. Light that passes through the light gates impinges on direct-print recording paper to record data waveforms with high fidelity and accuracy. The heart of the recorder is the array of light gates, composed of a number of PLZT plates containing vacuum-deposited electrodes spaced 0.0125 in. (0.318 mm) apart, thus providing high resolution. By using this type of fixed, digitally controlled solid-state

Figure 38 PLZT optical chart recorder manufactured by Bell and Howell, Inc. (now known as Transamerica Deval).

array, the data recorder has eliminated such problems as linearity, beam deflection, tangential error, overshoot, and inertia, which limit present galvanometer and CRT recording devices. The instrument has a frequency response from dc to 5-kHz sine wave or 10-kHz square wave, and a recording speed of 0.01 to 129 in. of paper per second.

7. Stereo-Viewing Devices

Various types of stereo three-dimensional viewing systems have been under development for a number of years by several companies. All these systems utilize a pair of lightweight PLZT viewing glasses, each lens of which is separately electronically controlled either remotely or through a small cable to a shirt-pocket power unit. Each lens is essentially an independently controlled transverse-mode shutter of the type described in Figure 26a. The PLZT shutters are synchronized to an alternating pair of displayed images so that only the left eye sees the left-eye view and the right eye the right-eye view. The views are alternated at a rate of more than 30 Hz for each lens, allowing the observer to perceive a single, stereoscopic image with a lifelike sensation of object depth. Some applications of this device are as flight simulators and trainers, air traffic control, medical imaging, scientific modeling, radar and sonar displays, and contour mapping.

8. Image Storage Devices

Research and development activities are continuing at Sandia National Laboratories on image storage devices utilizing PLZT ceramics. The Cerampic device has received extensive study for the past several years and continues to show promise for image storage applications of the future. It is a longitudinal-scattering-mode device, as described in Figure 26 (c). Early designs utilized a photoconductor layer which provided the spatial variations of switching voltage when exposed to spatial variations of light intensity (usually through a contact negative) needed to produce the image in the ceramic. This photoconductor was subsequently eliminated by exposing the image with near-ultraviolet light containing bandgap (3.35 eV) or higher-energy photons which produce a space charge field, thus aiding the domain switching process. Significant improvements in the sensitivity of the exposure and recording processes were reported by Peercy and Land [90] through the use of ion implantation into the surface of the PLZT. Reductions in exposure energy by as much as 10,000 times have recently been achieved through the coimplantation of argon and neon.

9. PLZT Segmented Displays

One of the most promising group of applications to open up for PLZT devices is in the area of segmented displays. These displays are now being offered commercially and were recently reported at the 1983 International Symposium on Applications of Ferroelectrics [91]. Although they compete in the marketplace with many other technologies, such as CRTs, LEDs, vacuum fluorescent displays, and electroluminescent displays, their appearance and principle of operation closely approximates that of LCDs. Like LCDs, they are not light emitters, but rather, light modulators (i.e., they do not generate light but control it). Their method of controlling light is by means of optical birefringence; hence polarizers and polarized light are employed. The PLZT displays offer some advantages over the LCDs, including (1) solid-state devices; (2) faster response times by three orders of magnitude; (3) wider viewing angle; (4) wider operating temperature range, from −40 to +80°C; (5) indefinite shelf life; and (6) less temperature-sensitive electrooptic properties. Their disadvantages are (1) higher operating voltages, of approximately

Figure 39 Four digit seven-segment PLZT reflective display manufactured by Motorola, Inc. Character size is 0.5 × 0.3 in. (12.7 × 7.6 mm). Photograph was taken at a 25° viewing angle.

150 V; and (2) a slightly reduced contrast ratio at normal viewing. The PLZT displays can be used in the reflective, transflective, transmissive, and projection modes. A typical example of a reflective display is given in Figure 39. Color displays are also possible with the use of special color polarizers.

REFERENCES

1. W. Cady, *Piezoelectricity*, McGraw–Hill, New York, 1946.
2. W. Mason, *Piezoelectric Crystals and Their Application to Ultrasonics*, D. Van Nostrand, New York, 1950.
3. "IRE Standards on Piezoelectric Crystals: The Electromechanical Coupling Factor," *Proc. IRE, 46*:764 (1958).
4. "IRE Standards on Piezoelectric Crystals: Measurements of Piezoelectric Ceramics, 1961," *Proc. IRE, 49*:1161 (1961).
5. B. Jaffe, W. Cook, Jr., and H. Jaffe, *Piezoelectric Ceramics*, Academic Press, New York, 1971.
6. R. Redin, G. Marks, and C. Antoniak, *J. Appl. Phys., 34*:600 (1963).
7. G. Haertling, *Am. Ceram. Soc. Bull., 43*:875 (1964).
8. G. Haertling, *Am. Ceram. Soc. Bull., 49*:564 (1970).
9. C. Land, *Sandia National Laboratories Rep. SC-R-67-1219*, Oct. (1967).
10. C. Land and P. Thacher, *Sandia National Laboratories Rep. SC-RR-68-866*, Dec. 1968.
11. C. Land and P. Thacher, *Proc. IEEE, 57*:751 (1969).
12. P. Thacher and C. Land, *IEEE Trans. Electron Devices, ED-16*: 515 (1969).
13. G. Haertling and C. Land, *J. Am. Ceram. Soc., 54*:1 (1971).
14. G. Haertling and C. Land, *Ferroelectrics*, 3:269 (1972).
15. J. Thompson, Jr., *Am. Ceram. Soc. Bull., 53*:421 (1974).
16. M. Murata, K. Wakino, K. Tanaka, and Y. Hamakawa, *Mater. Res. Bull., 11*:323 (1976).
17. G. Snow, *J. Am. Ceram. Soc., 56*:91 (1973).
18. R. Dungan and G. Snow, *Am. Ceram. Soc. Bull., 56*:781 (1977).
19. J. Pottharst and H. Schichl, *Ferroelectrics, 10*:191 (1976).
20. K. Okazaki, I. Ohtsubo, and K. Toda, *Ferroelectrics, 10*:195 (1976).
21. K. Hardtl, *Am. Ceram. Soc. Bull., 54*:201 (1975).
22. K. Nagata, H. Schmitt, K. Stathakis, and H. Muser, *Proc. 3rd Int. Meet. Mod. Ceram. (CIMTEC)*, Rimini, May 1976.
23. S. Hayashi, S. Waku, and H. Shibata, *Ferroelectrics, 10*:181 (1976).
24. M. Yokosuka, *Jpn. J. Appl. Phys., 16*:379 (1977).

25. P. Debely, P. Gunter, and H. Arend, *Am. Ceram. Soc. Bull.*, *58*:606 (1979).
26. A. Sternberg, V. Fritsberg, L. Shebanov, and A. Dobre, *Ferroelectrics*, *34*:153 (1981).
27. K. Nagata, H. Schmitt, K. Stathakis, and H. Muser, *Proc. First Meet. Ferroelectr. Mater. Appl.*, Kyoto, Nov. 1977.
28. K. Miyauchi and G. Toda, *J. Am. Ceram. Soc.*, *58*:157 (1975).
29. J. Cutchen and G. Haertling, *J. Am. Ceram. Soc.*, *56*:225 (1973).
30. A. Meitzler and H. O'Bryan, Jr., *Proc. IEEE*, *61*:959 (1973).
31. K. Carl and K. Geisen, *Proc. IEEE*, *61*:967 (1973).
32. Y. Fogel, N. Bar–Chaim, and A. Seidman, *Appl. Opt.*, *19*:1609 (1980).
33. C. Land, P. Thacher, and G. Haertling, in *Applied Solid State Science*, R. Wolfe, ed., Academic Press, New York, 1974.
34. E. Keve, *Ferroelectrics*, *10*:169 (1976).
35. C. Land, *Opt. Eng.*, *17*:317 (1978).
36. F. Micheron, J. Rouchon, and M. Vergnolle, *Appl. Phys. Lett.*, *24*:605 (1974).
37. C. Land and P. Peercy, *Appl. Phys. Lett.*, *37*:39 (1980).
38. J. Maldonado and A. Meitzler, *Proc. IEEE*, *59*:368 (1971).
39. W. Smith and C. Land, *Appl. Phys. Lett.*, *20*:169 (1972).
40. G. Haertling and C. McCampbell, *Proc. IEEE*, *60*:450 (1972).
41. J. Cutchen, J. Harris, Jr., and G. Laguna, *Appl. Opt.*, *14*:1866 (1975).
42. C. Land and W. Smith, *Appl. Phys. Lett.*, *23*:57 (1973).
43. A. Kumada, K. Suzuki, and G. Toda, *Ferroelectrics*, *10*:25 (1976).
44. J. Cutchen, *Ferroelectrics*, *27*:173 (1980).
45. F. Bloss, *An Introduction to the Methods of Optical Crystallography*, Holt, Rinehart and Winston, New York, 1961.
46. E. Wahlstrom, *Optical Crystallography*, Wiley, New York, 1951.
47. C. Land and P. Thacher, in *The Physics of Opto-electronic Materials*, W. Albers, Jr., ed., Plenum, New York, 1971.
48. C. Kirkby, *Ferroelectrics*, *37*:567 (1981).
49. H. Jaffe and D. Berlincourt, *Proc. IEEE*, *53*:1372 (1965).
50. G. Shirane and K. Suzuki, *J. Phys. Soc. Jap.*, *7*:333 (1952).
51. B. Jaffe, R. Roth, and S. Marzullo, *J. Appl. Phys.*, *25*:809 (1954).
52. K. Hardtl and D. Hennings, *J. Am. Ceram. Soc.*, *55*:230 (1972).
53. P. Lysne and C. Percival, *Ferroelectrics*, *10*:129 (1976).
54. S. Liu, J. Heaps, and O. Tufte, *Ferroelectrics*, *3*:281 (1972).
55. D. Appleby, S. Porter, and F. Ainger, *Ferroelectrics*, *14*:715 (1976).
56. Y. Ito, H. Takeuchi, S. Jyomura, K. Nagatsuma, and S. Ashida, *Appl. Phys. Lett.*, *35*:595 (1979).

57. J. Biggers and W. Schulze, *Am. Ceram. Soc. Bull.*, *53*:809 (1974).
58. D. McCarthy and R. Brooks, *Ferroelectrics*, *27*:183 (1980).
59. R. Newnham, D. Skinner, K. Klicker, A. Bhalla, B. Hardiman, and T. Gururaja, *Ferroelectrics*, *27*:49 (1980).
60. S. Lynn, R. Newnham, K. Klicker, K. Rittenmeyer, A. Safari, and W. Schulze, *Ferroelectrics*, *38*:955 (1981).
61. R. Dungan, *Sandia National Laboratories Rep. SC-DR-66-593*, Nov. 1966.
62. R. Pohanka, S. Freiman, and R. Rice, *Ferroelectrics*, *28*:337 (1980).
63. J. Maldonado and A. Meitzler, *Proc. IEEE*, *59*:368 (1971).
64. G. Haertling, *J. Am. Ceram. Soc.*, *54*:303 (1971).
65. J. Burfoot and G. Taylor, *Polar Dielectrics and Their Applications*, University of California Press, Berkeley, 1979.
66. G. Haertling, *IEEE WESCON Proc.*, Los Angeles, Aug. 1966.
67. K. Van Dyke, *Proc. IRE*, *16*:742 (1928).
68. D. Berlincourt, D. Curran, and H. Jaffe, in *Physical Acoustics*, W. Mason, ed., Academic Press, New York, 1964.
69. K. Hardtl, *Ferroelectrics*, *12*:9 (1976).
70. S. Liu, S. Pai, and J. Kyonka, *Ferroelectrics*, *22*:689 (1978).
71. D. Berlincourt, in *Ultrasonic Transducer Materials*, O. Mattiat, ed., Plenum Press, New York, 1971.
72. V. Isupov, *Sov. Phys.—Solid State*, *10*:989 (1968).
73. L. Benguigui, *Solid State Commun.*, *11*:825 (1972).
74. K. Okazaki, *Ferroelectrics*, *35*:173 (1981).
75. M. Multani, S. Gokarn, R. Vijayaraghavan, and V. Polkar, *Ferroelectrics*, *37*:652 (1981).
76. P. Thacher, *Appl. Opt.*, *16*:3210 (1977).
77. J. Cutchen, J. Harris, Jr., and G. Laguna, *IEEE WESCON Proc.*, San Francisco, Sept. 1973.
78. G. Jonker, *J. Am. Ceram. Soc.*, *55*:57 (1972).
79. K. Okazaki and K. Nagata, *J. Am. Ceram. Soc.*, *56*:82 (1973).
80. L. Cross, S. Jang, R. Newnham, S. Nomura, and K. Uchino, *Ferroelectrics*, *23*:187 (1980).
81. C. Germano, Clevite Corp. *Tech. Pap. TP-222*, July 1961.
82. C. Germano, Clevite Corp. *Tech. Pap. TP-237*, Mar. 1969.
83. J. van Randeraat and R. Setterington, eds., *Piezoelectric Ceramics*, Mullard, London, 1974.
84. H. Schafft, *J. Audio Eng. Soc.*, *14*:258 (1966).
85. H. Schafft, *Ferroelectrics*, *10*:121 (1976).
86. J. Bost, *J. Audio Eng. Soc.*, *28*:244 (1980).
87. J. Cutchen, J. Harris, Jr., and G. Laguna, *Proc. SPIE*, *88*:57 (1976).

88. J. Harris, Jr., J. Cyrus, and G. Laguna, *Proc. SPIE, 307*:53 (1981).
89. N. Samek and M. Munoz, *Proc. SPIE, 307*:60 (1981).
90. P. Peercy and C. Land, *IEEE Trans. Electron Devices, ED-28*: 756 (1981).
91. G. Haertling, *Int. Symp. Appl. Ferroelectrics*, Gaithersburg, Md. (NBS), June 1983; *Ferroelectrics, 50*:63 (1983).

4

Ferrite (Magnetic) Ceramics

THOMAS G. REYNOLDS III*
Amperex Electronic Corp., Saugerties, New York

RELVA C. BUCHANAN
University of Illinois at Urbana-Champaign, Urbana, Illinois

I. INTRODUCTION

All ferro- and ferrimagnetic materials exhibit the "hysteresis effect" (a nonlinear relationship between an applied magnetic field (H) and the magnetic induction (B) of the material. Consequently, they all have properties associated with hysteresis behavior (e.g., magnetic permeability, saturation magnetization, remanent magnetization, coercive forces, etc.). Many materials have important magnetic properties, including elemental metals, transition metal alloys, rare earth alloys, and ceramics.

*Present affiliation: Nederlandse Philips, Eindhoven, The Netherlands.

Among the magnetic ceramics, ferrites are the predominant class. These materials cover a wide range of crystal structures, compositions, and applications. However, they exhibit certain common properties:

1. They are all oxides.
2. They are all based on Fe_2O_3 as a major compositional component.
3. They exhibit a spontaneous magnetic induction in the absence of an external magnetic field.

The important properties that make these materials useful are their strong magnetic coupling, high resistivity, and low loss characteristics. Since its commercial beginning in the early 1940s, ferrimagnetic technology has advanced at a rapid rate, due in part to the ease of composition modification to meet specific applications.

Ferromagnetism in solids is due to the strong coupling or mutual attraction of magnetic dipole moments on the atoms in the solid. This results in a parallel mutual reorientation of the magnetic moments into a common direction, a condition defined as spontaneous magnetization. Ferrimagnetism, like antiferromagnetism, arises from the antiparallel alignment of the magnetic moments of the ions on different sublattices in the crystal. In ferrimagnetism, however, the oppositely directed magnetic moments do not exactly cancel, so that a net magnetic moment results.

The origin of these coupling behaviors may be related to the outer electron interactions of the neighboring ions. Although the bonding in magnetic ceramics (oxides) is predominantly ionic, secondary covalent bonding also occurs between p orbitals of the filled O^{2-} and the unfilled d orbitals of the transition metal cations. This results in a superexchange interaction between the outer electrons which is responsible for the coupling. This superexchange interaction is characteristic of coupled ions that both have half-filled or more than half-filled d orbitals and accounts for both antiferromagnetic and ferrimagnetic behavior. The strength of the coupling due to the superexchange interaction diminishes rapidly as the distance between the ions increases. Because of this, the dumbbell shape of the oxygen 2p orbital necessitates a cation-oxygen-cation bond angle of 180° for maximum interaction (or coupling) to occur.

A ferromagnetic or ferrimagnetic oxide exhibits its spontaneous magnetization in the form of Weiss domains, which are microscopic-sized regions of the material within which there is a single magnetic dipole. Weiss domains are separated by domain walls in which a gradual transition of magnetic moments occurs. Typically, this transition takes place over a 100 nm distance and has a wall energy of

$\approx 0.001\ J/m^2$. In the absence of an applied field, the domains orient themselves in a way that minimizes the energy of the system and which results in a net magnetization of zero. The energy of a magnetic ceramic is kept to a minimum when the magnetic flux (due to the domains) is kept mostly within its boundaries. This has an important bearing on the design of shapes for optimum magnetic applications.

Under an applied magnetic field, the Weiss domains of the ferrimagnetic material are responsible for the hysteresis behavior exhibited, the degree of magnetization being dependent on the strength and direction of the applied field. The hysteresis loop characteristic of a ferrite is obtained when it is subjected to an external field (Figure 1). In starting from a "virgin state" with zero magnetic induction (usually obtained by heating the material to a temperature above its Curie temperature, T_c), as the material is exposed to an increasing magnetic field of strength (H), the magnetic induction (B) increases linearly. The rate at which the induction increases defines the relative permeability (μ_i).

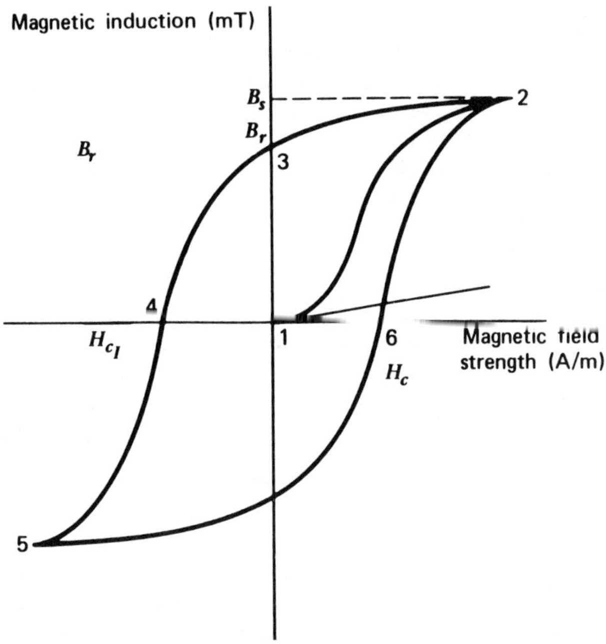

Figure 1 Hysteresis loop behavior for typical ferrites showing important properties. B_r, remanent polarization; B_s, saturation polarization; H_c, coercive field.

where

$$\mu_i = \frac{1}{\mu_0} \lim_{H \to 0} B/H \qquad (1)$$

$$\mu_0 = 4\pi \times 10^{-7} \text{ H/m}$$

B = magnetic induction, Wb/m^2 (1 Wb/m^2 = 1 tesla = 10^4 gauss)

H = magnetic field, A/m (1 A/m = 0.126 oersted)

As the field (H) continues to increase, magnetic domains within the ceramic grains rotate to align with it. When the rotational alignment is complete, near the knee of the curve, the domains that are aligned with the field now grow to maintain their lowest free energy. At point 2, the material has become "saturated"—any further increase in the magnetic field will not increase the magnetic induction. The value of induction obtained at the maximum field is the saturation magnetization (B_s).

Reducing the magnetic field from this value does not cause a drastic reduction in the induction initially because the energy stored in the domain walls must be overcome. Consequently, even when the external field has been reduced to zero, there remains a significant remanent magnetic induction (B_r)—point 3. Applying a magnetic field in the opposite sense causes the domains to switch their magnetic polarity to align themselves with the field. The magnetic field that causes this switching to take place is called the coercive field (H_c). Increasing the field further in the negative direction causes the domains to grow, as they store an increasing amount of magnetic energy until the material is once again saturated. At this point in the third quadrant of the hysteresis loop the material is in a state completely analogous to the first quadrant. Reversing the field causes the magnetic properties to continue around the hysteresis loop in a counterclockwise direction.

The magnetic properties described by the hysteresis loop (and these are by no means all the magnetic parameters of interest) are highly dependent on both the intrinsic properties of the materials themselves (crystal structure and composition) and extrinsic processing parameters (grain size, density, etc.). It is for this reason that the ferrimagnetic oxides commonly known as ferrites play such a key role as electronic materials.

For example, typical magnetic values for the cubic spinel ferrites and the hexagonal magnetoplumbite ferrites are dramatically different:

	H_c (Oe)	μ_i	B_r gauss
Spinel	0.1	100–20,000	400–1200
Magnetoplumbite	4000	1–10	4000

Consequently, these materials find considerably different applications. Spinel ferrites used as transformer cores in high-frequency power supplies are materials engineered to have low hysteresis losses (high resistivity) and high saturation magnetization in order to make the transformer as small and efficient as possible. The hexagonal ferrites on the other hand, are widely used as ceramic magnets. In this application the material must have very high coercive force to resist the demagnetizing forces encountered in motor applications. Both of the uses can be accomplished by ferrites using relatively cheap and abundant Fe_2O_3 as the major raw material.

Ferrites crystallize in a large variety of structures; these include spinel, magnetoplumbite, and garnet (Table 1). These three account for a majority of the technological applications. Common to all these materials is the simple ratio of Fe_2O_3 to the other oxide compound (Table 1). The crystal structure of these materials holds the key to their diverse magnetic properties because in each case it defines the interactions between the various ions on an atomic level.

II. SPINEL FERRITES

The spinel ferrites are isostructural with the naturally occurring spinel $MgAl_2O_4$ and conform to the general formula AB_2O_4. The

Table 1 Summary of Ferrite Structure Types, Typified by Changes in the Fe_2O_3:MeO (or Me_2O_3) Modifier Oxide Ratios

Spinel	1 Fe_2O_3:1 MeO	(where MeO = transition metal oxide)
Garnet	5 Fe_2O_3:3 Me_2O_3	(where Me_2O_3 = rare earth metal oxide)
Magnetoplumbite	6 Fe_2O_3:1 MeO	(where MeO = divalent metal oxide from group II A—BaO, CaO, SrO)

structure was first described by Bragg [1] and Nishikawa [2] in 1915. It is cubic (Figure 2), consisting of eight AB_2O_4 units. The relatively large oxygen anions are arranged in cubic close packing, with the octahedral and tetrahedral interstitial sites occupied by transition metal cations. Of the 64 available tetrahedral sites (A sites), 8 are occupied, and 16 of the 32 octahedral (B sites) are occupied.

Although to a first approximation the anions are considered to be rigid spheres, the presence of cations in the interstitial sites causes the oxygen sublattice to be distorted and the tetrahedrally coordinated sites to be slightly expanded, causing a slight displacement of the oxygen anions along the cell diagonals [3]. Consequently, the spherical radius of the tetrahedral and octahedral sites can be calculated in terms of the oxygen displacement parameter u:

$$r_{tet} = \left(u - \frac{1}{4}\right) a \cdot 3 - R_O \tag{2}$$

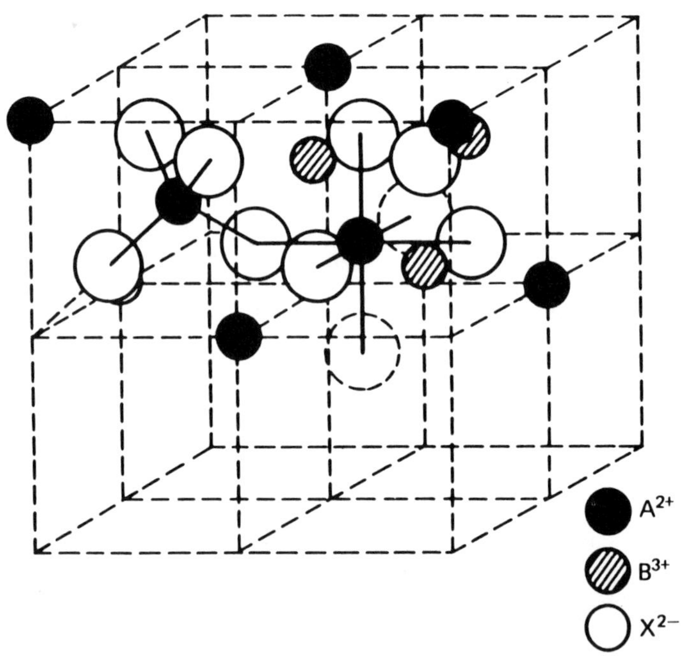

Figure 2 Unit cell for spinel structure with alternating tetrahedral (AO_4) and octahedral (BO_6) coordinated units (four each per unit cell).

$$r_{oct} = \left(\frac{5}{8} - u\right) a - R_O \tag{2}$$

where R_O is the ionic radius of oxygen. For materials typical of the spinel ferrites, this amounts to

$$\begin{aligned} 0.58 < r_{tet} < 0.67 \text{ Å} \\ 0.70 < r_{oct} < 0.75 \text{ Å} \end{aligned} \tag{3}$$

Within the transition metal series the d-electron structure of the cations changes in a regular and systematic way, whereas the ionic radius remains relatively constant (Table 2). This offers an excellent opportunity to engineer specific magnetic interactions in the crystal lattice by substituting various transition metal cations into the crystal structure, thereby altering macroscopic magnetic properties.

The AB_2O_4 structure offers many possible combinations of cations which could balance the −8 charge contribution of the oxygens. The most important of these are the 2−3 ferrospinels, in which Fe_2O_3 is a major component. In order to define the location of the cations on the octahedral and tetrahedral sites, the following convention has been adopted:

$$Me^{2+}[Fe_2^{3+}]O_4 \quad \text{normal spinel} \tag{4}$$

where the ions on the octahedral sites are enclosed in brackets. The ferric ion can occupy either tetrahedral or octahedral sites, depending on what other cations are present, and this results in the inverse spinel:

$$Fe^{3+}[Me^{2+} Fe^{3+}]O_4 \quad \text{inverse spinel} \tag{5}$$

As pointed out by Economos [4] and Broese Van Groenou [5], normal spinels are formed when Me^{2+} = Zn, Cd, and inverse spinels when Me^{2+} = Ni, Co, Fe, Mn, Cu.

The crystal structure can be considered as two magnetic sublattices: an A-site and a B-site lattice. Magnetic superexchange interactions between the octahedrally coordinated B sites and the A sites through the intervening oxygen anions cause these two sublattices to be antiferromagnetically oriented. Depending on the type

Table 2 Comparison of the d-Electron Structure, Coordination, and Cation Radii for First Transition-Series Elements

Element	d^0	d^1	d^2	d^3	d^4	d^5	d^6	d^7	d^8	d^9	d^{10}
Titanium	Ti^{6+}VI 0.0605	Ti^{3+}VI 0.067	Ti^{2+}VI 0.086								
Vanadium		V^{4+}VI 0.059	V^{3+}VI 0.0640	V^{2+}VI 0.079							
Tantalum		Ta^{4+}VI 0.066	Ta^{3+}VI 0.067								
Niobium		Nb^{4+}VI 0.069	Nb^{5+}VI 0.070	Nb^{2+}VI 0.071							
Chromium		Cr^{5+}IV 0.0350	Cr^{4+}IV 0.044 VI 0.055	Cr^{3+}VI 0.0615							
Molybde-num		Mo^{5+}VI 0.063	Mo^{4+}VI 0.0650	Mo^{3+}VI 0.067							
Manganese				Mn^{4+}VI 0.0540	Mn^{3+}VILS 0.058 HS 0.065	Mn^{2+}VILS 0.067 HS 0.0820					
Tungsten			W^{4+}VI 0.0650								

Iron	Fe^{3+}IVHS 0.049 VILS0.055 HS0.0645	Fe^{2+}IVHS 0.063 VILS0.061 LS0.0770				
Cobalt		Co^{3+}VILS 0.0525 HS0.061	Co^{2+}VILS 0.065 VIHS0.0735			
Nickel			Ni^{3+}VILS 0.056 VIHS0.060	Ni^{2+}VI 0.0700		
Copper					Cu^{2+}VI 0.073	
Cobalt						Co^{2+}IV 0.084 VI0.095
Zinc						Zn^{2+}IV 0.060 VI0.0745
Gallium						Ga^{3+}VI 0.0620

and site preferrence of the cations, a variety of magnetic properties are possible.

The saturation magnetization of the spinel ferrites is readily explained on this basis. Magnetite, Fe_3O_4, is an inverse spinel, and using the previous notation,

$$Fe^{3+}[Fe^{2+}Fe^{3+}]O_4 \tag{6}$$

The d-electron spins between the A and B sites are antiferromagnetically ordered,

$$d^5[d^4d^5] \tag{7}$$

and would indicate a net magnetic moment of four Bohr magnetons; there is good agreement between this value and the value of 4.1 observed at 0 K.

The manganese zinc ferrites and nickel zinc ferrites appear to have anomolous behavior of the saturation magnetization—the addition of nonmagnetic zinc ferrite *raises* the saturation magnetization. This behavior can be understood using the same model as for Fe_3O_4.

Zinc ferrite, $ZnFe_2O_4$, is a normal spinel, and as such the unit cell has no net magnetic moment:

$$ZnFe_2O_4$$

$$Zn^{2+}[Fe^{3+}Fe^{3+}]O_4 \tag{8}$$

$$d^0[d^5d^5]$$

Manganese ferrite is an inverse spinel and consequently the two magnetic sublattices are antiferromagnetically aligned:

$$MnFe_2O_4$$

$$Fe^{3+}[Mn^{2+}Fe^{3+}]O_4 \tag{9}$$

$$d^5[d^5d^5]$$

When the nonmagnetic zinc ion (d^{10}) is substituted into the manganese ferrite lattice, it has a stronger preference for the tetrahedral site than does the ferric ion and thus reduces the amount of Fe^{3+} on the A site. Because of the antiferromagnetic coupling the net result is an increase in magnetic moment on the B lattice and an increase in saturation magnetization (Figure 3).

$$Zn_x^{2+} Fe_{1-x}^{3+} [Mn^{2+} Fe^{3+}] O_4$$
$$d_x^{10} d_{1-x}^{5} [d^5 d^5] \qquad (10)$$

However, at large levels of zinc substitution, the A-site magnetic ion becomes so diluted that the coupling between the two lattices is lost and the saturation magnetization drops. An additional consequence of the weakening magnetic interactions is that they can more easily

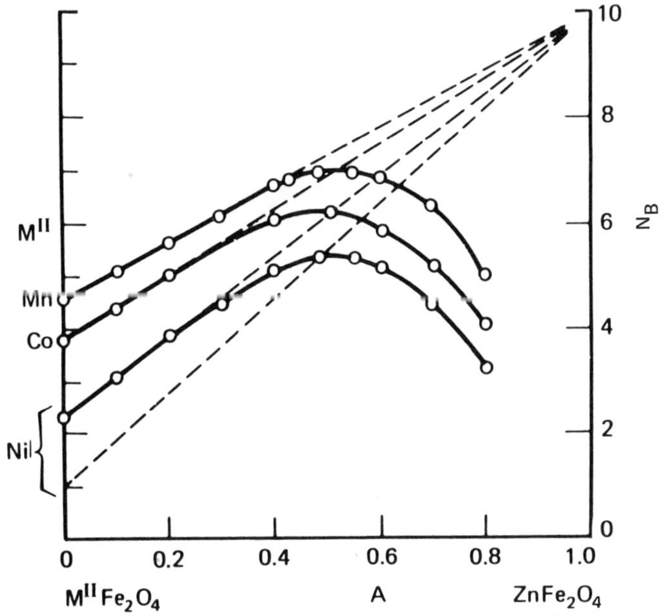

Figure 3 Plot of magnetization versus Zn additions to Ni, Co, and Mn ferrites showing increased magnetization and saturation effects.

be disrupted by thermal energy, so the temperature at which the magnetic properties disappear (the Curie temperature) decreases with increasing zinc content (Figure 4).

To achieve optimum magnetic properties it is necessary to control several other magnetic properties: crystalline anisotropy and magnetostriction. The magnetocrystalline anisotropy is that part of the crystal energy which is dependent on the direction of magnetization in the lattice and for a cubic material can be represented by an equation of the type [6]

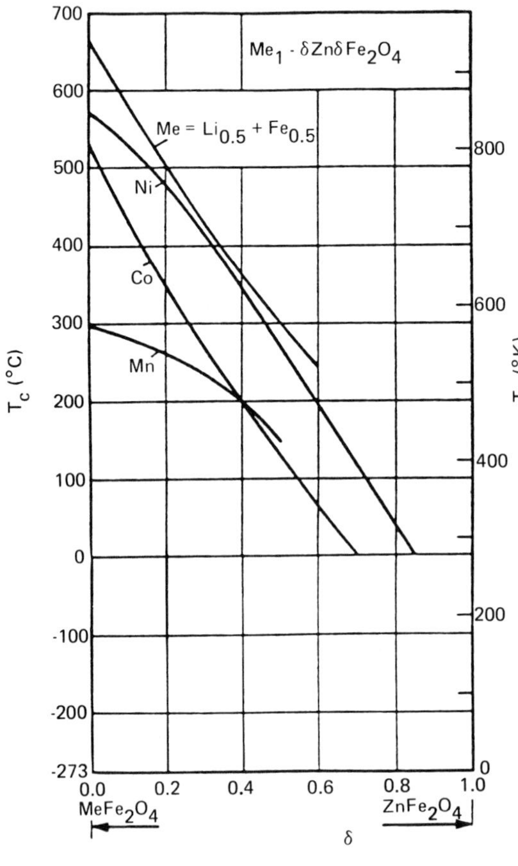

Figure 4 Depression of magnetic Curie point temperature with Zn additions to Mn, Co, Ni, and Li ferrites.

$$E_A = K_1(\partial_1^2 \partial_2^2 + \partial_2^2 \partial_3^2 + \partial_3^2 \partial_1^2) + K_2(\partial_1^2 \partial_2^2 \partial_3^2) + \cdots \qquad (11)$$

The anisotropy constants are different from one material to the other and are also temperature dependent. The magnetocrystalline anisotropy is predominantly dependent on K_1, the first-order anisotropy constant, and is strongly influenced by the cations present in the crystal structure [7]. In the manganese zinc ferrite system, for example, there is very little contribution to K_1 by Mn^{2+}, while the Fe^{3+} has a slight positive contribution to the anisotropy constant and Fe^{2+} has a strong positive contribution. Because of the temperature dependence of these effects, it is possible to have a net zero anisotropy value at some temperature. That temperature coincides with the well-known secondary maximum in the magnetic permeability.

A majority of the applications of the spinel ferrites depend on achieving high permeability, low coercive force, and low hysteresis losses. Consequently, the strong effect of divalent iron places stringent requirements on the composition and processing of modern-day spinel ferrites; these are discussed in more detail in Section V.

Ohata [8] has investigated the relationship tetween magnetocrystalline anisotropy and magnetic permeability in the $MnO\text{-}ZnO\text{-}Fe_2O_3$ system and effectively mapped out the regions where highest magnetic permeability can be expected to occur (Figure 5). The work of Stoppels et al. [9,10] extended these investigations to well-characterized single-crystal materials which were not available at the earlier time; this work has helped to better understand the role of the second-order constant, K_2, on permeability.

Magnetostriction is another anisotropy effect that determines the properties of ferrites. The application of a stress to a magnetostrictive material causes a change in the energy of the crystal lattice and hence the magnetic properties of that lattice [11,12]. The inverse effect is also possible whereby the application of a magnetic field causes a measurable strain to be induced in the material. However, the effect is very small in ferrites and has not found any practical application. The main consequence of magnetostriction in ferrites is to render them stress-sensitive. The magnetic permeability may drop 20–50% if they are handled roughly or dropped while being wound or assembled. Many ferrite components are coated to improve insulation resistance, and the shrinking of this coating can seriously alter the magnetic performance. Finally, the use of ferrite recording heads in such applications as audio- and videotape recorders requires a material with low magnetostriction to eliminate stress-induced noise in the signal. Fortunately, those areas where the magnetocrystalline anisotropy is low also correspond to low magnetostriction.

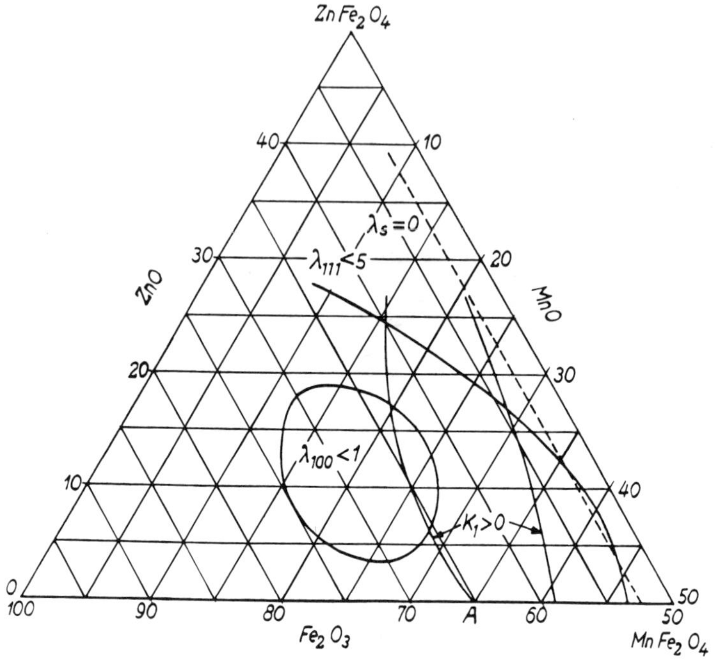

Figure 5 Composition field showing magnetocrystalline anisotropy and magnetic permeability regions in the MnO-ZnO-Fe_2O_3 ferrite system.

III. HEXAGONAL FERRITES

A number of hexagonal compounds exist in the general compositional field BaO-MeO-Fe_2O_3 (Figure 6). These are designated M, W, Y, and Z and correspond to (BaO + MeO)/Fe_2O_3 ratios of 1/6, 3/8, 4/6, and 5/12, respectively [13]. The compound of greatest technological interest is the M-hexagonal ferrite, $BaFe_{12}O_{19}$, which is isostructural with the naturally occurring magnetoplumbite, $PbFe_{7.5}Mn_{3.5}Al_{0.5}Ti_{0.5}O_{19}$ [14]. The structure is represented in Figure 7 and is considerably more complex than spinel, being built up of oxygen layers which have both cubic and hexagonal close packing [15]. The coordination environment of the cations is complicated and the competing, magnetically opposing moments of the Fe^{3+} ions result in a saturation magnetization of 4775 gauss [16].

The magnetocrystalline anisotropy of these materials is high, making them ideally suited as permanent magnets [17]. The high value also means that if the sintered particles are kept sufficiently small (<1.6 μm for barium ferrite), they will be single domain and quite difficult to demagnetize.

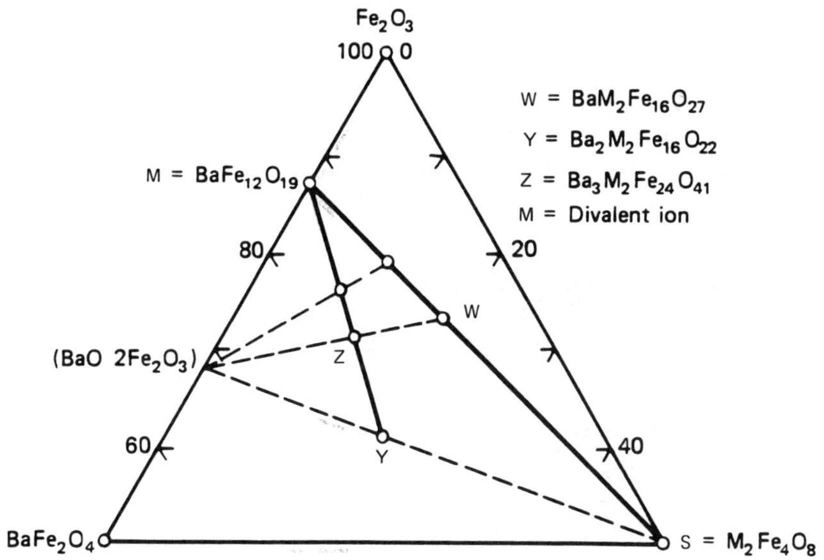

Figure 6 Hexagonal ferrite compositions in the field of the BaO-MeO-Fe_2O_3 system.

IV. GARNET

The rare earth yittrium iron garnet, $Y_3Fe_5O_{12}$ (YIG) is prototypical of the rare earth ferromagnetic insulators. The material is cubic, like spinel, and is isostructural with $Ca_3Al_2(SiO_4)_3$, the structure having been determined by Bertaut and Forrat [18] and Geller and Gilleo [19] (see Figure 8). The lattice constant is considerably larger (12.376 Å) than that of spinel (8.41 Å is typical for nickel zinc ferrite) and the unit cell contains eight formula units or 160 atoms. The coordination of the cations is considerably more complex, with the metal cation located in tetrahedral, octahedral, and dodecahedral sites. The structure has been investigated in detail and a number of workers have reviewed known compounds and the crystal chemistry [20,21]. The tetrahedral sites, of which there are 24, are the smallest and are normally occupied by Fe^{3+}; the octahedral sites number 16 per unit cell and are occupied by Fe^{3+}; the 24 dodecahedral sites in the structure are the largest and are occupied by Y^{3+} ions:

$\{Re_3^{3+}\}$	$[Fe_2^{3+}]$	$(Fe_3^{3+})O_{12}$	(12)
(c)	(a)	(d)	Crystal site positions
Dodecahedral	Octahedral	Tetrahedral	

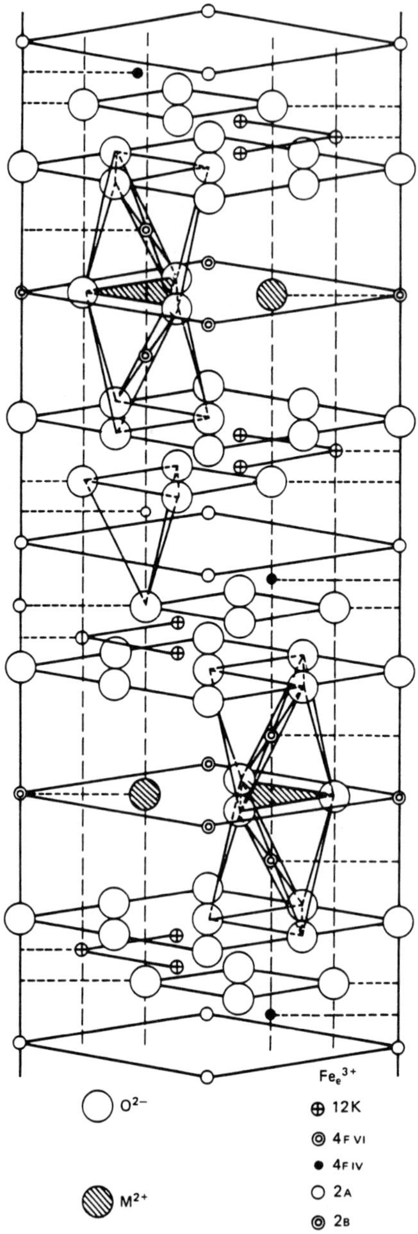

Figure 7 Unit cell structure of M-hexagonal ferrite ($M^{2+}Fe_{12}O_{19}$). Stacking of oxygen layers in the C (magnetically active) direction shows both cubic and hexagonal close packing.

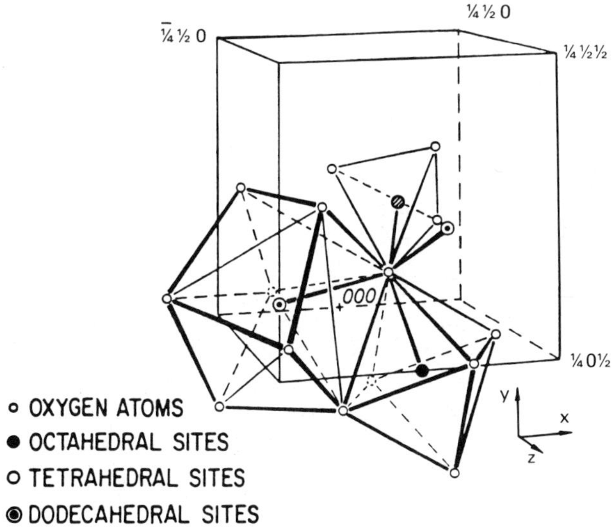

Figure 8 Unit cell structure of the garnet ferrite ($Y_3Fe_5O_{12}$) showing cation coordination structures.

As was the case with the spinels, a wide range of transition metal cation substitutions are possible and rare earth ions may be substituted on the octahedral and dodecahedral sites [22]. Consequently, the saturation magnetization, magnetocrystalline anisotropy, and other magnetic properties may be altered over a wide range.

V. PROCESSING

A. Powder Preparation

There are a wide variety of preparative methods that can be used to prepare ferrite powders. Economos [4], Takada [23], Robbins [24], and Goldman [25] have described a variety of chemical precipitation methods. The usual synthetic approach is to start with a solution containing the appropriate metal ion sulfates and precipitate a ferrite intermediate using hydroxide or carbonate as the precipitating agent. The precipitation is carried out at pH 10–12 at an elevated solution temperature, and often, air is bubbled through the solution and the precipitate to encourage oxidation. Repeated washing of the precipitate may be necessary to remove trace amount of sulfate, which could have detrimental effects on the sinterability and magnetic properties of the powder. The resulting powder is finely divided 0.05–1.0 μm and chemically homogeneous. However, this method has found application only in certain specialized areas because of higher manufacturing costs and the necessity of disposing of large quantities of aqueous solutions.

Schnettler and Johnson [26] of Bell Laboratories have investigated the possibility of preparing finely divided, highly reactive powders by a freeze-drying technique whereby an aqueous salt solution containing the proper metal cations is sprayed into chilled hexane. The frozen pellets are then freeze dried and the resultant anhydrous salt is reacted in a furnace to yield a ferrite powder. The anhydrous powder is highly reactive and forms ferrite at temperatures substantially below those required for solid-state reaction. The method, while yielding high-purity reactive material, does not appear to be suitable for scale-up to commercial quantities. These and other preparative methods are quite useful in the preparation of very pure materials which are instrumental in the study of the structure and magnetic properties of ferrites: however, *standard ceramic processing* (see Figure 9) is the predominant technology for most of the ferrite industry.

Initially, the raw materials are assayed for purity. For the spinel ferrites specific attention is paid to the silica content of the Fe_2O_3 (it should be in the range <200 ppm and for the highest-grade materials, <100 ppm) because Fe_2O_3 constitutes approximately 65% of the composition. The usual source of iron oxide is chemically precipitated powder prepared from sulfate solutions or spray-roasted iron oxides prepared from HCl-based pickle liquors. These spray-roasted oxides are considerably less expensive ($0.25/kg) but contain a higher level of SiO_2 (approximately 200 ppm) because of silica being in the steel strip being pickled. Typically, spray-roasted oxides find application in entertainment ferrites, ferrites for transformers and power supplies, and the hexagonal ferrites used in ceramic magnets.

Following weighing, the mixture of oxides or carbonates is mixed to yield a homogeneous powder. Mixing is either done dry or, more usually, by wet milling in a steel mill with steel balls. This process is commonly referred to as red milling because the oxide mixture has a characteristic red color. Two types of ball mills are used in ferrite processing: ball mills and attritor mills. Ball mills are large cylindrical mills which are filled with steel balls, powder, and water and rotated about their cylindrical axis so that the balls fall by gravity and impact on the powder particles. These mills are inexpensive to operate and maintain and are available in a wide variety of sizes, from 1-kg mills which rest on rotating rollers to mills that will accommodate a 2500-kg charge of powder.

Attritor mills are filled with smaller steel balls and the milling action is caused by a steel rod with "lifters" which stirs the balls,

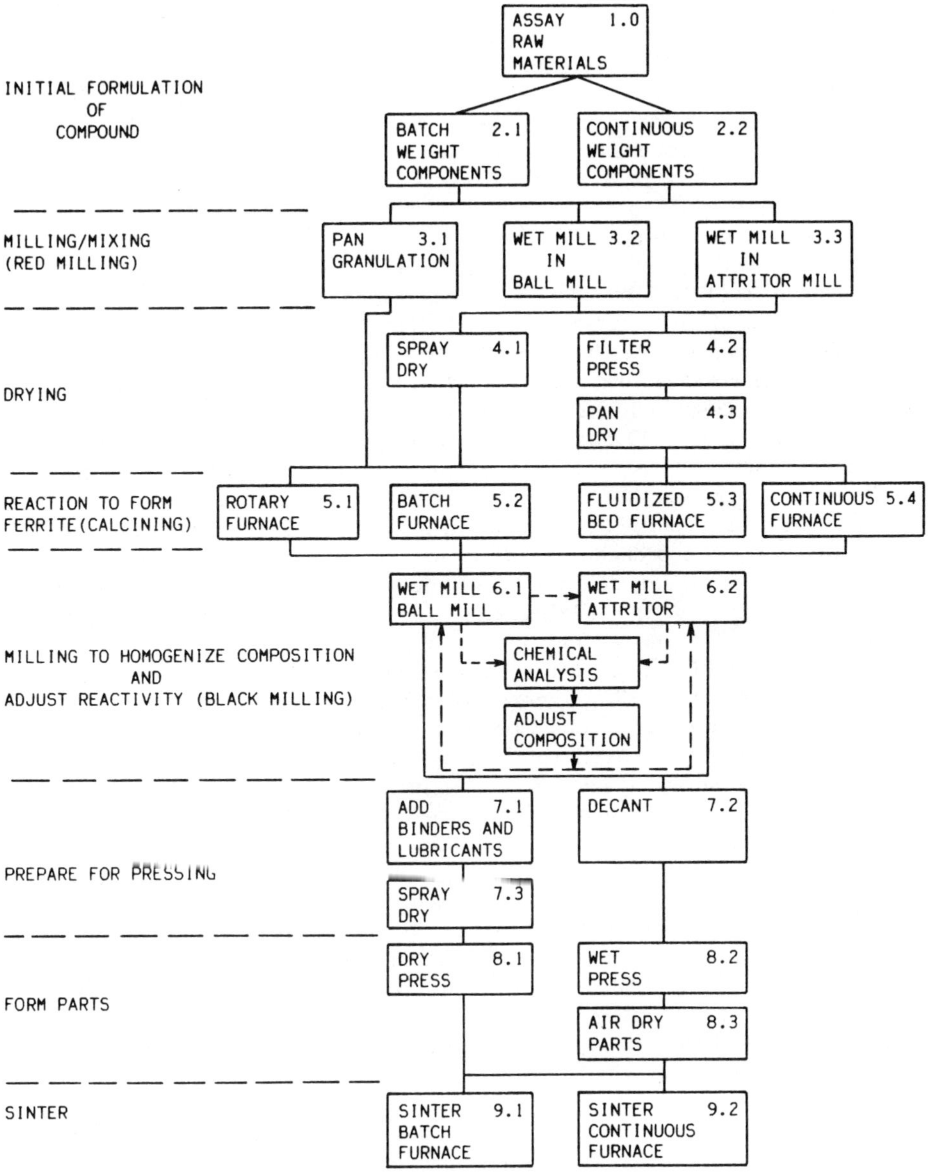

Figure 9 Flowchart for standard ceramic processing of ferrite components.

causing them to lift up and fall back. The milling action is much more intensive because the force restoring the media downward is the weight of all the media above it. Because of the higher impact energy it is possible to use smaller-diameter media and it is usual that attritor mills produce a powder with a smaller particle size distribution. Also, because the mill itself does not rotate, it is possible to use the attritor in a semicontinuous fashion by pumping slurry and an oxide mixture through it. Attritor mills, however, require more power and are more complicated to run and maintain. Following this first (red) milling the water is removed from the slurry by a high-capacity filter press or by spray drying.

Because of the energy costs associated with red milling and the subsequent removal of the water, for less demanding technical grades of ferrite the red milling step is omitted completely. In that case the raw materials are weighed and mixed dry, or with a very low percentage of water. Mixing can be done by simple pan granulation, or if it is desired that the material be compacted slightly to enhance the solid-state reaction, it can be passed between rollers to form thin sheets or extruded to form small pellets. The hexagonal ferrites are weighed on a semicontinuous basis and dry mixed, granulated, and fed by conveyor belt or bucket belt to direct-fired rotary kilns.

Following this first mixing step the raw materials are reacted to form the ferrite phase, this operation is called calcining or presintering. The reaction temperature is typically 700–1000°C in air. As was the case with milling, there are a variety of ways in which calcining can take place. The most common is for the semidry powder to be fed into a gas-fired rotating cylindrical kiln. Smaller quantities can be reacted in ceramic boxes in batch kilns. The calcining reaction itself is critical to the future properties of the ferrite because it determines the reactivity of the powder and the final shrinkage in the subsequent sintering step. While nickel zinc ferrites are reacted almost 100% to the spinel phase during this step, the manganese zinc ferrites are not fully converted to spinel (40–65% spinel).

The presintering of hexagonal materials takes place at a higher temperature (1000–1250°C) than that for the spinels. In this case it is important to convert the oxides 100% to the hexagonal phase so that there is little, if any, reaction-sintering with subsequent grain growth taking place when the parts are fired. It is necessary to keep the residence time in the high-heat portion of the kiln as short as possible so that the hexagonal particles remain as small as possible, to minimize the energy required to mill them. To further enhance millability, the hot pellets exiting the kiln are quenched to induce stresses and microcracks in them.

The result of the calcining process is a powder that is lightly sintered into friable lumps. This material must be milled again to further homogenize its composition and improve reactivity by having a uniform small particle size. Milling is done again in ball or attritor mills with water as the carrier. The black milling step, as it is called, is monitored or controlled by measuring the particle size or surface area of the powder. Extended milling beyond the desired particle size should be avoided because it introduces a small amount of ultrafine particles which can nucleate nonuniform grain growth during the sintering cycle and a degradation of magnetic properties. This is especially important in the M-hexagonal materials, where the final grain size target is 1–2 μm.

During the final milling the chemical composition is carefully analyzed and additions are made to bring it to specified values. For example, in microwave ferrites of the type $Y_3Fe_{5(1-\Delta)}O_{12}$, it has been shown that Δ must be controlled to ±0.002 to achieve optimum properties [27]. At this time, permanent-magnet materials are doped with small amounts of MeO and SiO_2. These additives combine to form a secondary phase during sintering and retard grain growth [28].

Near the end of the black milling cycle of the soft ferrites, one or several organic binders are added to the mill [29]. This multicomponent binder system serves several purposes: (1) there is a dispersant which enhances milling and improves the solids-to-liquid ratio, thereby reducing costs and increasing throughput; (2) as a binder and plasticizer to increase green strength in the pressed parts; and (3) as a lubricant to help reduce die wall friction during pressing. It has been pointed out by Ghate [28] that the binder system itself can play a major role in determining the final magnetic properties, especially of high-permeability materials. One such system that is in use is the following [30]:

0.4 wt % polyethylene imine (Polymin)
0.4–1.6 wt % polyvinyl alcohol
0.2–1.0 wt % polyvinylpyrrolidone
0.03–0.1 wt % methyl cellulose

This binder offers the advantage over more traditional ceramic binders which include gum arabic of being clean burning, leaving no residue in the ceramic and allowing high permeability levels to be obtained. The ceramic slurry, including binders, is spray dried to yield a free-flowing powder which can be dry pressed into a wide variety of standard shapes: pot cores, E cores, U cores, toroids, and so on (see Figure 10).

Although some ceramic magnets are dry pressed in an orienting magnetic field, their properties are rather low, so to optimize the shape and magnetocrystalline anisotropy, most ceramic magnets are wet pressed in the presence of a field. Rather than have a binder added and then be spray dried, the black milled ceramic magnet slurry is allowed to stand in decanting tanks for several days and the excess water is drawn off. The resultant slurry, with about

Figure 10 Ferrite component parts (E cores, U cores, pot cores, toroids) prepared by dry pressing and sintering of spray-dried powders.

the consistency of yogurt, is pumped directly to the presses and injected into the die cavity. Once the die cavity is closed, a magnetic field is applied and the particles align themselves with the field because of the water present in the slurry. After alignment is complete the water is removed by vacuum filtration through holes in the die itself (the inside of the die cavity is lined with filter paper) and the parts are removed. Since there is no binder to impart strength to these wet-pressed parts, they must be handled and stored gently. The parts must be dried for a period of time ranging from 2–3 days to several weeks to reduce the moisture content from 10–12% to 2–4% before they can be sintered.

B. Firing

The firing of the ferrites can be divided into two cases:

1. Materials that are prepared in the fully oxidized state [e.g., nickel zinc ferrites, hexagonal (magnetoplumbite) ferrites, and garnets]

2. Materials in which it is necessary to control the valence state of the cations, typically the manganese zinc ferrites

The nickel zinc ferrite, $Ni_{1-x}Zn_xFe_{2-y}O_4$, is formulated with a slight defficiency of iron ($0 < y < 0.025$) and the firing is done in air or oxygen at a temperature of 1100–1350°C to ensure that all the cations are in their highest oxidation state. These compounds are characterized by very high resistivity (>10^7 Ω-cm), relatively low permeability, and low saturation magnetization. Because of their high resistivity they are used primarily in high-frequency applications. One additional feature of nickel zinc ferrite is that the material can be sintered to a dense (>99.8%), pore-free ceramic. This makes it ideally suited for the fabrication of magnetic recording heads for computer and audio applications. During the sintering cycle the initial ceramic grains grow from 0.5–1.0 μm to 10–25 μm.

The ferromagnetic garnets must also be sintered in air or oxygen to ensure that all iron is in the trivalent state; otherwise, operating at frequencies in the range 100 MHz to 500 GHz they would exhibit excessive dielectric losses.

The hexagonal M-type ferrites, $SrFe_{12}O_{19}$, are also fired in an oxidizing atmosphere. However, in this case, the sintering cycle is optimized to yield a dense material with minimal grain growth. For use as a ceramic magnet it is necessary to keep the grain size as small as possible (<1–2 μm) so that they are single domain and the high coercive force necessary to prevent demagnetization is not degraded. The firing of hexagonal ferrites usually takes place in large gas-fired car kilns, with the ceramic parts stacked on ceramic tiles or placed in ceramic boxes. Although the sintering cycle may be as long as 24 hr, the time at peak temperature (1350°C) may be as short as 30 min to minimize grain growth.

The manganese zinc ferrites, $Mn_{1-x}Zn_xFe_{2+y}O_4$, are formulated with an excess of iron ($0.05 < y < 0.20$). This additional iron permits compensation of the magnetocrystalline anisotropy to a net value of zero by controlling the amount of Fe^{2+} present in the spinel structure. As a result the resistivity is much lower (caused by the presence of Fe^{2+}-Fe^{3+}), but the permeability and saturation magnetization are much higher than in the nickel zinc series (see Table 3). Consequently, the majority of the applications of manganese zinc ferrites occur in the frequency range 100 KHz to 3 MHz.

The precise control of the Fe^{2+}/Fe^{3+} ratio (typically 0.05/2.00) during the sintering of large quantities of ceramic parts represents a major technological challenge in the production of ferrites. A number of workers have studied the gas-solid equilibrium [31,32] between ferrites and the atmosphere in which they are fired [33–36] and the existence boundaries of the various phases are well defined (see Figure 11). Another way to visualize the interaction with the atmosphere is through an Arrhenius-type plot of the relationship

Table 3 Properties of Commercial (Ferroxcube) Ferrite Formulations

Characteristic	Unit	Ferroxcube grade					
		4C4	3D3	3B9	3B7	3C8	3E2A
Initial permeability at 298 K	μ_i	125	750	1800	2300	2700	5000
Saturation flux density at 298 K (mT)	B_s	300	380	320	380	440	360
Coercive force (mA/m)	H_c	37.7	12.6	3.8	2.5	2.5	1.0
Residual flux density (mT)	B_r					100	
Loss factor at:							
4 kHz	$\tan\delta/\mu_i$			$\leqslant 1 \times 10^{-6}$	$\leqslant 1 \times 10^{-6}$		
100 kHz	$\tan\delta/\mu_i$	35×10^{-6}		$\leqslant 5 \times 10^{-6}$	$\leqslant 5 \times 10^{-6}$	$\leqslant 10 \times 10^{-6}$	$\leqslant 10 \times 10^{-6}$
500 kHz	$\tan\delta/\mu_i$	35×10^{-6}	$\leqslant 14 \times 10^{-6}$	25×10^{-6}	25×10^{-6}		
1 MHz	$\tan\delta/\mu_i$	$\leqslant 40 \times 10^{-6}$	$\leqslant 30 \times 10^{-6}$	120×10^{-6}	120×10^{-6}		
5 MHz	$\tan\delta/\mu_i$	60×10^{-6}					
Temperature factor	$\dfrac{\Delta\mu_i}{\mu_i^2 \Delta T}$	-6.0×10^{-6} min.	$+1.0 \times 10^{-6}$ min.	$+0.9 \times 10^{-6}$ min.	-0.6×10^{-6} min.		-2×10^{-6} min.

		$+6.0 \times 10^{-6}$ max 278–328 K	$+3.0 \times 10^{-6}$ max. 243–343 K	$+1.9 \times 10^{-6}$ max. 243–343 K	$+0.6 \times 10^{-6}$ max. 253–343 K		$+2 \times 10^{-6}$ max. 293–343 K
Power losses at 16 kHz 200 mT, at:	mW/cm^3						
25°C						$\leqslant 110$	
100°C						$\leqslant 100$	
Disaccommodation	$\frac{\Delta \mu_i}{\Delta \mu_i^2 t_1 \log t_2/t_1}$	$< 15 \times 10^{-6}$	12×10^{-6}	$\leqslant 2.5 \times 10^{-6}$	$\leqslant 3.5 \times 10^{-6}$		
Hysteresis losses at kHz, (t^{-1})	ηB		$\leqslant 1.8 \times 10^{-3}$	$\leqslant 1.1 \times 10^{-3}$	$\leqslant 1.1 \times 10^{-3}$		
Curie temperature	K (°C)	>573 (>300)	>423 (>150)	>418 (>145)	>443 (>170)	>483 (>210)	>443 (>170)
Uses		1–20 MHz, filter coil applications, high-frequency, wide-band, and pulse transformers	200 kHz–2.5 MHz, matches polystyrene capacitors, inductors, pulse transformers	Audio to 300 MHz, inductors and tuned transformers	Audio to 300 kHz, inductors used with low-temperature coefficient capacitors, pulse and power transformers	High-flux-density applications, switched-mode power supplies	Wide-band and pulse transformers, inductors

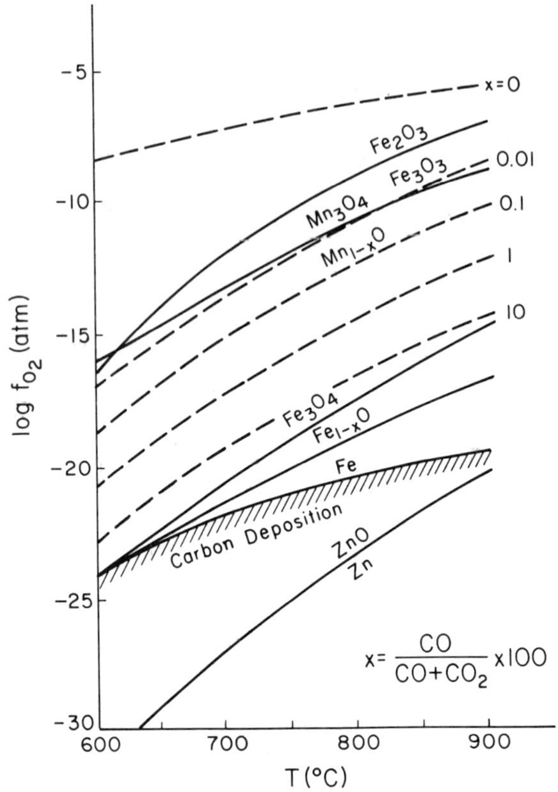

Figure 11 Gas-solid equilibrium phases as a function of oxygen partial pressure (P_{O_2}) and temperature for ferrite component oxides.

between temperature, atmosphere, and the boundaries between the various phases (see Figure 12). In this instance the phase boundaries approximate straight lines and the atmosphere-temperature relationship can be expressed as

$$\log(\%O_2) = \frac{-C_1}{T} + C_2 \qquad (13)$$

In theory, then, to control the amount of divalent iron, and hence the temperature at which there is zero magnetocrystalline anisotropy, it is necessary for the sintering to follow a specific equilibrium line (defined by C_1 and C_2) throughout the firing and cooling down of the product. In practice, however, ferrites are fired in continuous kilns (tunnel kilns). These kilns are rendered relatively gastight by doors or gates at each end which open and close to permit one ceramic tile load of parts to enter at a time. Atmosphere regulation is accomplished by introducing nitrogen gas into the cooling portion of the kiln through inlet ports. The result is a stepwise reduction in the amount of oxygen from the 5–21% present at the sintering temperature (1250–1450°C) to 100–50 ppm at an anneal temperature of 1050–1200°C (see Figure 13). At this elevated temperature, equilibrium with the ambient atmosphere is rather quick, which means that the anneal zone can be relatively short and then the parts are pushed out through the cooling portion of the kiln. It is fortunate that as the temperature drops, the reaction kinetics between the ferrite and the atmosphere drop rapidly because the equilibrium atmosphere required in this low-temperature portion of the kiln can not be achieved in practice.

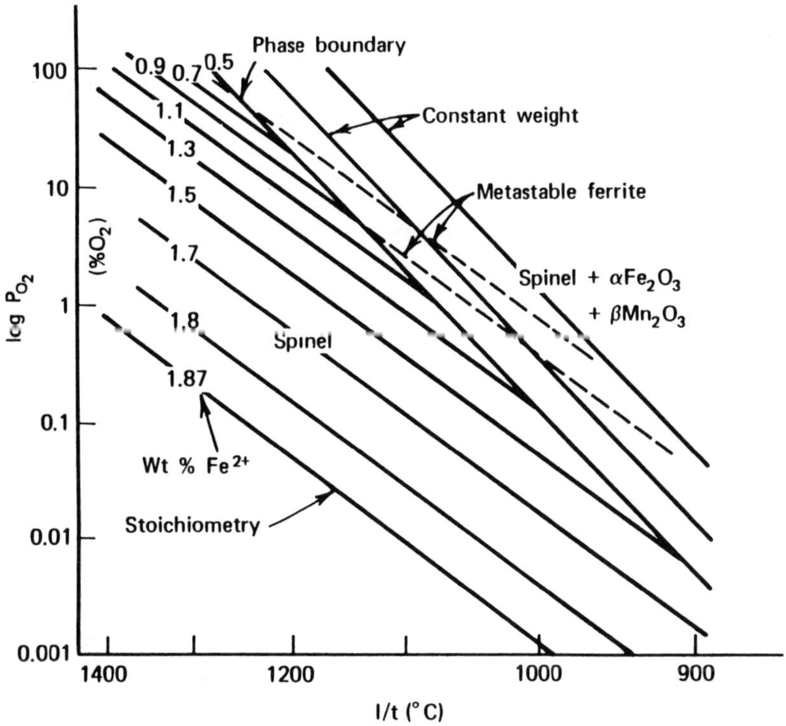

Figure 12 Relationships between oxygen partial pressure (P_{O_2}), temperature, and phase stability in spinel ferrites.

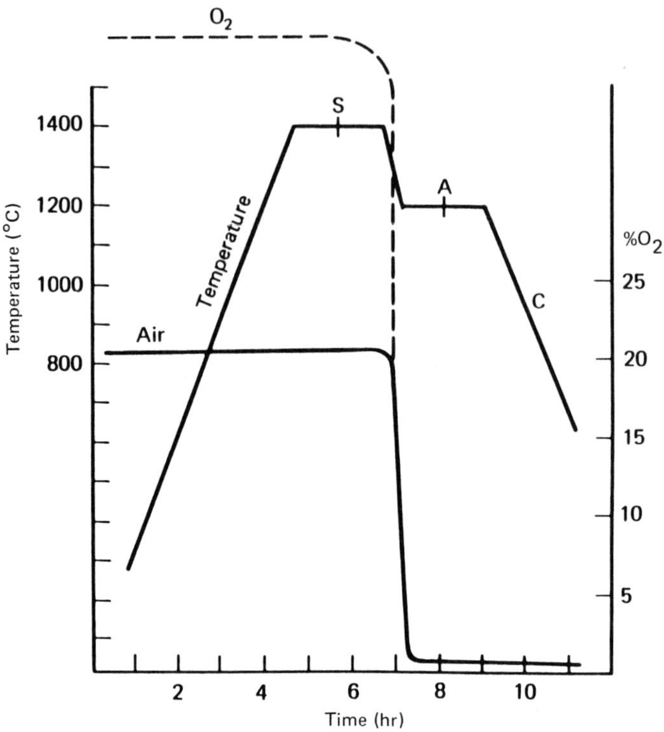

Figure 13 Sintering profile as a function of temperature and atmosphere for MnZn ferrite. S, Soak; A, anneal; C, cooling temperatures.

Small laboratory and production batch kilns with working volumes of 0.7–1.0 m^3 have made it possible to approximate atmosphere control along a C_1–C_2 curve using computer-controlled mixing of O_2-N_2 gases. With a total cycle time of 8–24 hr, these offer the potential of firings individually tailored for specific properties. It is also possible, as pointed out by Plaskett et al. [36], to use a dynamically varying gas-gas buffer system (CO-CO_2) to control the atmosphere; however, this has not found an application in production equipment. It is also possible to "pack" the parts to be fired in a coarse, fully reacted powder of the same composition, the high-surface-area powder acting as a shield or atmosphere source for the parts being fired.

C. Hot Pressing

For most applications the densities obtained by conventional sintering are adequate; however, applications such as magnetic recording require dense, pore-free materials which have the highest saturation and magnetic permeability. The manganese zinc ferrites are ideal, having permeabilities of 5000–6000 (Ni-Zn = 2000) and saturation mannetization 5000–6000 (Ni–Zn = 3500). However, the sintered density is in the range 94–97% and some other densification technique must be used before they are suitable.

Uniaxial hot pressing was the first method employed to obtain fully dense Mn-Zn material. The usual method is to press a preform in the shape of a flat circular disk and then to place that disk in a die constructed of graphite, silicon carbide, or silicon nitride [37]. Densification takes place by heating the die assembly at 1000–1380 C for periods of 1/2–2 hr. Following the hot-pressing step it is usual to do an anneal at a lower temperature to optimize magnetic properties but not induce further grain growth [38]. This method, while yielding dense parts, has the disadvantages that (1) pressure is limited by die wall strength at elevated temperatures, (2) thermal shock characteristics of the die place a lower limit on cycle time, (3) the pressed volume of each disk is relatively small, and (4) each individual piece sees a separate hot-pressing cycle and thus piece-to-piece reproducibility may be poor. For these reasons the hot isotatic pressing (HIP) technique has been developed for ferrites.

Hardtl [39] demonstrated that ceramic materials could be densified in a HIP using inert gas under high pressure (N_2, 1250–1350°C, 2000 psi, 2–10 hr) without the need of an impervious metal envelope to transmit the forces. The key to this method is to sinter the pressed part at a low temperature prior to hipping [40]. This sintering step must be carefully controlled to minimize grain growth yet assure that there is no interconnected porosity through which the gas could permeate the sample during hipping.

VI. SINGLE-CRYSTAL FERRITE

Initial attempts to fabricate single-crystal ferrite during the late 1960s concentrated on the Vernueil method, in which a ferrite powder was melted in a flame and fell on a cooler rod of ferrite, causing the single crystal to grow upward. The results of this method were less than suitable because the crystal was highly stressed and deficient in ZnO because of volatilization in the flame. The properties of these materials were greatly reduced.

Currently Mn-Zn single crystals are grown for fabrication into video recording heads by the Bridgeman method. In this technique a small single-crystal seed of ferrite is placed in a capillary-shaped extension at the bottom of a cylindrical platinum-iridium crucible. The area above the single crystal is filled with ferrite powder or a preform of sintered ferrite. A narrow high-temperature (1500°C) zone passes from the seed to the upper end of the crucible as it is lowered slowly through the furnace (∿1 mm/hr), causing the ceramic to melt and resolidify as a single crystal. The platinum crucible is then cut off the crystal ingot, to be reclaimed, and the single-crystal rod, 100 mm × 800 mm, is fabricated into video recording heads.

The use of the nonmagnetic gadolinium gallium garnet, $Gd_3Ga_5O_{12}$ (GGG), as a substrate for the epitaxial growth of substituted yittrium iron garnet films in bubble memory device fabrication has resulted in the necessity of growing large, well-characterized single crystals of GGG. Fortunately, the rare earth gallium garnets melt without decomposition (congruently melting) and can be pulled from the melt as single-crystal boules (Czochralski method) [41]. Because of the high melting temperature of the materials, 1500–1700°C, iridiumn crucibles are used and heating is done by radio-frequency (RF) coupling to the crucible. A suitable seed crystal is attached to the end of a cooled rod and is slowly brought down to the molten interface and then raised at a rate of 0.5–5.0 mm/hr. To enhance stoichiometry, the crucible and the growing crystal boule are rotated in opposite directions while growth is taking place. After growth is complete, the crystal is sawed into thin disks and polished. It then becomes the host lattice upon which a thin layer of magnetic garnet is grown for the fabrication of magnetic bubble memories [42].

VII. APPLICATIONS: LINEAR FERRITES

Slick [43] has detailed the wide variety of applications in which ferrites are used. These cover the frequency range from 60 Hz to 500 MHz, with the majority of the applications being served by two materials, Mn-Zn and Ni-Zn ferrites. Within these systems a variety of compositions have been developed with specific properties (see Table 4). The general applications can be divided into two types: signal handling (or low-level) and power handling.

The low-level uses are those in which the applied field (or excitation current) and the resulting magnetic flux in the ferrite are linear; the permeability of the material is thus constant (see Figure 14). This means that driving fields are low and that core and winding losses can be neglected. The application of linear design principles

Table 4 Summary of Nonmicrowave Ferrite Compositions and Applications

Ferrite chemistry	Device	Device function	Frequencies	Desired ferrite properties
		Linear B/H, low-flux density		
MnZn, NiZn	Inductor	Frequency selection network Filtering and resonant circuits	$\leqslant$1 MHz (MnZn) $\sim$1–100 MHz (NiZn)	High μ, high μQ, high stability of μ with temperature and time
MnZn, NiZn	Transformer (pulse and wide band)	V and I transformation Impedance matching	Up to 500 MHz	High μ, low hysteresis losses
NiZn	Antenna rod	Electromagnetic wave receival	Up to 15 MHz	High μQ, high resistivity
MnZn	Loading coil	Impedance loading	Audio	High μ, high B_s, high stability of μ with temp., time, and dc bias
		Nonlinear B/H, medium to high-flux density		
MnZn, NiZn	Flyback transformer	Power converter	<100 kHz	High μ, high B_s, low hysteresis losses
MnZn	Deflection yoke	Electron beam deflection	<100 kHz	High μ, high B_s
MnZn, NiZn	Suppression bead	Block unwanted ac signals	Up to 250 MHz	Mod. high μ, high B_s, high hysteresis losses

Table 4 (Continued)

Ferrite chemistry	Device	Device function	Frequencies	Desired ferrite properties
MnZn, NiZn	Choke coil	Separate ac from dc signals	Up to 250 MHz	Mod. high μ, high B_s, high hysteresis losses
MnZn, NiZn	Recording head	Information recording	Up to 10 MHz	High μ, high density, high μQ, high wear resistant
MnZn	Power transformer	Power converter	<60 kHz	High B_s, low hysteresis losses
		Nonlinear B/H, rectangular loop		
MnMg, MnMg-Zn, MnCu, MnLi, etc.	Memory cores	Information storage	Pulse	High squareness, low switching coefficient, and controlled coercive force
MnMgZn, MnMgCd	Switch cores	Memory access transformer	Pulse	High squareness, controlled coercive force
MnZn	Magnetic amplifiers			

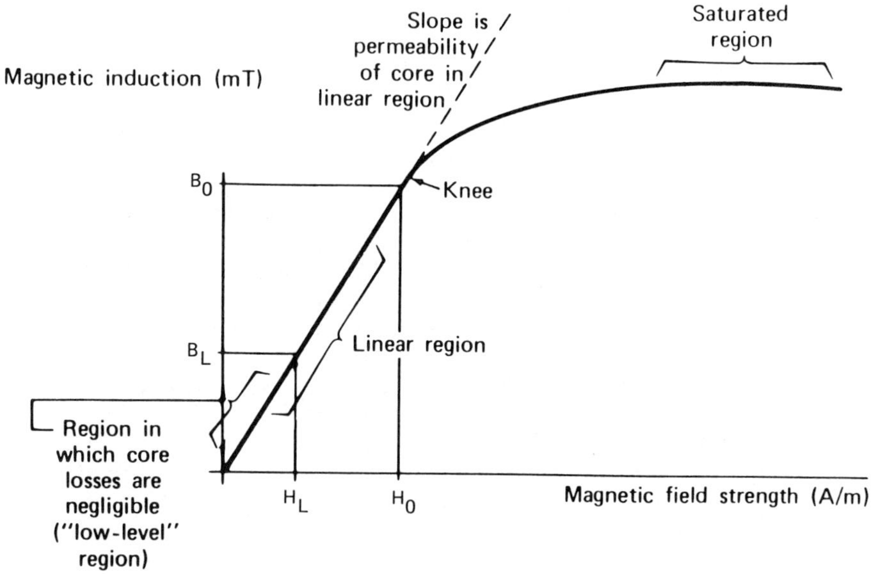

Figure 14 Plot of permeability versus magnetic field strength for a linear ferrite.

can be found in low-level transformers, inductors, telecommunications filters, and so on, and are usually limited to a magnetic induction level of less than 250 gauss [44].

Manganese zinc ferrites having high saturation magnetization and low power losses (e.g., Ferroxcube 3C8, Siemens N27, TDK H7C1) offer considerable advantages over laminated transformer steels at frequencies above 1 kHz. At these frequencies the low resistance of metal laminates ($\sim$50 $\mu\Omega$-cm) causes eddy current losses and transformer heating to be significant.

Although ferrite has a lower operating flux density (3300 gauss) than that of transformer steel (18,000 gauss), the higher operating frequency decreases the flux density in the transformer:

$$B = \frac{E \times 10^8}{4.44 \times f \times N_p \times A_e} \qquad (14)$$

where

B = flux density

E = volts (rms)

f = frequency, Hz

A_e = effective area of the transformer, cm^2

N_p = primary turns

Because of the inverse relationship between flux density and frequency, high-frequency operation reduces the flux density to within acceptable limits. It is also apparent from this relationship that the transformer can be made smaller (reduce A_e) and the number of turns (N_p) reduced as the frequency increases. Ferrite does, however, have a much lower thermal conductivity than metallic transformer cores, so for applications greater than several kilowatts it becomes difficult to cool the transformer effectively.

A major application of high-frequency power components is in the switched-mode power supply (SMPS) [45]. These devices rectify and smooth the primary voltage and switch it through the transformer at high frequencies (15–100 kHz). The output of the transformer is then highly regulated, filtered dc voltage. Offering high efficiencies (60–90%) and better performance in a smaller package, SMPS are finding wide usage in the instrument and computer power supply markets.

Many of these principles have been applied to fluorescent lighting ballasts. Rather than operating at line frequencies using metal transformers, new more efficient ballasts and self-contained fluorescent bulbs are used in incandescent fixtures.

Except for certain military and other specialized applications, the use of ferrite memory cores in computer mainframes has all but disappeared because of advances in semiconductor memory technology. However, spinel ferrites are used extensively in the recording of both analog (audio and video recording) and digital signals. In 1966, IBM introduced the 2314 series of peripheral memories using a Ni-Zn ferrite recording head. This ceramic recording head was easier to construct than one using thin metal laminations because it could be machined to close tolerances from bulk ceramic and glass bonded in air to form a precise, durable gap. The track density was low (100 tracks per inch) and the system storage density was less than 1/4 megabit per square inch. Storage densities of digital systems have increased steadily at 27% per year since that time and current systems offer storage densities in excess of 12 megabits per square inch. With improvements in the processing of Mn-Zn ferrites (e.g., hot pressing), it is now possible to use these materials with their higher permeability and saturation in applications where the track width is less than 0.001 in. (25 μm).

The disappearance of the magnetic properties of ferrites at the Curie temperature can be used to construct a fixed-point temperature

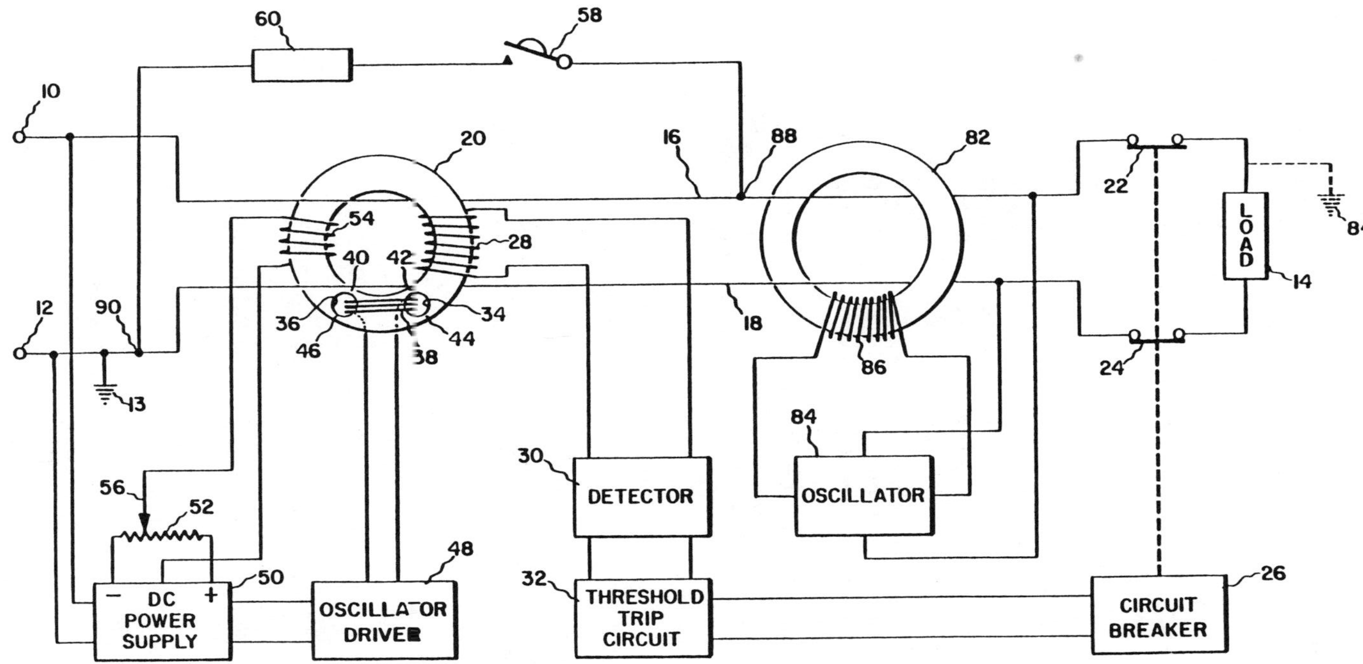

Figure 15 Circuit block diagram for magnetic field detector utilizing magnetic ferrite core for flux control and sensing.

sensor. The Curie temperature can be controlled precisely by substituting varying amounts of zinc ferrite into manganese ferrite. After it has been sintered the Curie temperature is fixed and it is possible to fabricate a miniature fixed-point temperature sensor either by using the material as a signal transformer (above the Curie temperature a signal on the primary will not appear on the secondary) or as the inductive component in a tuned circuit.

The high-frequency properties have been used by Tanka and Reynolds [46] to fabricate a very sensitive magnetic field detector by gating the flux through a sense coil (Figure 15). A controlling coil is wound on a narrowed portion of the magnetic path and then this coil is energized with a square wave driving that portion of the circuit into magnetic saturation. Saturation of this portion of the magnetic patch prevents magnetic flux from circulating through the sense coil. As the square wave decays, the narrowed portion desaturates abruptly, allowing the flux to circulate. The high-frequency flux gating through the sense coil induces a voltage in the sense coil. Of particular advantage in such applications as the sensing of ground fault currents is the fact that the output is essentially independent of the frequency of the signal being measured and the device is active for a very brief period, making it immune to spurious noise effects.

REFERENCES

1. W. H. Bragg, *Philos. Mag.*, *30*, 315 (1915).
2. S. Nishikawa, *Proc. Tokyo Math. Phys. Soc.*, *8*, 199 (1915).
3. J. Smit and H. P. J. Wijn, *Ferrites*, Wiley, New York, 1959, p. 139.
4. G. Economos, *J. Am. Ceram. Soc.*, *38*(7) 241 (1955); *38*(8) 292 (1955); *38*(9), 335 (1955).
5. A. Broese Van Groenou et al. *Mater. Sci. Eng.*, *3*, 317 (1968–1969).
6. A. Broese Van Groenou et al., *Mater. Sci. Eng.*, *3*, 317 (1968 to 1969).
7. Y. Watanabe et al., *Phys. Status Solidi (b)*, *90*, 697 (1978).
8. K. Ohta, *J. Phys. Soc. J.*, *18*, 685 (1963).
9. D. Stoppels, *J. Appl. Phys.*, *51*, 2789 (1980).
10. D. Stoppels, U. Enz, and J. P. M. Damen, *J. Magn. Magn. Mater.*, *20*, 231 (1980).
11. R. V. Jones, *IEEE Trans. Son. Ultrason*, *SU-13*, 86 (1966).
12. M. O'Connor and H. Belson, *J. Appl. Phys.*, *41*, 1028 (1970).
13. J. Smit and H. P. J. Wijn, *Ferrites*, Wiley, New York, 1959, p. 177.
14. V. Adelslöld, *Ark. Kem. Min. Geol.* *12A*(29), 1 (1938).

15. G. Albanese, *J. Phys.*, Colloque Cl, Suppl. 4, *38*, C1–85 (1977).
16. J. Smit and H. P. J. Wijn, *Ferrites*, Wiley, New York, 1959, p. 194.
17. A. L. Stuijts et al., *Philips Tech. Rev.*, *16*, 141 (1954).
18. Bertaut and Forrat, *Compt. Rend.*, *242*, 383 (1956).
19. S. Geller, *Z. Kristallogr.*, *125* (1967).
20. R. H. Langley and G. D. Sturgeon, *J. Solid State Chem.*, *30*, 79 (1979).
21. F. C. Hawthorne, *J. Solid State Chem.*, *37*, 157 (1981).
22. M. A. Gilleo, in *Ferromagnetic Materials*, Vol. 2, E. P. Wohlfarth ed., North–Holland, New York, 1980, p. 1.
23. T. Takada, *Ferrites: ICF 3*, 19 , p. 3.
24. H. Robbins, *Ferrites: ICF 3*, 19 , p. 7.
25. Bu–Fan B. Yu and A. Goldman, *Ferrites: ICF3*, 19, p. 52.
26. F. J. Schnettler and D. W. Johnson, *Ferrites: ICF 3*, 19 , p. 105.
27. J. Nicolas, in *Ferromagnetic Materials*, Vol. 2, E. P. Wohlfarth ed., North–Holland, New York, 1980, p. 243.
28. B. B. Ghate, in *Grain Boundary Phenomena in Electronic Ceramics*, The American Ceramic Society, Columbus, Ohio, 1981, p. 477.
29. G. Y. Onoda, Jr., in *Ceramic Processing Before Firing*, G. Y. Onoda, Jr., and L. L. Hench, eds., Wiley, New York, 1978, p. 235.
30. M. Dixon et al., "Fabrication of Ferrite Material," U.S. Patent 4,247,500, Jan. 27, 1979.
31. W. G. Ernst, *J. Am. Sci.*, *264*, 37 (1966).
32. H. P. Eugster and G. B. Skippen, *Res. Geochem.*, *2*, 492 (1967).
33. J. M. Blank, *J. Appl. Phys.*, *32*, Suppl. 3, 378S (1961).
34. E. D. Macklen, *J. Appl. Phys.*, *36*, 1022 (1965).
35. P. I. Slick, in *Ferrites (Proceedings of the International Conference July 1970, Japan)*, Y. Hoshino, S. Iida, and M. Sugimoto, eds., University Park Press, Baltimore, Md., 1971, p. 81.
36. T. S. Plaskett et al., *J. Appl. Phys.*, *53*, 2428 (1982).
37. A. Ikeda et al., in *Ferrites (Proceedings of the International Conference July 1970, Japan)*, Y. Hoshino, S. Iida, and M. Sugimoto, eds., University Park Press, Baltimore, Md., 1971, p. 337.
38. E. Hirota et al. in *Ferrites (Proceedings of the International Conference, Sept.–Oct. 1980, Japan)*, H. Watanabe et al., eds., Center for Academic Publications, Japan, 1981, p. 667.
39. K. H. Hardtl, *Am. Ceram. Soc. Bull.*, *54*, 201 (1975).
40. C. Buthker and Th. Berben, *J. Phys.*, Colloque C1, Suppl. 4, *38*, C1–341, 1977.
41. C. D. Brandle and A. J. Valentino, *J. Cryst. Growth*, *12*, 3 (1972).
42. Hsu Chang, *Magnetic Bubble Technology: Integrated-Circuit Magnetics for Digital Storage and Processing*, IEEE Press, New York, 1975.

43. P. I. Slick, in *Ferromagnetic Materials*, Vol. 2, E. P. Wohlfarth, ed., North–Holland, New York, 1980, p. 189.
44. *Linear Ferrite Magnetic Design Manual,* Ferroxcube Division of Amperex Electronic Corp., Saugerties, N.Y.
45. *SMPS Transformer Design Manual*, Ferroxcube Division of Amperex Electronic Corp., Saugerties, N.Y.
46. D. Tanka and T. Reynolds, "Ground Fault Interrupter," U.S. Patent 4,280,162, July 21, 1981.

5

Ceramic Sensors: Theory and Practice

DAVID C. HILL
MLC Technologies, Inc., Attleboro, Massachusetts

HARRY L. TULLER
Massachusetts Institute of Technology, Cambridge, Massachusetts

I. INTRODUCTION

The operation of our industrialized world has depended, in large part, on our ability to measure and control a variety of physical and chemical parameters, including pressure, temperature, and position. In recent decades, development of electrical measurement and control systems that couple input sensor and output transducer via ever more sophisticated signal processing units has continued at a rapid pace. Furthermore, due to the recent proliferation of inexpensive and compact microprocessors, the scope of potential applications of such systems has expanded enormously, rapid growth in robotization being a notable example. Comparable improvements in the input and output functions provided by the sensors and actuators have not, however, kept pace and now often limit the feasibility of the overall system. Increasing attention to developments in these areas can, therefore, be expected in the near future.

In this chapter the authors focus on ceramic sensors, which, because of their well-known stability, find wide application in areas unsuitable for other types of sensors. To provide a basis for understanding and predicting the operation of the most common sensors, we begin by reviewing the fundamental relationships between the electrical properties of ceramics and the important variables, such as composition, temperature, atmosphere, and microstructure. Because the operation and range of applicability of specific sensors are so intimately tied to the manner in which they are fabricated, much attention has been focused on the practical difficulties that must be addressed during the development of new and improved sensors. Detailed case studies of representative sensors are provided to illustrate the fundamentals of device operation. Extension of these concepts to other sensor types is indicated.

For each type of sensor, the following features will be analyzed: theory of operation, device manufacture, applications, and trends and directions. In this manner it is hoped that this chapter will serve the needs of the applications engineer, the processing specialist, and the research scientist concerned with developing new or improved materials and sensor devices.

II. THEORY

Oxide ceramics exhibit a number of special characteristics which make them particularly attractive for sensor applications. These include marked sensitivity to oxidizing and reducing gases or to temperature excursions. Additionally, high ionic conductivities render some oxides unique choices in electrochemical-based sensors.

Each type of sensor must satisfy certain criteria. Semiconducting gas sensors, while responding readily to modulations in gas composition, should be insensitive to temperature. Thermistors, on the other hand, rely on their sensitivity to temperature excursions but must do so while exhibiting little or no sensitivity to atmosphere. Sensors operating on electrochemical principles cannot tolerate electronic conduction but must exhibit exceptionally high ionic conduction.

Furthermore, consideration of only bulk electrical characteristics is often insufficient to predict sensor response. Device behavior may be dominated, for example, by transport across grain boundaries.

It therefore follows from the discussion above that an understanding of charge carrier generation and transport, both in bulk materials and at interfaces, is necessary to provide a basis for predicting the thermal and chemical characteristics of various sensor elements. In the following review we emphasize the salient features of charge transport relevant to sensor response.

In Chapter 8 the principles of electronic conduction in oxides are reviewed in some detail. In this section we highlight those aspects particularly relevant to the operation of ceramic sensors and begin with consideration of their temperature dependence.

It is well known that all matter exhibits a temperature-dependent resistivity. The nature of this dependence, however, differs widely among the different materials. Thus whereas the resistivity of metals increases weakly with temperature, the resistivity of semiconductors and insulators, in general, decreases exponentially with temperature. The latter feature (i.e., a large temperature coefficient of resistance, TCR) is the basis of the most common oxide sensor—thermistors (thermally sensitive resistors)—in use today. The term "negative temperature coefficient" (NTC), usually associated with these devices, reminds the user that the thermistor resistance decreases with "negative coefficient" increasing temperature.

In contrast to semiconductors with relatively few carriers (10^{10}–10^{15} cm^{-3}), metals already possess an optimum number of carriers ($\sim 10^{22}$ cm^{-3}) at 0 K, and so respond to temperature variations only through their carrier mobilities, which decrease due to thermally induced scattering. This is reflected in a nearly linear increase in resistance with temperature (>100 K), that is, a positive temperature coefficient of resistance (PTCR). In the 1950s a more useful form of PTC thermistor based on ceramic barium titanate was discovered with enhanced TCR (10–150% K^{-1}) values markedly above that for metals. This phenomenon is now understood to be associated with grain boundary effects in this material.

In the following, transport in intrinsic, extrinsic, and nonstoichiometric semiconducting metal oxides, as well as the influence of interfaces on these characteristics, are examined.

A. Principles of Electrical Conduction Relevant to Sensor Operation

The electrical conductivity of a solid is composed of the sum of the partial conductivities associated with each type of charge carrier and is given by

$$\sigma = \sum_j \sigma_j \tag{1}$$

Each partial conductivity σ_j, which may represent transport of either ionic or electronic carriers, is defined by

$$\sigma_j = C_j Z_j q \mu_j \tag{2}$$

in which C_j is the carrier density (per m^3), Z_j the effective charge (coulombs), and μ_j the mobility (m^2/V-sec) of the jth species. For semiconductors, the C_j term generally represents nearly the entire temperature dependence of the partial conductivity (small polarons may be the exception), whereas for ionic conductors the mobility term μ_j can be strongly temperature dependent.

Three mechanisms for carrier generation need to be considered: (1) thermally induced intrinsic electronic and ionic disorder, (2) impurity, and (3) redox-induced defects. In the following we discuss the manner in which C_j is influenced by temperature and atmosphere in a number of different defect regimes.

1. Intrinsic Conductors

Thermal electron-hole generation across the semiconductor band gap, represented by the reaction

$$O \longrightarrow e + h \tag{3}$$

takes on the familiar mass action form

$$np = AT^3 \exp\left(-\frac{E_g}{kT}\right) \tag{4}$$

Since each electron is compensated by a hole,

$$n_i = p_i = (AT^3)^{\frac{1}{2}} \exp\left(-\frac{E_g}{2kT}\right) \tag{5}$$

in which the subscripts $_i$ represent intrinsic carrier densities, A is a constant dependent on carrier effective masses, and E_g is the band gap energy. Choosing a relatively small band gap of 1 eV (as in Si), one obtains a carrier density of about 10^{12} cm^{-3} at 500 K, a factor of 10^{10} smaller than in metals. Because band gaps in semiconductors range from tenths of an electron volt (InSb) to 10 eV (SiO_2), this in principle provides for a wide choice of both bulk resistivities and temperature sensitivities.

Intrinsic ionic disorder, a consequence of either Schottky disorder,

$$O \longrightarrow V'_m + V^{\bullet}_O \tag{6}$$

or Frenkel disorder,

$$O \longrightarrow V'_m + M^{\bullet}_i \tag{7}$$

results in the formation of charged ionic defects.* The mass action relations that follow are represented by

$$[V'_m]\ [V^{\bullet}_O] = K_s = N^2 \exp\left(-\frac{\Delta H_s}{kT}\right) \tag{8}$$

$$[V'_m]\ [M^{\bullet}_i] = K_f = NN_i \exp\left(-\frac{\Delta H_f}{kT}\right) \tag{9}$$

in which N represents the number of normal lattice sites, N_i the number of interstitial sites, and ΔH_s and ΔH_f the enthalpies for Schottky and Frenkel disorder, respectively. Since these defects are also generated in pairs, the temperature dependence of the individual defects is given by

$$[V'_m] = V^{\bullet}_O] = N \exp\left(-\frac{\Delta H_s}{2kT}\right) \tag{10}$$

for Schottky disorder and

$$[V'_m] = [M^{\bullet}_i] = (NN_i)^{\frac{1}{2}} \exp\left(-\frac{\Delta H_f}{2kT}\right) \tag{11}$$

---

*For a detailed description of defect thermodynamics and notation, see Ref. 1.

for Frenkel disorder. Since ΔH_s and ΔH_f are typically on the order of several electron volts and more, thermistors based on intrinsic ionic disorder would have exceptionally large negative TCRs.

2. Extrinsic Conductors

Altervalent impurities induce the formation of either charged electronic or ionic defects in order to maintain charge balance in the crystal. The energies required to dissociate these carriers from the impurity centers fall far below the intrinsic defect formation energies, such as E_g, ΔH_s and ΔH_f. Consequently, the concentration of such extrinsic defects generally lies far above that of intrinsic carriers at reduced temperatures.

The temperature dependence of the extrinsic electron carriers in the low-temperature region is given by [2]

$$n = BT^{3/4} \exp\left(-\frac{E_d}{2kT}\right) \tag{12}$$

$$p = CT^{3/4} \exp\left(-\frac{E_a}{2kT}\right) \tag{13}$$

in which E_d and E_a are the donor and acceptor ionization energies and B and C are constants proportional to the square root of the donor and acceptor densities, respectively.

At sufficiently high temperatures (i.e., $\frac{1}{2}kT \sim E_d$, E_a) full ionization of the impurities are achieved and the carrier densities become independent of temperature. This region, called the "exhaustion" or "saturation" region in semiconductor terminology, separates the thermally activated intrinsic region at elevated temperatures from the much weaker thermally activated region at low temperature. All three regimes are included in Figure 1, in which log carrier density is plotted against reciprocal temperature for a typical semiconductor.

In the oxide literature, doped oxides, which normally lie in the saturation region due to low-impurity ionization energies, are often referred to as "controlled valence semiconductors" [3]. This serves to emphasize the fact that each free carrier may be visualized as changing the valence of one of the host cations, partially where one is dealing with localized carriers in the conduction or valence bands. A well-known example is the compensation of a Li^+ ion in NiO by a Ni^{3+} ion, which may alternatively be viewed as a hole in the valence band with Ni 3d character.

Controlled-valency semiconductors exhibit low TCR values, for one of two reasons. First, as mentioned above, the impurity ionization energy is low. This applies, however, only to temperatures

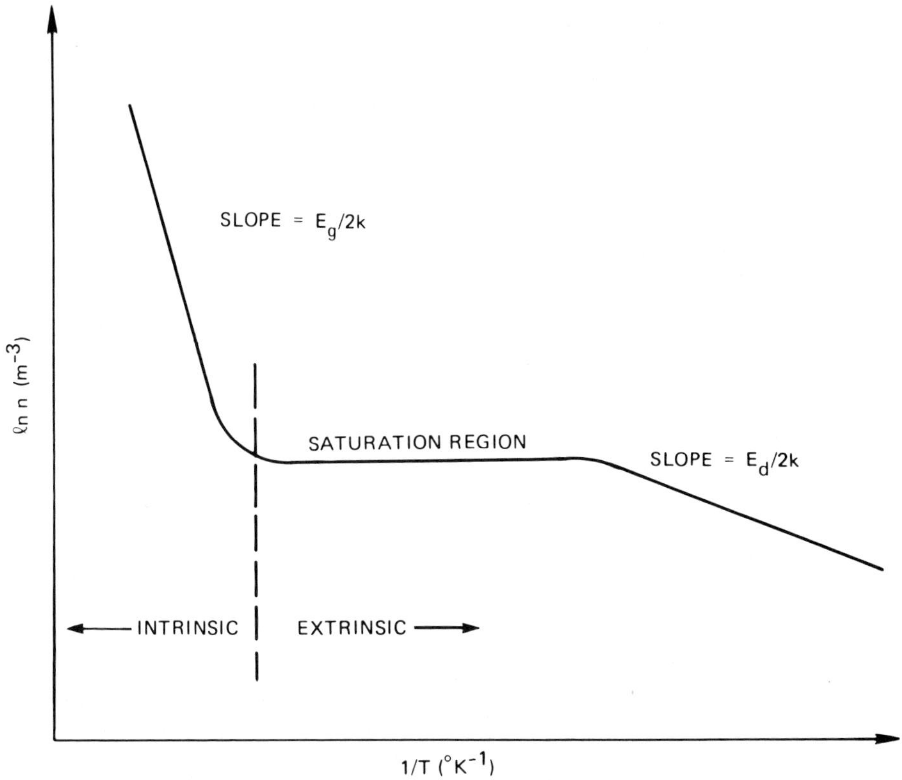

Figure 1 Variation of carrier density with temperature in an n-type semiconductor. Note the three different segments of the curve corresponding to different defect regimes.

below the saturation region. Second, the localized nature of carriers in many transition and rare earth metal oxides results in activated mobilities. The latter feature is probably the primary source of the TCR in many oxide thermistors. The mobility factor is treated explicitly below.

In a parallel manner, altervalent impurities sometimes result in the enhancement of certain ionic defects. These become mobile only after breaking away from the impurity of opposite charge. Again an expression of the form (for example)

$$[V_m'] = (NN_a)^{\frac{1}{2}} \exp\left(-\frac{E_b}{2kT}\right) \tag{14}$$

is obtained in which N_a is the number of altervalent impurities and E_b is the binding energy of the defect pair. Similar equations hold for the other ionic defects. Furthermore, as in the semiconductor case, a saturation region is also found at higher temperature. Here, however, a strong temperature dependence is maintained, given the activated nature of ion mobility.

When solubility limits of altervalent ions are high (e.g., Y_2O_3 in ZrO_2), anomalously high ionic conductivities may result. Such materials find use as solid electrolytes in electrochemical sensors. Further details concerning their electrical properties may be found in Chapter 8.

3. Nonstoichiometry

Since thermistors must often operate at elevated temperatures, their stability under these conditions is of great concern. In contrast to other types of semiconductors, oxides can react, often reversibly, with their gaseous environments, for example:

$$MO \longrightarrow MO_{1-x} + \frac{x}{2} O_2 \uparrow \tag{15}$$

This resultant change in composition or stoichiometry can introduce an additional source of temperature sensitivity over and above those already discussed in Section II.A.2.

Oxidation-reduction reactions contain a number of important implications for thermistor operation. These include the potential for nonreproducible operation, sensitivity to changes in ambient atmosphere, and potentially large TCR values at elevated temperatures. These features will become apparent after discussing one such potential reaction in some detail.

Consider the reduction reaction given in Eq. (15). If we rewrite this reaction in terms of the defects that are probably formed during this reaction, we obtain

$$O_O \longrightarrow V_O^{\bullet\bullet} + 2e' + O_2 \tag{16}$$

in which an oxygen ion is observed to leave the solid for the gas phase, leaving behind a new positively charged vacant site and two excess electrons. By comparison with a conventional semiconductor, the vacancy can be viewed as a donor defect which may recombine with the excess electrons at reduced temperatures [4].

The temperature dependence of reaction (16) may be obtained from the mass action relation given by

$$[V_O^{\cdot\cdot}] \times n^2 \times P_{O_2}^{\frac{1}{2}} = K(T) = K_O \exp\left(-\frac{E_R}{kT}\right) \tag{17}$$

where $[V_O^{\cdot\cdot}]$ and n are the vacancy and electron density, respectively, and K(T) is the equilibrium constant. Assuming that the temperature is high enough to ensure full ionization and further that charge neutrality is satisfied.

$$n = 2[V_O^{\cdot\cdot}] \tag{18}$$

then the charge carrier densities are given by

$$n = 2V_O^{\cdot\cdot} = 2K_O^{1/3} \exp\left(-\frac{E_R}{3kT}\right) P_{O_2}^{-1/6} \tag{19}$$

Since E_R commonly varies between approximately 5 and 10 eV, this reaction can result in a rather steep temperature dependence. Note further that the carriers are also sensitive to the oxygen partial pressure which would result in isothermal resistivity changes during atmosphere transients. These observations clarify the latter two of our above remarks concerning large TCRs and atmosphere sensitivity.

The solution for n and $[V_O^{\cdot\cdot}]$ given in Eq. (19) assumes the device to be in equilibrium with its surroundings. Upon change of temperature or atmosphere, oxygen must diffuse out of the element during reduction or into the element from the gas phase during oxidation. If this diffusive process is slow, as it must be at lower temperatures, it will result in a time-dependent and generally nonreproducible operation of the device. Practical devices are thus designed to operate at temperatures where ionic motion is not significant. Under these conditions, a given defect is, in effect, frozen in, leaving only the electronic ionization processes and mobility factors as sources of temperature dependence for the resistivity. This clearly places a practical upper operating temperature limit on standard thermistor devices.

For the reasons just discussed, a controlled valency approach, rather than quenching in carriers produced by nonstoichiometry, would appear to be the most reliable means for obtaining time-independent electrical properties. Nevertheless, even for controlled valence solids, remnants of other defect regimes which occur under high-temperature processing conditions are rarely entirely removed during specimen cooling [5,6].

Since the manner in which thermistor devices are fabricated implies, to some degree, operation under nonequilibrium conditions, it must be expected that some device aging will occur over extended periods. Whether the rate of aging is acceptable will depend in large part on the mobilities of the ionic species, the extent that the frozen-in composition deviates from the equilibrium value, and the temperature of operation. In well-designed thermistors these features have been factored in and are maintained under control. In Section IV. A. 5 some practical aspects of device stability are discussed.

4. Phase Transformations

The extremely large changes in resistance that occur over a relatively narrow temperature range in the vicinity of a phase transformation are sometimes used in switching a circuit from one state to another at the critical temperature. The change in resistivity may either be due to a change in the bulk transport mechanism [4] or related to a grain boundary phenomenon as discussed shortly for PTC devices based on $BaTiO_3$. In the former case, care must often be taken in the fabrication of the device to ensure that the lattice parameter changes associated with the transitions do not result in sufficiently high stresses to damage the device during manufacture or destroy it during subsequent thermal cycling.

5. Mobility Factors

Since all presently utilized thermistors are semiconductors, we begin by considering the temperature dependence of electronic carrier mobility. For pure, nonpolar, broad-band semiconductors, electronic drift mobilities are limited by acoustic phonon scattering [7] and take the general form

$$\mu = \mu_0 (m^*)^{-5/2} T^{-3/2} \tag{20}$$

Values for μ in such systems vary from approximately 500 (ZnTe) to 65,000 (InSb) $cm^2/V\text{-sec}$ at room temperature. At sufficiently low temperatures, ionized impurity scattering, given by

$$\mu = \mu_0^* (m^*)^{-1/2} N_i^{-1} T^{3/2} \tag{21}$$

may take over.

In ionic compounds such as oxides, the electron mobility is determined in large part by its interaction with the polar modes of the crystal. Here the electron induces a deformation in the surrounding lattice, thus tending to self-trap itself in a potential well of its own making. The electron and its surrounding polarization cloud are commonly referred to as a polaron.

Two types of polarons are distinguished. When the radius of the well is many atomic spacings, the electron continues to move in a band but with an enhanced effective mass. The mobility of a large polaron [8], above the Debye temperature ϕ, is given by

$$\mu = \mu_0 T^{-1/2} \tag{22}$$

and is expected to be $\sim$1–100 cm^2/V-sec. If, instead, the well radius approaches one atom spacing, the electron becomes self-trapped at a given lattice site and can move to an adjacent site only by an activated hopping process similar to that exhibited by ionic diffusion. Small polaron mobility [9] is described by

$$\mu = \left[\frac{(1 - c)ea^2\nu_0}{kT}\right] \exp - \frac{E_H}{kT} \tag{23}$$

in which E_H is the hopping energy, ν_0 the attempt frequency, and $(1 - c)$ is the fraction of sites not already occupied by other electrons. Small polaron mobilities are generally reported to be on the order of 10^{-4}–10^{-2} cm^2/V-sec at elevated temperatures. Since many thermistor formulations include large fractions of transition metal cations, small polaron transport can often be expected given the narrow 3d bands characteristic of these elements. Hopping energies typically add 0.1–0.4 eV to the overall activation energy of conduction [8].

Ionic motion is well known to be an activated process and is described satisfactorily by Eq. (23). Until recently, ionic migration energies were believed to be at least of the order of 0.5 eV and typically much higher in oxides [e.g., E_m (Mg in MgO) $\sim$2.1 eV]. With the discovery of fast ion conductors, however, it has become clear that much lower ionic migration energies are possible [e.g., E_m (Ag in $RbAg_4I_5$) $\sim$0.09 eV] [10]. Although some early attempts at utilizing ionic conductors were made [11], difficulties associated with electrolysis-induced changes in composition and properties greatly hampered their utility. Nevertheless, as mentioned above, ionic transport still remains an important parameter in establishing the aging characteristics of thermistors even in oxide semiconductors.

In summarizing the conclusions of this section, the TCR value of material in which the bulk transport properties are controlling is related to (1) carrier generation and (2) carrier mobility. In general, at least one of these factors is thermally activated, and so the resistivity takes the form

$$\rho = \rho_0 \exp\left(\frac{E}{kT}\right) \tag{24}$$

the exceptions being the nonpolar semiconductor in saturation and metals. Table 1 summarizes the values that E takes on for the various types of conductors. The terms E_H and E_m correspond to electron hopping and ionic migration energies, respectively.

When discussing practical devices, resistance R is generally used and Eq. (24) is conventionally rewritten in the form

$$R = A \exp\left(\frac{B}{T}\right) \tag{25}$$

where the material constant, B value or beta factor, is simply E/k, and A is a constant related to device dimensions and material resistivity. Typical values of B fall around 3500 K, which corresponds to an activation energy of ∿0.3 eV. It should be noted that B does not always remain constant over the full range of interest but may vary due to a transition between transport regimes as discussed above.

Another constant already mentioned (i.e., the TCR) is often designated by the symbol α and is defined more precisely by

$$\alpha = \frac{1}{R}\frac{dR}{dT} \tag{26}$$

It may be easily shown by differentiating Eq. (25) with respect to temperature that the α and B are related to each other by

$$\alpha = \left(-\frac{B}{T^2}\right) \tag{27}$$

B. Boundary Effects on Electrical Properties

1. NTC Devices

Under thermal equilibrium the electron and hole densities in a semiconductor (nondegenerate) or insulator are tied to the Fermi energy [7] by the expressions

Table 1 Activation Energies for Resistivity

Mobility type	Carriers[a]			
	Intrinsic	Saturation	Extrinsic	Nonstoichiometric
Bandlike	$E_{g/2}$	0	$E_{D/2}$[b]	$E_{R/a}$
Small polaron	$E_{g/2} + E_H$	E_H	$E_{D/2}$[b] $+ E_H$	$E_{R/a} + E_H$
Ionic	$E_{f/2}$[c] $+ E_M$	E_M	$E_{b/2} + E_M$	$E_{R/a} + E_M$

[a] "a" represents ionization state of the defect and E_b the binding energy of a defect pair.

[b] Or $E_{A/2}$ for acceptor impurities.

[c] Or $E_{S/2}$ for Schottky defects.

$$n = N_c \exp\left(-\frac{E_g - E_f}{kT}\right) \tag{28}$$

$$p = N_v \exp\left(-\frac{E_f}{kT}\right) \tag{29}$$

in which N_c and N_V are the electron and hole effective density of states, respectively, and the top of the valence band has been taken as the zero of energy.

When a compositional inhomogeneity is introduced into the material (e.g., the donor dopant density, N_D, a gradient in Fermi energy or chemical driving force is initially generated. A net flow of charge is thereby induced to eliminate the gradient. Under steady-state conditions, local electrical neutrality is accommodated by the formation of a space charge in the close vicinity of the inhomogeneity, thus resulting in electric fields which tend to repel further flow of charge in that direction. In the familiar case of the p-n junction, the space charge region is depleted of mobile carriers, due to the massive recombination of electrons and holes in its vicinity.

For an abrupt p-n junction in which the n and p regions are in saturation, the electrostatic barrier formed at the junction or the built-in potential V_{bi} [7] is given by

$$qV_{bi} = kT \ln \frac{N_A N_D}{n_i^2} \tag{30}$$

Note that since n (n-region) $\sim N_D$ and n (p-region) is equal to $n_i^2/p = n_i^2/N_A$, thus

$$V_{bi} \sim \ln \frac{n\ (\text{n-region})}{p\ (\text{p-region})} \tag{31}$$

which demonstrates that the larger the initial gradient in the carrier density of a given carrier, the larger the built-in potential barrier.

By their nature, grain boundaries represent inhomogeneities in structure and composition. Compositional inhomogeneities can be marked due to processes such as impurity segregation and/or kinetically limited reactions in the vicinity of the boundaries. These grain boundary defects and impurities give rise to a distribution in energy of a planar array of localized interface states. If, as above, we assign different individual Fermi energies E_{FB} and E_{FG} to the boundary regions and to the grains prior to their effective contact and equilibration with each other, we can expect the formation of space charge regions and potential barriers subsequent to such equilibrium. Figure 2a shows the resultant spatial energy distribution for two semiconducting grains and their boundary. Given the symmetry of this arrangement, one now has, in effect, two back-to-back diodes with barrier heights ϕ_{BO}, space charge layer widths $d_l = d_r = d$, and boundary width W, which for real materials is typically much smaller than d_l and d_r. The symbol ζ represents the energy difference between the Fermi energy and the conduction-band edge.

Application of the one-dimensional form of Poisson's equation to such junctions gives the following relationship between the built-in barrier height ϕ_{BO} and the depletion layer width d [7]:

$$\phi_{BO} = \frac{qN_D d^2}{2\varepsilon\varepsilon_0} \tag{32}$$

where N_D is the donor density in the bulk and ε the relative dielectric constant. If we take N_b to be the two-dimensional charge trapped in the boundary states, then

$$N_b = 2dN_D \tag{33}$$

to satisfy charge neutrality for the junction. Therefore, ϕ_{BO} may be rewritten as

$$\phi_{BO} = \frac{qN_b^2}{8\varepsilon\varepsilon_0 N_D} \tag{34}$$

The reader should note the inverse relationship between the barrier height and the relative dielectric constant. This relationship plays a crucial role in ferroelectrics in which ε is a strong function of temperature in the vicinity of the Curie temperature. We will return to this feature in our discussion of PTC thermistors.

When a bias voltage V is applied across the boundary, the spatial energy distribution changes to that shown in Figure 2b, with the left side going into forward bias and the right side into reverse bias. Beginning with the assumption that the only important mechanism for current flow across the barrier is thermionic emission, Pike and Seager [12] derived the following current-voltage relationship:

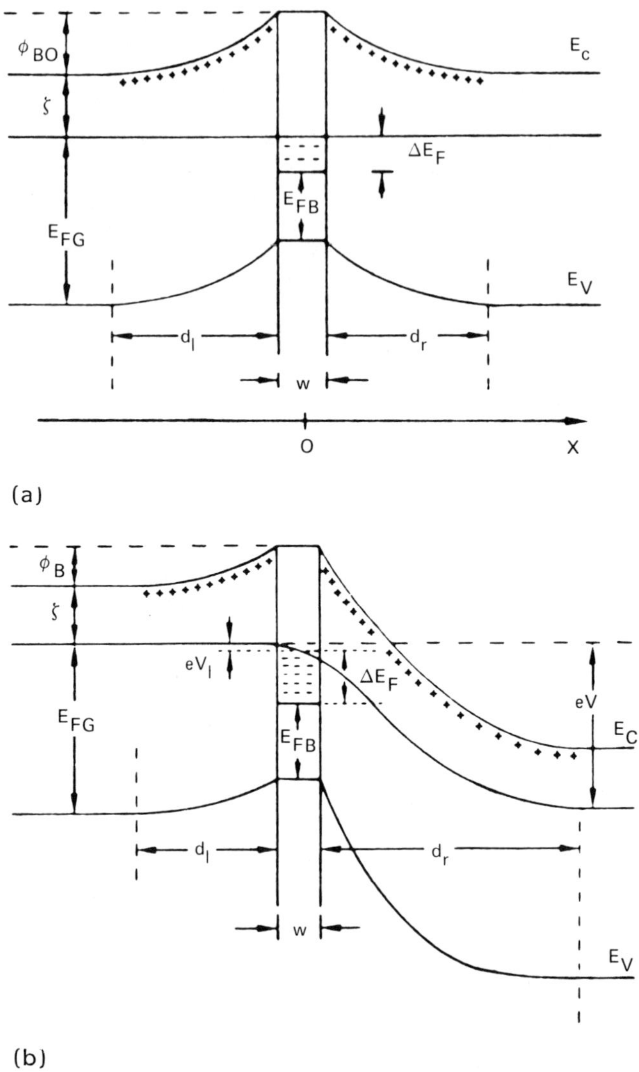

Figure 2 (a) Spatial energy distribution in the vicinity of a grain boundary. The two back-to-back diodes are characterized by barrier heights ϕ_B, space charge layer widths l_r and l_l, boundary width w, and the energy difference between conduction-band edge and Fermi energy ζ; (b) spatial energy distribution as modified by bias field (eV). (From Ref. 12.)

$$J = A \exp\left(-\frac{\zeta + \phi_B}{kT}\right)\left[1 - \exp - \frac{eV}{kT}\right] \qquad (35)$$

in which J is the current density, A the Shottky–Richardson prefactor, ϕ_B the band bending, and V the applied voltage.

Pike and Seager also show the near-zero bias resistance to be

$$R_O = C \exp\left[\left(\phi_{BO} - T\,\frac{d\phi BO}{dT}\right)/kT\right] \qquad (36)$$

where $\phi_{BO} = \phi_B(V = O)$. Clearly, R_O is a sensitive function of the built-in barrier height ϕ_{BO}.

For a large percentage of polycrystalline oxide semiconductors, space charge barriers at the grain boundaries result from segregation of dopants, more rapid redox equilibrium at the grain boundaries due to enhanced grain boundary diffusion, and/or intrinsic interface states. Under all of these conditions, a negative temperature coefficient (NTCR) with a B value roughly equal to ϕ_{BO}/k results in addition to any temperature dependence resulting from bulk effects.

Metal electrode-semiconductor contacts are often also rectifying. Analysis of such Schottky barriers proceeds in a fashion similar to that of p-n junctions except for the fact that the space charge is now entirely localized on the semiconductor side. In developing thermistors and related devices, one is often concerned with identifying electrodes which are "ohmic" and of low resistance, thereby adding little to the temperature dependence of the device. In general, however, the TCR of a material is made up of (1) bulk, (2) grain boundary, and (3) electrode contributions. In developing a device with well-defined characteristics, each of these factors must be taken into account and controlled.

2. PTC Devices

Space charge barriers at grain boundaries were shown above to result in NTCR behavior. An exception to this rule may be found in the perovskite structured family of ferroelectric semiconductors based on $BaTiO_3$. Here, as first pointed out by Heywang [13], given that the dielectric constant, which obeys the Curie–Weiss law

$$\varepsilon = \frac{C}{T - \phi} \qquad (37)$$

becomes anomously large near the Curie temperature ϕ, it follows that the built-in potential ϕ_B will decrease drastically as the temperature is lowered toward ϕ [see Eq. (34)]. Given the rapidly increasing barrier height with temperature above ϕ, the source of the PTCR in these materials becomes immediately clear [see Eq. (36)]. Typical resistance-temperature plots for PTC thermistors prepared from solid solutions of $BaTiO_3$ with $SrTiO_3$ or $PbTiO_3$ are illustrated in Figure 3. A close correspondence between the magnitude of the switch temperature and the Curie temperature is obvious.

An overview of PTC chemistry is discussed in Section IV. B. 5. Some general comments can be made, however, which indicate how a meaningful value for ϕ_B can be induced at the boundaries by proper control of the chemistry. One begins by doping the titanate n type with a donor impurity such as $La^{\bullet}_{Ba}$. At elevated temperature the La impurity is fully compensated by electrons and thus the neutrality relation

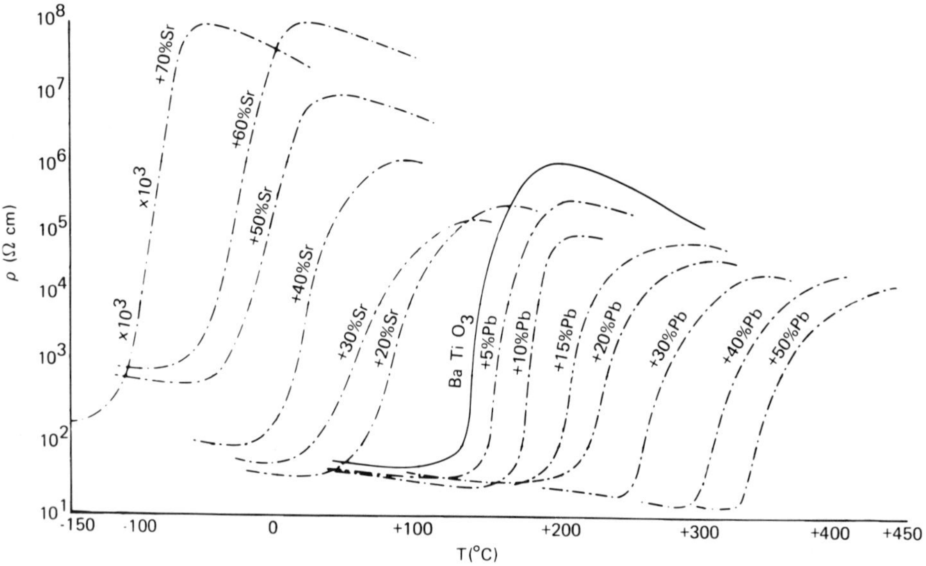

Figure 3 Resistivity-temperature curves for $BaTiO_3$ PTC thermistors containing isovalent substitution for barium. The solid line represents $BaTiO_3$ and the dashed lines show the effect of strontium (shift to left) and lead (shift to right) substitution in the proportions indicated. (From Ref. 43.)

$$[La^{\bullet}_{Ba}] = n \quad (38)$$

This renders the material highly n-type semiconducting. At lower temperatures, Daniels et al. [14] have reported that the compensation mechanism changes from electronic, as in Eq. (38), to ionic given by

$$[La^{\bullet}_{Ba}] = 2[V''_{Ba}] \quad (39)$$

This means that as the temperature is lowered, acceptorlike barium vacancies diffuse inward from the boundaries and annihilate the free carriers generated at higher temperature, causing the region to become insulating. In PTC thermistor materials the kinetics are controlled in such a manner as to freeze in the conducting regime at all but a thin section of material bounding on the grain boundaries. This is possible given the low diffusivity of barium vacancies in the bulk of the grains [14]. Figure 4 illustrates the frozen-in defect distributions and the corresponding potential barrier for $BaTiO_3$ ceramic materials.

Small concentrations of dopants (later referred to as modifiers), typically transition metal impurities, are also added to such devices to further enhance the height of the barriers. These elements are

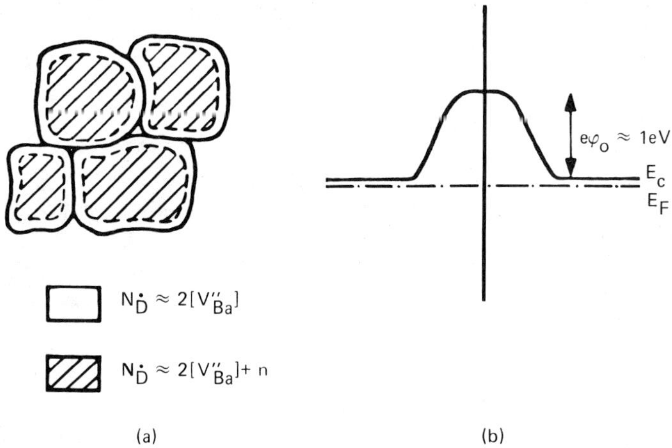

Figure 4 (a) Schematic of defect distributions in grains and grain boundary regions; (b) resultant potential barriers at grain boundaries. (From Ref. 61.)

believed to segregate to the boundaries and form acceptorlike interface states, resulting in increased depletion of free carriers [15] in the vicinity of the grain boundary.

Free carrier depletion to achieve maximum PTC behavior is accomplished after the microstructure is fully developed. Controlled oxidation of the grain boundary region occurs during the cooling cycle of the PTC material. Daniels et al. [14] have demonstrated the importance of controlling this oxidation step and emphasized the interdependence of microstructure, chemistry, and P_{O_2} to achieve desired PTC performance. It is also interesting to note that further oxidation of similar compositions results in the production of boundary layer capacitors with exceptionally high apparent dielectric constants ($\varepsilon > 50{,}000$).

3. Summary

To summarize, the electrical response of a material to temperature and atmosphere excursions depends on the manner in which the electronic and ionic carriers are generated and the manner in which they move. In oxides, both of these factors are complicated by the high-temperature nonstoichiometric effects, on one hand, and the existence of a number of mobility mechanisms, on the other. Furthermore, in contrast to conventional semiconducting devices, which are single crystalline, most sensor elements are polycrystalline. Consequently, electrical phenomena occurring at grain boundaries must be considered. For the reasons given above, only a systematic approach, which takes such factors into account, can be expected to result in proper operation of sensing devices.

C. Basics for Chemical Gas Sensors

Two basic approaches for detecting the composition of gases with electrical ceramics exist. The first relies on the atmospheric dependence of electrical conductivity exhibited by some ceramic semiconductors. The second utilizes ionic conductors as solid electrolyte membranes in electrochemical concentration cells. In the following we summarize briefly the principal characteristics of each approach.

1. Semiconducting Gas Sensors

The basis for semiconducting gas sensors was alluded to in Section II. A. 3. There we observed that changes in stoichiometry due to redox reactions result in the isothermal generation of donor or acceptor states together with excess electrons or holes, respectively.

Under such circumstances the electrical conductivity is found to follow a power law dependence on the partial pressure of the relevant gas. In the example for doubly ionized vacancies derived above, a $\sigma \propto P_{O_2}^{-1/6}$ dependence was obtained.

A number of factors go into determining the exact power law dependence. First, the sign of the power law depends on the sign of the major carrier. Thus a $P_{O_2}^{-1/x}$ dependence is obtained for n-type while a $P_{O_2}^{+1/x}$ dependence is obtained for p-type semiconductors. Second, the exponent $1/x$ is a function of the nature and ionization state of the predominant ionic defect and the electroneutrality relation. For example, in the doubly ionized oxygen vacancy case under conditions defined by Eq. (18), the exponent would show $-1/6$ dependence. The same defects under conditions where the vacancy density is fixed by impurities I" (e.g., $[V_O^{\cdot\cdot}] = [I'']$) will, following Eq. (17), result instead in an exponent of $-1/4$. In general, a series of simultaneous defect equations need be solved to establish which power law dependence is likely to hold when a sensing device is subjected to a range of temperature and gas partial pressure conditions. The general approach taken is outlined for barium-strontium titanate in Chapter 8. Figures 11 and 12 of Chapter 8 illustrate the types of P_{O_2} dependencies one can expect for oxide semiconductors. We use Figure 12 to illustrate some intrinsic difficulties in applying such materials as sensors.

Consider attempting to utilize this titanate as P_{O_2} sensor at reduced P_{O_2} values. One observes immediately the strong temperature dependence of the conductance of the material at a given P_{O_2} level [see Eq. (19)]. Thus while $\sigma \sim 10^{-1}$ (ohm-cm)$^{-1}$ at 800°C corresponds to $P_{O_2} = 10^{-19}$ atm, it corresponds to roughly 10^{10} times higher P_{O_2} at 1100°C. Clearly, a highly effective temperature stabilization mechanism must be built into such a sensor.

An additional potential problem is related to the minimum in σ observed at, for example, 10^{-7} atm at 800°C. Here a given conductance can be a two-valued function of atmosphere [e.g., 10^{-3} (ohm-cm)$^{-1}$ at $P_{O_2} \cong 10^{-11}$ and 10^{-1} atm]. This problem can in principle be minimized by inducing a shift in the conductivity minima to higher P_{O_2} values. Based on the defect analysis presented in Chapter 8, this can be accomplished by making the material more n-type by either doping with donors or reducing the acceptor density. One can readily show that the P_{O_2} at the minimum increases two orders of magnitude for every order of magnitude decrease in [A'] [16].

Equilibration of ceramics with atmosphere and temperature was implicitly assumed above. At reduced temperatures, bulk diffusion and surface reactions kinetically limit the bulk response of such devices. Under these circumstances ceramics may still be used to sense atmosphere variations due to stoichiometry-induced changes at the surfaces or grain boundaries in highly porous ceramics [17]. The results of recent molecular orbital calculations on ZnO surfaces [18], for example, demonstrate that absorbed oxygen results in the depression of the surface cluster Fermi energy from the conduction-band edge toward the valence-band edge. This results in the formation of a space charge barrier at the interface and a consequent decrease in conductance of the device with increasing P_{O_2}. Catalytic electrodes are often employed to catalyze specific reactions to improve selective sensor response to one of a number of competing reactions [19]. Applications of similar devices as auto exhaust sensors are described in Section IV.C.1.

2. Electrochemical Sensors

One may show that for a material maintained in, for example, an oxygen potential gradient under open-circuit conditions an electromotive force (emf) given by

$$E = \frac{kT}{4q} \int_{I}^{II} t_{ion} \; d \ln P_{O_2} \tag{40}$$

is produced across the material [16]. In Eq. (40), t_{ion} corresponds to the fraction of the total conductivity which is ionic in nature. A simplifying assumption is often made (i.e., that t_{ion} is nearly constant across the material). Then Eq. (40) simplifies to

$$E = t_{ion} \frac{kT}{4q} \ln \frac{P''_{O_2}}{P'_{O_2}} \tag{41}$$

For conditions where t_{ion} is equal to unity, this emf is commonly known as the Nernst emf. In a sensor application P'_{O_2} is maintained at a well-defined reference value (e.g., air where $P_{O_2} \cong 0.2$ atm), while E is measured at a given temperature for unknown values of P''_{O_2}. To ensure meaningful response, t_{ion} must be maintained at ≥ 0.99 over the full range of operating temperature and atmospheres.

In Section II.A.2. we mentioned that it is often possible to produce a solid electrolyte by heavily doping some oxides with altervalent ions (e.g., Y_2O_3 in ZrO_2). Even under these circumstances, a shift to highly reducing or oxidizing conditions may induce a sufficiently large shift from stoichiometry in the oxide to produce large numbers of electronic carriers and thus reduce t_{ion} below unity.

In Figure 11 of Chapter 8, the bounds of the electrolytic region are designated by P_p and P_n, the P_{O_2} values where t_{ion} falls to 0.5 at high and low P_{O_2} values. The electrolytic domain typically shrinks rapidly with increasing temperature given a strong electronic carrier dependence on temperature. As an illustration, domain boundaries of stabilized zirconia, a widely used oxygen ion electrolyte, are shown in Figure 5. Note that whereas at 1000 K, the material may

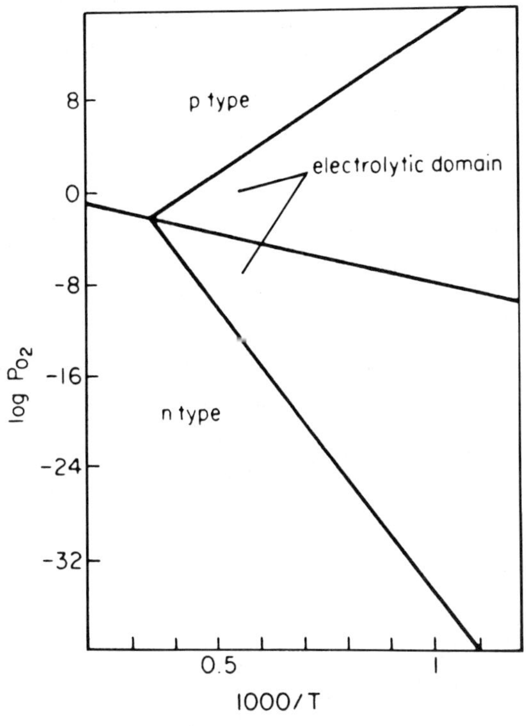

Figure 5 Domain boundaries of stabilized zirconia as projected on the log P_{O_2}−1000/T plane. (From Ref. 59.)

be used over an extensive P_{O_2} range, by 2000 K it has become semiconducting at nearly all atmospheres. It is thus essential to map out the electrolytic domain of a material before considering it for use in a gaseous sensor cell. Techniques for optimizing the electrolytic domain of oxygen ion conductors have recently been reviewed by one of the authors [16]. Further comments concerning these devices may be found in a later section concerned with applications.

III. TRANSDUCER CLASSIFICATIONS

The term "sensor" is generally perceived to mean some form of input transducer. Some common examples are thermocouples, photodetectors, and pressure cells that convert thermal, optical, or mechanical energy into electrical energy. Output transducers operate in a similar fashion, converting one form of energy to another. Given the recent progress made in a number of technologies, the number of different types of transducers that can be conceived of has increased substantially over the last decade or so. To assist in visualizing the different types of alternatives it is useful to follow the approach recently proposed by Middelhoek and Noorlag [20].

One begins by examining those energy forms that are most relevant in the analysis of signal processing and instrumentation. These include (1) radiant, (2) mechanical, (3) thermal, (4) electrical, (5) magnetic, and (6) chemical energies. Signals based on these energy forms can be applied to the input of a transducer or may appear at the output. Some examples of transducers for which the same form of energy appears at both output and input include optical filter (rad, rad), transistor (electrical, electrical), and catalyst (chem, chem).

Upon considering the different types of transducers, it appears that two formats exist: those for which an auxiliary source of energy is required for operation and those which generate a useful output without need for such an auxiliary source. The first type, exemplified by a phototransistor, is called a modulating transducer; the second type, exemplified by a thermocouple, is called a self-generating transducer.

For most effective analysis of transducers it is convenient to arrange them in a three-dimensional grid as shown in Figure 6. Here the x-axis gives the form of auxiliary energy, the y-axis the output signal, and the z-axis (not labeled) the input modulating signal. Some examples with their three-dimensional notation are included in Table 2. In all, 216 points exist in this three-dimensional array of sensor possibilities, some of which cannot be related to a specific known transducer. This figure demonstrates that much may

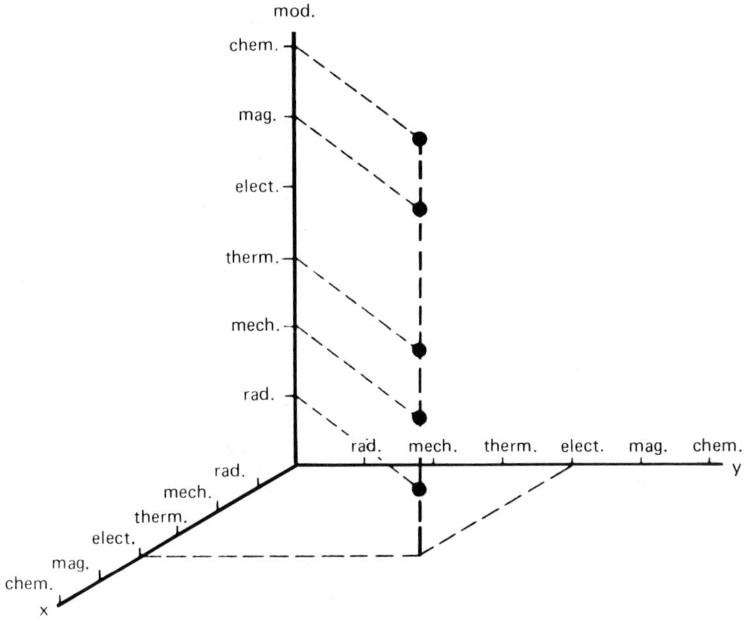

Figure 6 Three-dimensional transducer diagram showing the five possible modulating input transucers based on photoconductance, piezoresistance, thermoresistance, magnetoresistance, and electrolytic conductance effects. (From Ref. 20.)

Table 2 Transducers and Notation

Thermocouple	[Thermal, elect, 0]
Thermistor	[Elect, elect, thermal]
LED	[Elect, rad, 0]
Oxygen sensor, galvanic	[Chem, elect, 0]
Oxygen sensor, resistive	[Elect, elect, chem]

Table 3 Popular Ceramic Sensors and Transducers

Stimulus	Type	Materials
Thermal	NTC thermistor PTC heaters	Transition metal spinels, barium titanate
Chemical	Gas humidity	Cubic ZrO_2, TiO_2, SnO_2, Al_2O_3, $MgCr_2O_4$
Mechanical	Piezoelectric optical	PZT ceramics, glass fibers
Electrical	Varistor phosphor	ZnO, YAG, Y_2O_3
Optical	Laser filter	YAG:Nd, photochromic glass
Magnetic	Memory	Ferrites, YIG

yet be done in developing new sensor devices. Given progress in instrumentation and packaging—future efforts in this area are to be expected. Given the scope of this chapter and limitation of space, only a number of the alternatives will be examined in some detail. These include thermal sensors (NTC and PTC thermistors) and gas sensors (oxygen, in particular). A summary of popular ceramic sensors is included in Table 3.

IV. TRANSITION FROM THEORY TO PRACTICE

Section II described the underlying thermodynamic and semiconductor theory necessary for describing the manner in which energy stimuli influence the electrical response of sensor devices. This theoretical foundation has evolved since Faraday discovered the first semiconductor material in 1833. Semiconductor technology has progressed dramatically since the 1950s. Practical sensor devices [21] introduced in the early 1930s were made using processing methods based on empirical observations. Sensor device improvements have occurred as a result of process technology improvements and more recently aided by interdisciplinary approach to sensor systems design. The integration of physics, chemistry, materials science, and manufacturing disciplines has resulted in the production of sensor devices of improved quality. This section attempts to bridge the gap between theory and practical device utilization.

Solids respond to energy stimuli in a predictable manner if certain important parameters are maintained under control. These include:

Geometry (size, shape, etc.)
Chemistry (composition, reactivity)
Thermophysical properties
Electrical response
Mechanical properties
Performance (reliability, stability)
Environmental concerns

The interdependence of these considerations becomes obvious as one becomes more deeply involved with sensor R&D and manufacture. Each of these considerations is examined below for a number of representative devices. Subsequent sections deal with fabrication, testing, and application of devices.

In the following we deal with three major areas of ceramic sensor activity: NTC thermistors, PTC resistors, and gas sensors (in particular, oxygen). The general history of each type of sensor is reviewed to obtain a perspective on its development and operation. Device theory, materials construction, manufacture, and applications are then presented. An overview on trends, challenges, and needs sums up each sensor category.

A. NTC Thermistors: Device Manufacture and Testing

1. History and Future Developments

Faraday discovered and reported on the semiconducting behavior of Ag_2S in 1833. For the next 100 years the negative temperature behavior of semiconductors remained a curiosity, due to the difficulty in reproducibly making useful devices. Commercial utilization of ceramic sensors began in the 1930s and early 1940s with pioneering work at Siemens, Philips, and later Bell Telephone Laboratories. Major applications of these early devices were telecommunications, circuit temperature compensation, and temperature measurement.

More sophisticated applications were introduced and sensor manufacturing expansion occurred in the postwar era (1950–1960) when aerospace and cryogenic devices were introduced. Greater emphasis on sensor stability and performance resulted in better sensor housing designs and materials capabilities. From 1960 to 1970, accuracy and interchangeability were emphasized for the newer applications. Significant development efforts for medical applications were made as a result of government support programs. Thermistor manufacturers began to focus on cost-effective manufacture for volume products. High-accuracy "chip" and bead thermistors were introduced by several manufacturers.

2. NTC Device Construction

Typical NTC device constructions are shown in Figure 7. The semiconducting ceramic sensor (top left) has two (sometimes more) electrode contacts. The geometric configuration of the device, the sensor electrical properties, and the heat flow characteristics to and from the sensor device determine performance. A sensor's electrical character is determined by the formula

$$\rho = \frac{RA}{l} \tag{42}$$

where the material resistivity ρ and configurational parameters (effective area A and the path length l) govern the electrical resistance R of the device. Table 4 illustrates how dimensional parameters vary for commercial NTC sensor devices.

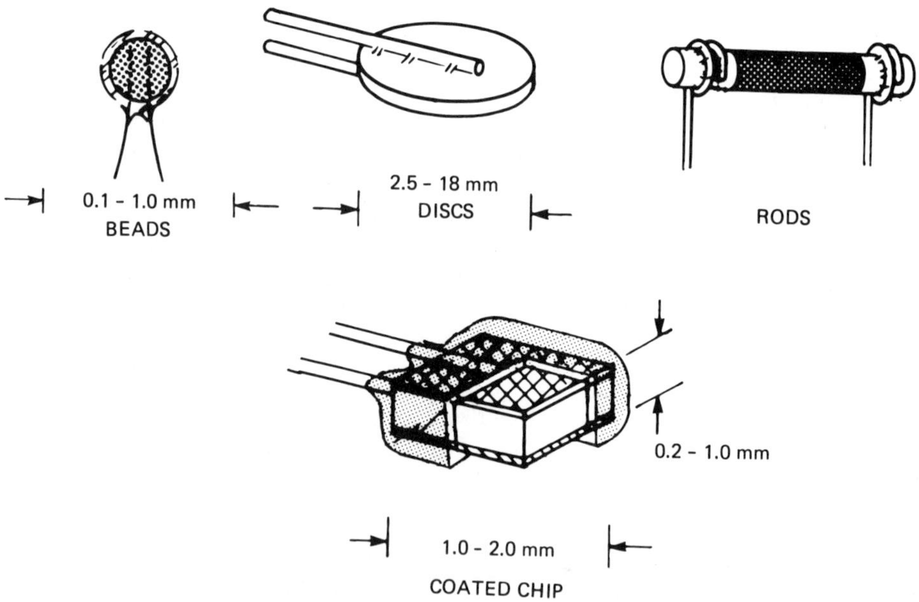

Figure 7 Some typical NTC thermistor device configurations.

Table 4 Commercial NTC Sensor Device Sizes

Device type	Diameter or width (cm)	Length or thickness (cm)	A/L (cm)
Rod	0.5–1.5	0.5–4.0	0.2–3.0
Bead	0.005–0.1	0.010–0.10	1.0–100
Disk/chip	0.01–3.0	0.005–50	0.01–100
Thick-film	Variable	0.002–0.02	10^{-4}–10^{3}

Some actual NTC devices are shown in Figures 8 to 10. Each device is carefully designed, coated, or placed in a protective housing. Several such probe designs are illustrated in Figures 11 to 13. Materials and configuration selections are generally based on designs found to operate successfully after years of field trials.

3. Electrical Properties

a. Resistance-Temperature Response. Materials respond to energy stimuli in a number of unique ways. Consider the resistance-temperature behavior of the four materials shown in Figure 14. Each

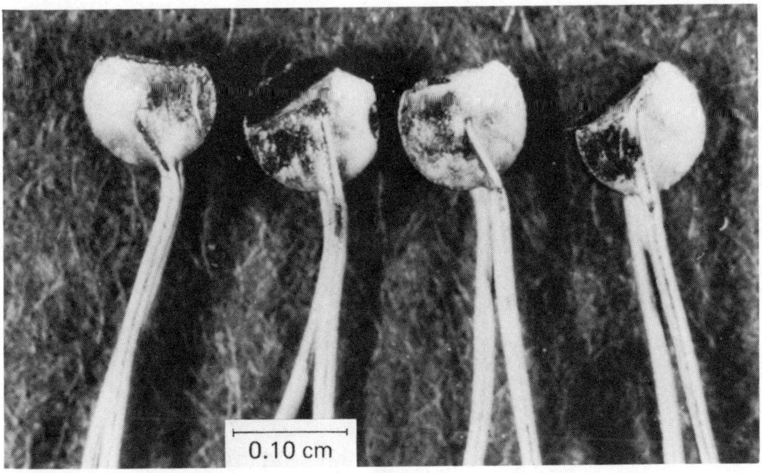

Figure 8 Precision NTC thermistors with abrasion resistance adjustment.

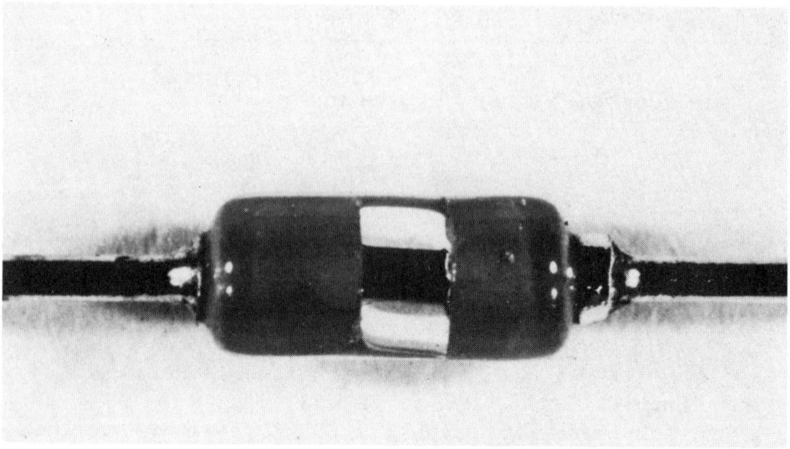

Figure 9 Chip thermistor with glass diode encapsulation.

Figure 10 Miniature glass bead NTC thermistor and typical molded probe assembly.

Figure 11 Automotive water temperature NTC sensor probe.

Figure 12 Surface temperature monitor NTC probe assembly.

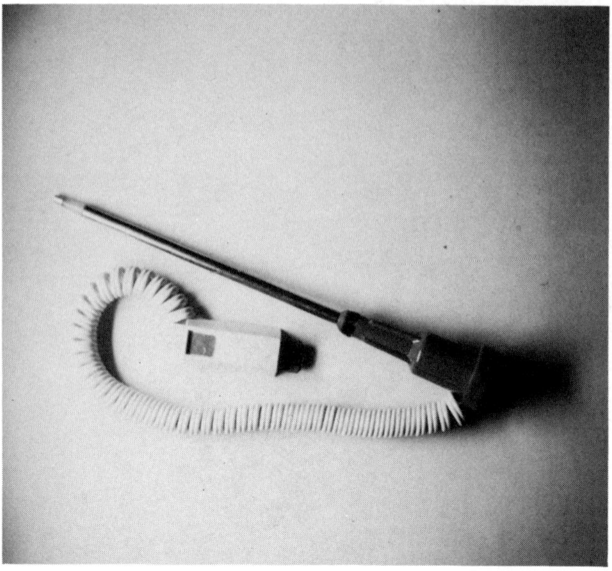

Figure 13 Microwave oven NTC temperature probe.

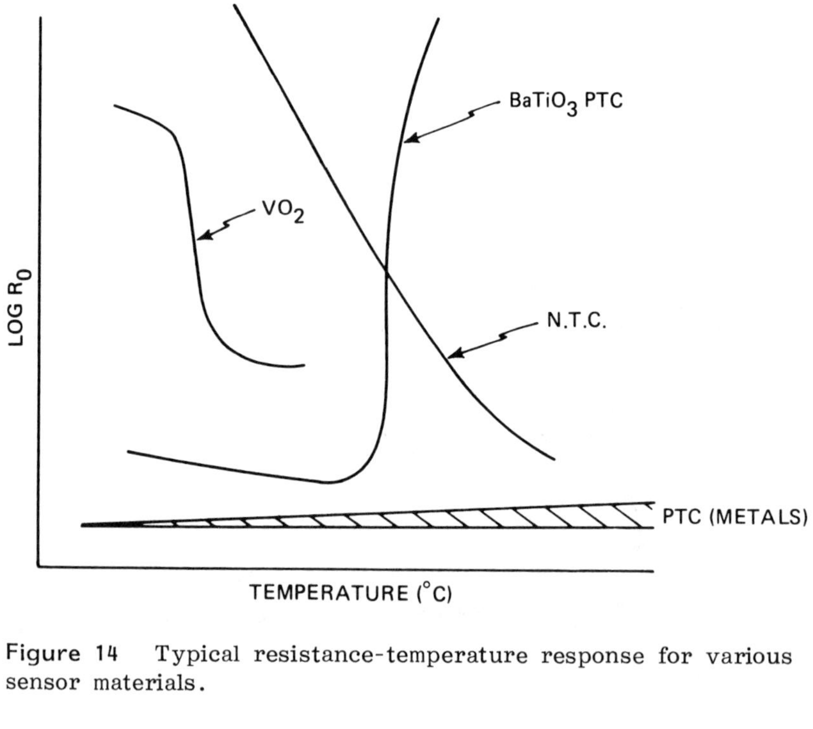

Figure 14 Typical resistance-temperature response for various sensor materials.

material and its resistance response is governed by one of a number of phenomena covered in Section II. All but the metals exhibit a negative coefficient of resistance over some temperature range. The particular choice of material will of course depend on the intended application.

Semiconductors that exhibit a reasonably constant activation energy over a broad temperature range (designated NTC) are most desirable from the standpoint of developing a highly sensitive thermal detector. The VO_2 and $BaTiO_3$ materials, although exhibiting a strong sensitivity to temperature excursions, are limited for this use due to a number of sharp discontinuities in their log R-T slopes. Metals, on the other hand, possess a well-defined temperature dependence but suffer from limited sensitivity.

Many applications exist in which an alarm or switching process need be initiated. Under these conditions the discontinuities in the resistance-temperature curves exhibited by materials such as VO_2 [22] and $BaTiO_3$ [23] are highly desirable. The PTC behavior of $BaTiO_3$ can further be used to self-limit current flow through circuit elements. PTC devices are examined in detail in a subsequent section.

Table 5 summarizes stability, relative temperature sensitivity, temperature range, and relative cost of sensors commonly used for thermal measurement applications. It should be noted that each sensor has its own distinct advantage and application ranges. For example, the ready applicability of metallic sensors to both high and low temperature must be viewed in the context of their rather poor temperature sensitivity.

We next concentrate on the temperature dependence of conventional NTC devices. The resistance-temperature response for some NTC compositions is shown in Figure 15. Deviation from Eq. (25) is obvious, and Eqs. (26) to (27), at times, approximate real NTC

Table 5 Stability, Sensitivity, and Relative Cost for Thermal Sensors

Sensor material	Resistance change dR/Rod T (%/°C)	Stability	Temperature range (°C)	System relative cost
Metals	+0.1 to 0.5	Fair to good	100–1000	1–5
Silicon	+0.1	Fair to good	150	1
$BaTiO_3$ PTC	+10 to 100	Good at T_c	200+	1–3
NTC	−6.0 to −1.0	Good to excellent	300	1–3

behavior only over limited temperature intervals. The basic equation (25) may, however, be expanded as the polynomial

$$R_T = A\left(\frac{B}{T} + \frac{C}{T^2} + \frac{D}{T^3} + \cdots\right) \tag{43}$$

This new expression lends itself to microprocesser control and can be effectively used for sensor applications if the accuracy and stability

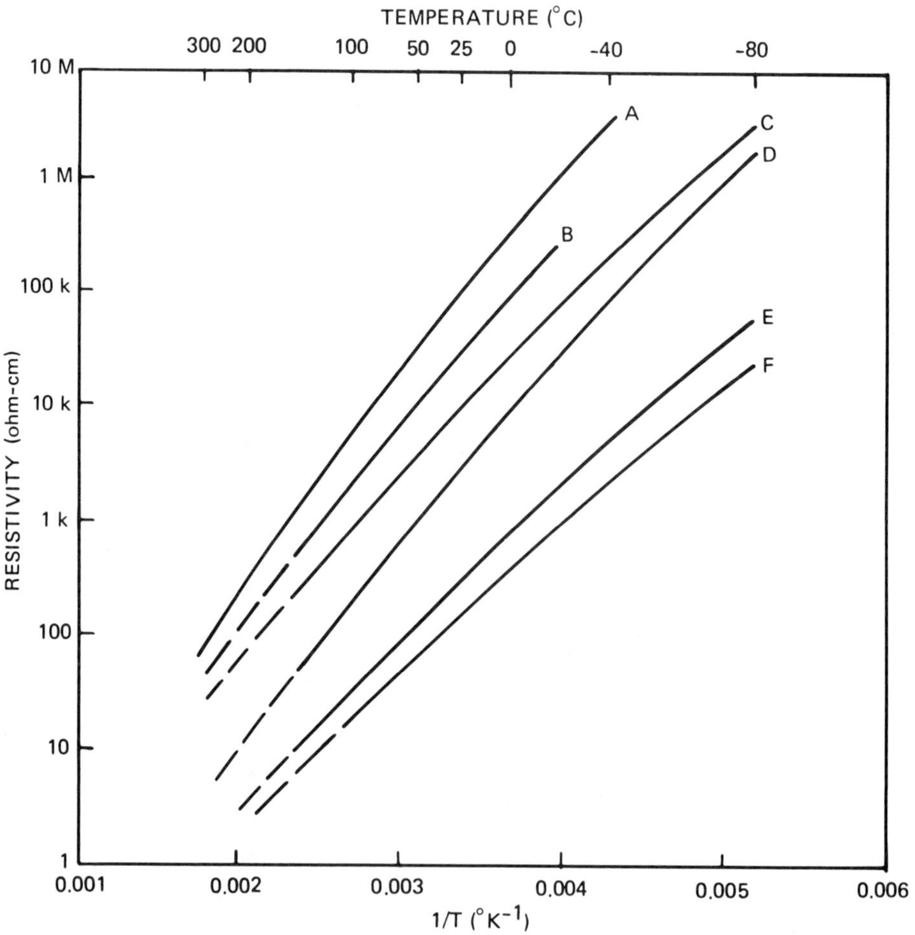

Figure 15 Resistivity-temperature behavior of some commercial NTC thermistor materials.

of sensor devices can be adequately maintained. Circuit designs that take into account the nonlinear resistance behavior of NTC devices have been reported [21,24] and allow for high-accuracy analog measurement applications.

b. Resistance-Temperature Accuracy. Curve C of Figure 15 has been expanded and replotted as shown in Figure 16. Beta tolerance bands are placed on the original curve to demonstrate slope deviation. This technique is often used by both customers and manufacturers to specify product acceptability and define maximum temperature deviation of a NTC sensor device.

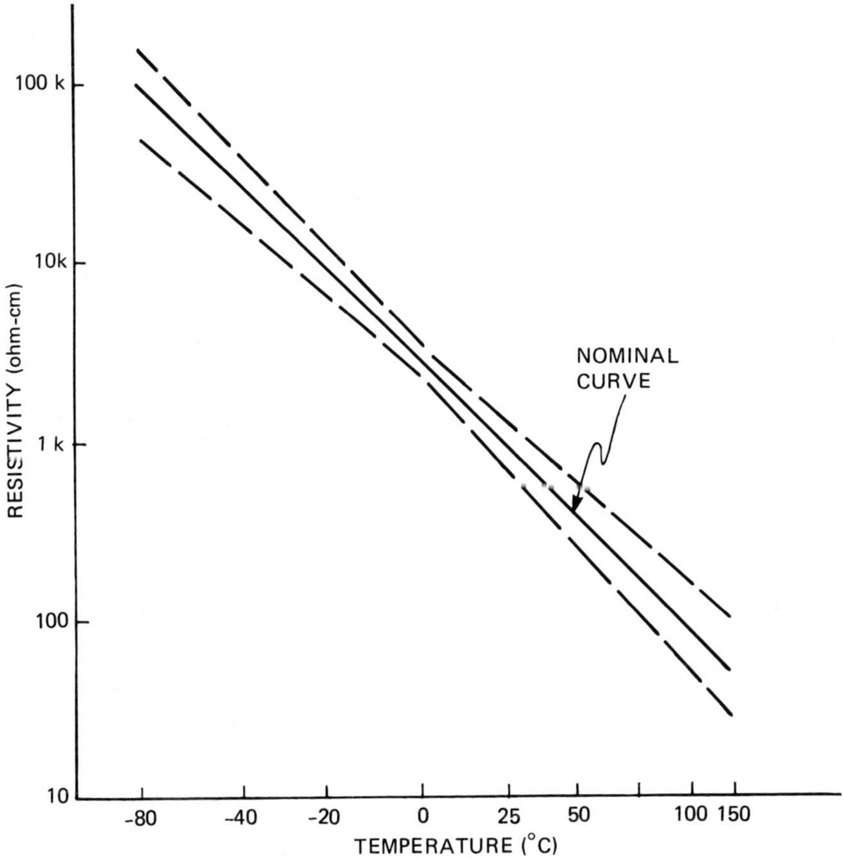

Figure 16 Resistivity-temperature envelope for thermistor material C. ρ at 25°C = 2800 (Ω-cm ± 20%; B = 3850 K ± 10%).

NTC manufacturers produce sensors which have resistance limits that vary depending on processing, resistance accuracy sorting measurement, and overall resistance drift after contacting. The curve bands (Figure 16) are further expanded, as illustrated in Figure 17a, to account for sort resistance tolerance limits. These resistance tolerances over the full temperature range translate into bands of temperature uncertainty as indicated in Figure 17b. Practical resistance tolerance bands depend on the inherent beta control, material stability, and controls the NTC producer imposes on the manufacturing process.

c. *Time Constant—Dissipation Constant.* A number of sensor applications require close control of the thermal-time response of thermistor devices. The physical size of a sensor, its thermophysical properties, and the coupling to its surrounding environment determines the heat exchange kinetics.

Detailed relations between the thermal time constant t_{TC} of the device and the geometrical parameters, the bulk thermal conductivity K_s, the interfacial heat transfer coefficient h, and the heat capacity C_p are derived in the Appendix for both low current and self-heating conditions. In Table 6 we list typical t_{TC} values for both bare and housed NTC thermistor devices.

4. Device Chemistry

The resistance-temperature behavior of NTC devices can vary significantly with chemical composition and crystal structure. A fairly complete disclosure of past patent and technical literature pertaining to composition-processing-electrical property characterization for NTC materials is presented elsewhere [21,24]. In this section a few systems will be selected to illustrate important features.

The two materials parameters, beta and resistivity, determine the conditions under which a given thermistor material may be utilized. High-beta and high-resistivity materials are generally used at elevated temperature to ensure adequate sensitivity and optimum match to measurement circuitry. The low-beta, low-resistivity materials find application in circuit temperature compensation and cryogenic measurements.

The beta-resistivity relation for a number of commercial thermistor compositions have been combined and plotted as bands in Figure 18. Cryogenic sensor formulations based on Ti^{4+}-doped hematite (Fe_2O_3) [21] appear to extend general trends observed for spinel-structured compositions. Devices based on the cubic Li^+-doped transition metal monoxides [3], on the other hand, exhibit a shift to higher beta values for equivalent resistivity levels. Similar shifts would be expected for nonoxide NTC semiconductors (e.g., SiC, Si, diamond).

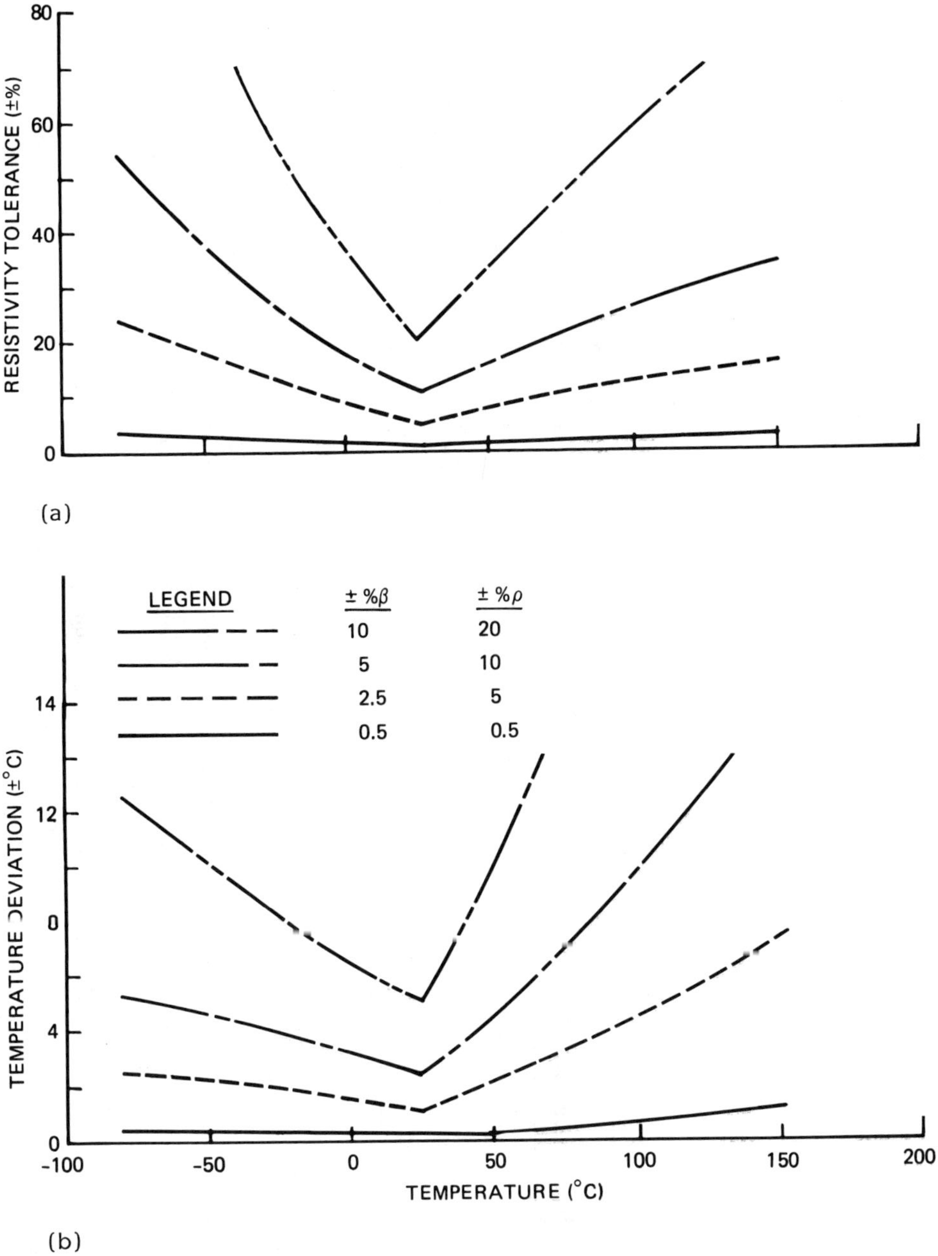

Figure 17 Effect of sensing temperature on maximum variance of (a) resistivity, and (b) temperature deviation from the nominal thermistor curve (material C). Curves display variances for NTC thermistors with beta and resistivity (at 25°C) envelopes as shown in the table.

Table 6 Typical Time and Dissipation Constant Values for NTC Devices and Probe Assemblies

	Physical size of device probe			Device/probe thermal behavior	
Device probe description	Diameter (cm)	Thickness (cm)	Volume (cm^3)	Dissipation constant (MW/K)	Time constant (sec)
NTC devices					
Thin flake (bare)	0.125	0.002	0.000031	0.20–0.40	0.004–0.008
Bead					
Micro (glass coated)	0.050	0.050	0.000098	0.05–0.30	0.2–3.0
Glass probe	0.150	0.050	0.000884	0.20–2.0	2.0–15
Chip					
Bare	0.125	0.025	0.000307	∼1.0	5.0
Epoxy coated				∼1.0	<10

Disk (leaded)	0.250	0.125	0.00614	3.5	9.0–12
	0.500	0.125	0.0245	7.0	25
	1.250	0.125	0.1534	15	70
	1.250	0.250	0.3068	15	100
	2.500	0.125	0.6136	25	150
Rod (leaded)	1.000	0.300	0.2356	2.8	35
Probe assemblies					
Fluid probe, bead, stainless steel housing	0.400	NA	0.0700	3.0	100
	0.09[illegible]	NA	0.0100	0.7	16
Surface probe, chip on aluminum plate	0.625	NA	0.0040	2.3	25

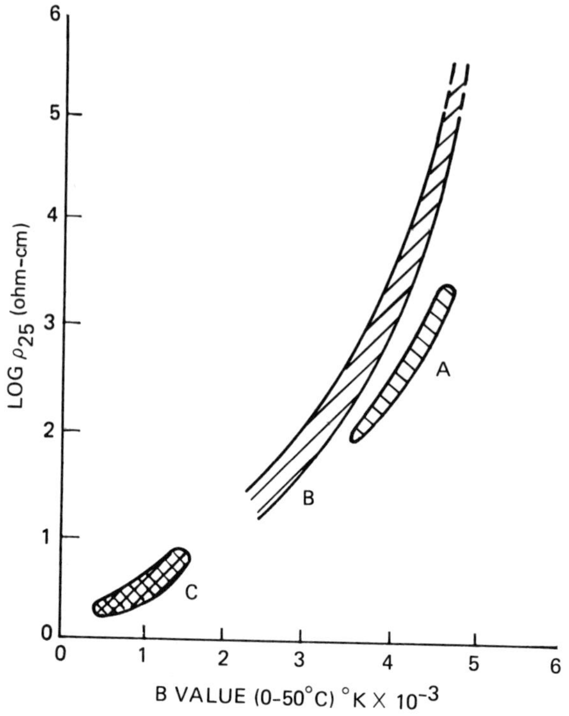

Figure 18 Relationships between resistivity and beta value for selected NTC thermistor materials classes: (A) Li-doped (Mn, Ni, Co) O; (B) complex spinels AB_2O_4: $(NiMn)_3O_4$, $(NiMnCo)_3O_4$, $(NiMnFeCo)_3O_4$; (C) hematite $(Fe, Ti)_2O_3$.

The resistivity-temperature behavior of various compositions was presented earlier. Room-temperature resistivities of compositions in the three-component Ni-Mn-Fe-oxide system prepared under the same conditions are shown in Figure 19, thus illustrating the influence of even small variations in composition on electrical properties. A similar beta versus composition triangle plot (not shown) can also be expected.

More extensive compositional studies have been performed by Futaki [25] and Sheftel et al. [26] in an attempt to fully characterize potential NTC thermistor compositions for commercial use. It is important to appreciate that significant shifts in electrical properties and stability of such devices occur with minor shifts in processing conditions and sintering cycles. Materials technologists must therefore be concerned with each step of the processing cycle, including starting materials, formulation, particle distributions, homogeneity, and so on, as discussed below.

Consider the impact of processing conditions on the manufacturability of NTC formulations. The 80 mol % MnO—20 mol % NiO composition [27] may be used to illustrate this point. This composition falls at the resistivity minimum for the two-component system, and minor processing variations do not significantly affect product yields or quality. For this reason this formulation has been used extensively by the industry.

High-performance NTC applications require much tighter property controls. NTC producers tend to develop special formulations and use processing methods to optimize device performance. Such details are considered proprietary and not usually disclosed.

Regarding processing/compositional complexities, most commercial NTC compositions are comprised of multivalent transition metal oxides (Ti, Fe, Mn, Ni, Co, Cu, etc.). Examination of the phase equilibria for these systems show that complex oxidation-reduction phenomena occur throughout the processing temperatures normally used for NTC manufacture. The Mn-Co-O system shown in Figure 20 is illustrative of this feature.

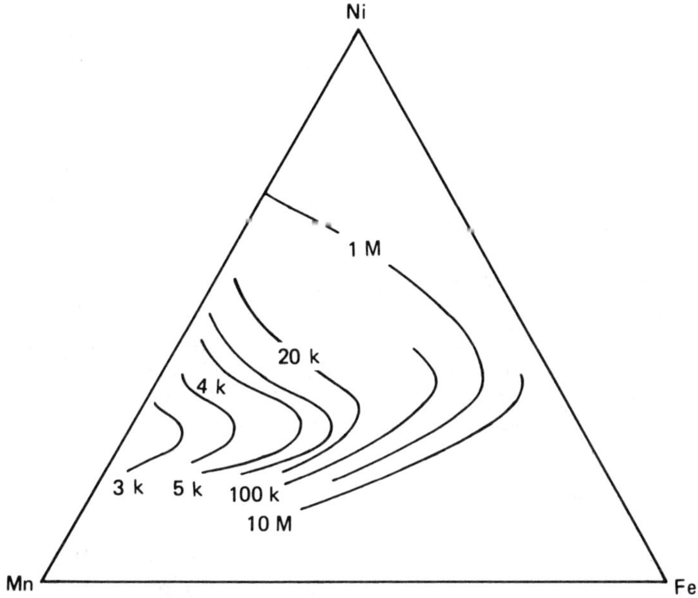

Figure 19 NTC thermistor resistivity contours at 25°C for the compositional system Ni-Fe-Mn oxides.

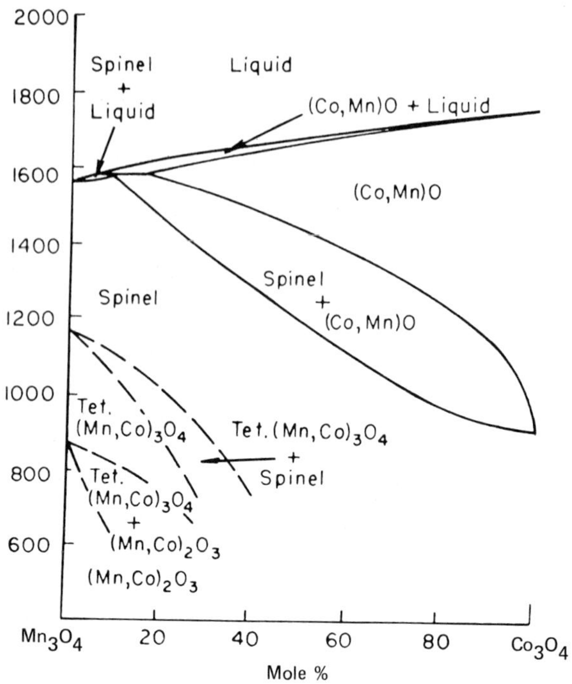

Figure 20 Phase diagram for the system Mn-Co oxides in air.

A spinel single-phase region exists over a wide compositional range for the $Mn_xCo_{3-x}O_4$ system when sintered in air at typical temperatures used for NTC device manufacture (1100–1300°C). Mn/Co oxide compositions tend to oxidize at lower temperature (400–1100°C). If oxidation is allowed to proceed uncontrolled, chemical inhomogeneities and phase changes must occur. Such changes have been known to cause significant electrical and mechanical uncertainties in NTC devices. One must therefore be aware of the processing-composition-oxidation interactions in order to produce reliable NTC devices for the intended sensor application.

5. Stability and Reliability

Device performance characteristics, particularly those pertaining to the utilization, life expectancy, and stability of the sensor, are represented in Figure 21. It is obvious that the mechanical, chemical, and electrical aspects must be controlled in such a manner as to minimize all potential inhomogenities to the extent possible.

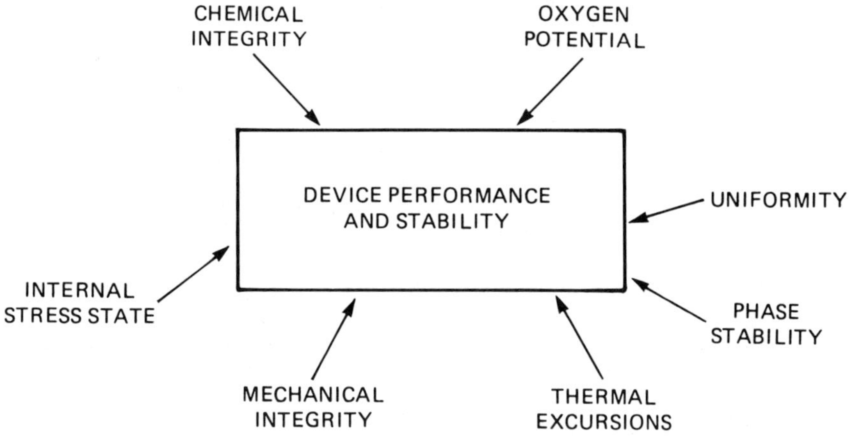

Figure 21 Factors affecting ceramic sensor performance, stability, and life.

There are few references regarding NTC stability in the technical literature [28,29]. The best information reported is for disk or chip sensor devices used for temperature measurement and control. Resistance shifts (dR) measured at room temperature after interval aging at elevated temperature is the typical experimental approach. The most interpretive results are obtained when initial measurements (R_0 at t_0) are taken subsequent to sintering and electroding. Resistance changes for such devices usually follow the general expression

$$dR = R - R_0 = R_0 \ln (t - t_0) \tag{44}$$

Figure 22 demonstrates some early stability data published by Becker et al. [28]. Reference to standard thermistor texts [21,24] indicates that chemical composition, processing, chemical attack, elevated-temperature aging, contacting variables, and other variables can all influence aging behavior. The resistance shift-time plots shown in Figure 22 can be extremely helpful in predicting life expectancy and measurement accuracy for NTC devices when reliable and statistically valid curves are obtained.

Thermistor manufacturers describe procedures wherein NTC devices are subjected to burn-in at elevated temperatures to improve time-dependent resistance change and resistance accuracy. Two types of aging behavior can be obscured by this practice, as shown in Figure 23.

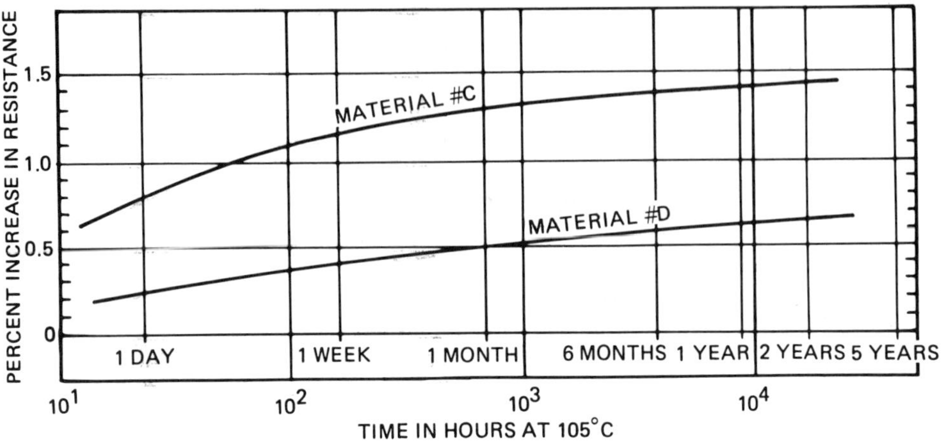

Figure 22 Aging characteristics of two thermistor materials, oven aged at 105°C.

The resistance change shown in curve A of Figure 23 is considered intrinsic for a thermistor composition when aged at elevated temperature. A continuous resistance change with predictable dR/dln t slope has been measured over time spans of years. Consider reestablishing initial R_0 and t_0 values at time t_{100} and plot the remaining resistance values from t_{100} to t_{end} as virgin aging data (curve B of Figure 23). The plot shows a plateau of minimal resistance change followed by a discontinuity and then a curve paralleling the original curve A.

The uninitated observer might conclude that the initial aging behavior (dotted line, curve B) is exceptional compared to that of curve A and projects a long life based on the initial slope measurements. Continuing the aging measurements beyond the discontinuity, however, would show that an R_0 shift specification limit of 0.2% would be exceeded after look hours.

Subtle resistance shifts can occur when aging conditions are changed. The data shown in curve C of Figure 23 are used to illustrate this point. Shifts, up or down depending on the thermal history of the device, have been observed. The subsequent resistance shifts observed for aging at the new temperature environment tend to recover the shift caused by the original thermal change. Shifts of this type confuse stability assessment and temperature uncertainty.

In the remainder of this section we touch on a variety of features to illustrate some facts and pitfalls of stability assessment.

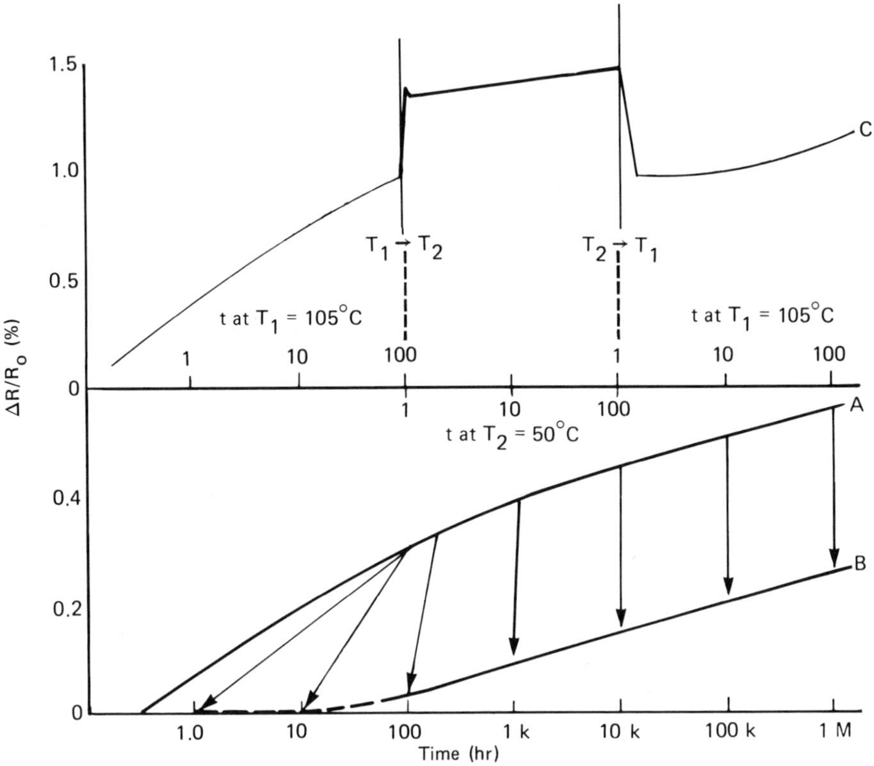

Figure 23 Resistance-aging responses from various NTC thermistors subjected to elevated temperature. A, Virgin thermistor device; B, unit A aged 100 hr; C, unstable unit under cycle aging.

a. Mechanical. A NTC sensor body may encounter mechanical stresses that induce microscopic cracks during its manufacture. Resistance shifts can occur if crack propagation proceeds after initial resistance measurement. These show up as a resistance shift discontinuities equivalent to changes in A/l ratio.

Most commercial NTC formulations have relatively small piezoresistive coefficients. Permanent resistance shifts due to stress relaxation or induced mechanical stresses up to fracture loading have not been observed. The low piezoresistive coefficients for NTC resistors give them a distinct advantage over elemental silicon devices which exhibit significant resistance shifts (temperature inaccuracy) under induced stress conditions.

The inherent strength and mechanical performance of sensor devices can be depicted by flexural strength. Ceramic sensor strength measurements are alluded to in Table 7 and are considered to be intermediate to metals and encapsulant materials. This assessment can be misleading, owing to the brittle nature (fracture toughness) of ceramic sensors. Careful design of sensors and probes is required to minimize stresses induced during assembly and subsequent thermal excursions. Sensor designers give particular attention to thermal expansion mismatch of materials and stress relief at sensor interfaces. Sensor construction and subsequent processing are optimized to achieve the highest mechanical integrity. Attainment of high-density, controlled grain size microstructures and appropriate dimensional designs are important factors in good sensor design.

b. *Chemical.* Any basic chemical reactions that result in a change in electron flow between the electrode contacts affects device resistance. The following concerns are important for high-accuracy applications:

Electrode-sensor material reactions
Oxidation-reduction reactions
Grain boundary segregation
Gas penetration and reaction in open porosity
Chemical gradients and inhomogeneities from processing
Encapsulant reaction and penetration
Electrode deterioration

Most of the concerns listed above are thermally activated and can be studied kinetically as functions of temperature, atmosphere, composition, microstructure, and electrode material.

Chapter 4 describes how minor adjustments in electronic/ionic structure influence magnetic properties (initial permeability) of ferrite spinels. Trolander et al. [30] follow a similar approach to explain bulk stability in Mn-Ni oxide compositions. Phase transitions and oxidation gradients also contribute to instability, as pointed out by Larson et al. [31].

Earlier in this chapter, segregation of impurities and charge compensation at grain boundaries was discussed. Such effects influence both device performance and stability, due to corresponding time-dependent changes in interfacial characteristics.

Physical, electrical, and chemical imbalances (diffusion, surface reaction, phase changes, vaporization/condensation phenomena, etc.) all influence device performance. A more rigorous treatment of these concerns has been presented elsewhere [24]. It is sufficient to note that reliable sensors are often protected from such environmental changes (glass, seals, potting compounds, polymer coatings, etc.).

These are designed to provide suitable life expectancy during sensor utilization, but such protection schemes also affect sensor time response.

Chemical, structural, and electrode deterioration mechanisms all appear to be driven by rate processes consistent with the behavior of Eq. (44). Stability control of NTC devices, however, has followed an empirical approach since the 1940s with varying degrees of success. An extension of NTC device usage to temperatures greater than the 300–400°C limit presently agreed upon by users and producers might succeed if a more measured approach were taken.

Although much information concerning the relations between stability of devices and composition remains unavailable to the public for proprietary reasons, considerable technical information on ceramic sensors and related ceramic materials is available in the technical literature. Table 8 illustrates general materials formulas, metallic content, crystalline phases, semiconduction type, and electrode materials that comprise various ceramic electrical sensors. Future progress in this area will be assisted by the development of new or modified material systems which have more clearly defined structure-composition-property relations and which are less sensitive to processing variables.

6. Device Manufacture

NTC sensor components are usually small, less than 1 cm^{-3}, but are complex in construction and materials content (see Figure 7). For high-volume, tight-tolerance manufacture, innovative machinery and more elaborate process controls have to be employed.

NTC sensor manufacturing methods have a firm foundation in ceramic processing technology. Unit operations typical of chip thermistor manufacture are shown in Figure 24. This manufacturing process might be used for producing temperature sensors for medical probes to be described later.

Consider the following basic steps in NTC sensor manufacture:

1. Thermistor materials synthesis
   a. Raw materials selection
   b. Powder synthesis
   c. Device forming
   d. Sintering and chemistry microstructure development
   e. Materials properties definition

2. Finishing operations
   a. Electroding or electrical contact attachment
   b. Resistance measurement adjustment

Table 7 Typical Thermophysical Properties of Materials Used for Ceramic Sensor Devices

Material description	Density (g/cm^3)	Spec. heat, C_p (cal/gm)	Therm. cond. K (cal/sec-cm k)	TCE × 10^6 (K^{-1})	Yield or flex (kpsi × 10^3)	Elastic mod., E × 10^3 (kpsi)	Use temp. max. (°C)
Sensor materials							
Spinels (NTC)	5.50	0.20	0.0150	10.0	10–20	15–30	400
Stabilized zirconia	6.00	0.13	0.0001	10.0	20–50	20	1000
Rutile	4.20	0.18	0.0006	10.0	10–15	30	400
Barium titanate (PTC)	5.70	0.12	0.0005	12.0	7–20	15–30	300
Electrode materials							
Platinum	21.00	0.03	0.165	8.9	2–10	25	1000

Silver	10.40	0.06	1.00	19.0	8–15	11	150
Gold	19.30	0.03	0.71	14.2	nil	12	800
Aluminum	2.70	0.22	0.53	23.6	10	10	300
Nickel iron alloys	8.00	0.12	0.04	4.0–12.0	50	23	600
Copper	8.92	0.09	0.94	16.5	10	17	300
Metal-filled polymers	2.0–4.0	0.15	0.015	20–50	6–15	6	<300
Encapsulant/housing materials							
Organic polymers	1.3–2.0	0.19	0.0005	30–70	10–15	1	<300
Glasses	2.5–6.0	0.18	0.003	3.0–10.0	4–10	10	300–600

Table 8 Ceramic Sensors: Classification of Sensors, Electrodes, and Crystal Chemistry

Sensor type	Semi-conductor type	General formula	Typical composition	Major crystalline phase	Typical electrode used
NTC thermistor	p	$M^{2+}M_2^{3+}O_4$	$Ni_{0.6}Mn_{2.4}O_4$	Spinel	Ag, Pt, or noble metals
PTC titanates	n	$M^{2+}M^{4+}O_3$	$Ba_{0.005}Y_{0.005}Ti_{1.02}O_{3+}$	Perovskite	Al, electroless Ni, In
Oxygen	Ionic	$M^{2+}_xM^{4+}_{1-x}O_{2-x}$	$Y_{0.1}Zr_{0.9}O_{1.95}$	Fluorite	Porous Pt
Humidity	p	$M^{2+}M_2^{3+}O_4$	Ti doped $MgCr_2O_4$	Spinel (porous)	Porous RuO_2
Hydrocarbon gas	n	MO_2	Doped SnO_2	Rutile	Porous Pt

3. Probe Assembly

   a. Coating protection
   b. Sensor containment

The materials technologist generally focuses on the first two aspects when developing the necessary operations for acceptable device manufacture. Interaction between the materials and systems specialists helps to define sensor construction, manufacturing approach, and final probe assembly design.

a. Raw Materials Selection and Powder Synthesis. The choice of chemical precursors as raw materials sources for NTC device manufacture is very broad. The raw materials options are initially defined by the materials technologist's experiences and training. The unit processes are then defined and checked for compatability with the raw materials selections. After an initial development trial, processing ground rules and precursor choice are more firmly established. Major considerations of concern during this effort are:

Consistency upon processing
Cost
Precursor chemistry
Impurity levels
Reactivity
Sensor synthesis approach
Physical parameters
Size, packing efficiency
Mixing ease/mix homogeneity

Two general synthesis approaches have been used to prepare NTC thermistor devices from available high-purity raw materials sources.

(1) NTC sensors by reactive sintering of oxide-mixture compacts. Oxide raw materials are usually selected for chemical purity, particle size (reactivity), and lot-to-lot consistency. Thermistor formulations are weighed, intimately mixed, compacted into the desired shape, and subjected to the prescribed time-temperature-atmosphere sintering cycle. This results in the required chemical uniformity, microstructure, and electrical and mechanical properties needed for the application. This reactive sintering approach has generally been used for volume disk manufacture where low-cost and wider resistance-slope tolerances are expected. General performance and reliability of devices made by this approach are considered fair to good.

(2) NTC sensors by sintering synthesized powder compacts. This powder synthesis approach has been used for higher-quality, close-tolerance NTC sensors and parallels the flowchart of Figure 24. The selection of chemical precursors can be rather broad. For example, soluble salts, salt precipitates, powder mixtures (carbonates, hydroxides, oxides), and even organically derived formulations have been used.

Intimate powder mixtures of NTC formulations in fluid suspension are consolidated, dried, and granulated. The resultant powders are then calcined for 2 to 8 hr at temperatures below normal sintering conditions (600–900°C). This heat treatment partially reacts the precursor mixture to form an oxide agglomerate that approximates the final NTC chemistry and crystal structure. Calcination results in significant physical shrinkage of the precursor mix and chemical homogenization. The oxide powder agglomerates thus produced generally have well-controlled crystallite size (reactivity)

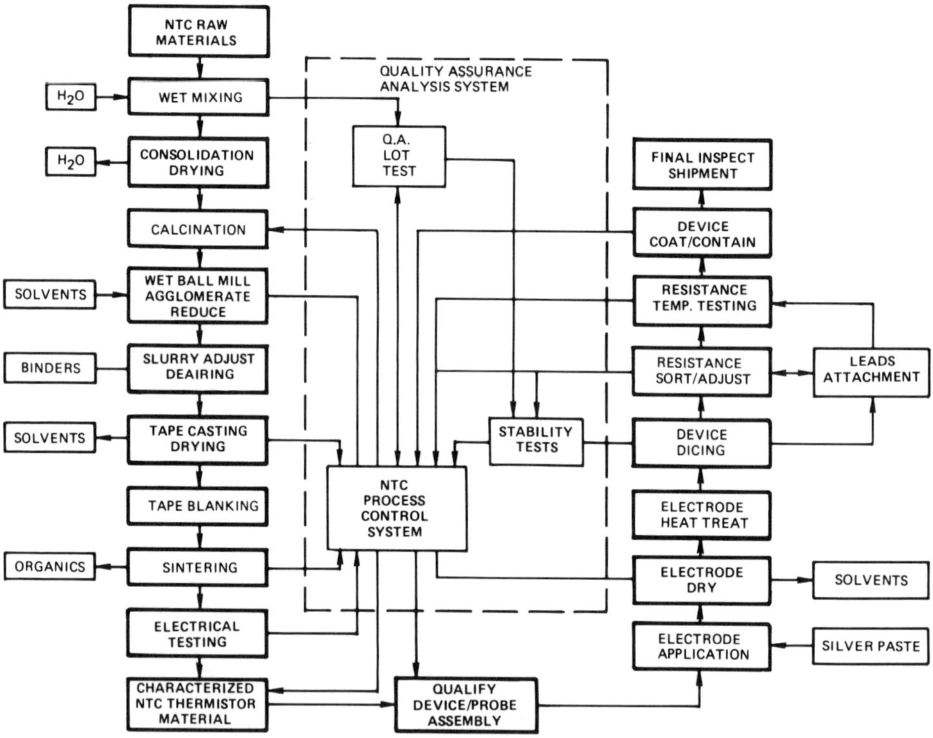

Figure 24 Typical materials inputs, unit operations, process flow, and quality assurance interrelationships for NTC thermistor manufacture.

and predictable packing efficiency once agglomerates have been reduced.

The calcined agglomerates are broken down toward the ultimate crystallite size in an effort to achieve controlled powder compaction, better chemical homogeneity, and predictable sintering reactivity. Comminution using rotary, impact, vibratory, and colloid milling equipment is common. Organic binders may be introduced at the final stages of milling to assist in the NTC device-forming operation.

A sample lot of the NTC devices is usually processed using a standard manufacturing procedure. A statistical sample of processed units is measured for critical electrical and dimensional parameters. This establishes the potential manufacturing yields and processibility for the formulation lot. Once accepted, the batch is committed to a device order and processed accordingly. A number of process control checks would be used to monitor the actual production lot and assure product and processing conformance.

b. *Forming*. NTC products come in a variety of sizes and shapes to satisfy the device-sensing requirements demanded by the user. The major forming methods employed include:

Device type	Prime forming method
Bead	Slurry cast on wires
Disk washer	Uniaxial compaction
Rod	Dough extrusion
Chip	Slurry tape casting
Film	Screen print, sputtering

Bead thermistors are usually produced by hand and require operator dexterity and judgment. The standard method employs tensioning two parallel precious-metal wires in a precision frame. Wire separations of 0.05–0.50 mm are typical and precisely controlled. A drop of oxide slurry is placed on the wires using a fine-pointed glass or metal tool. The size of the bead placed on the wires is controlled by drop size and drawing the tool across the wires. Proper control of slurry solids content, viscosity and surface tension helps the operator to control the bead size, length, and concentricity on the wires. Additional beads are placed along the wires at desired interfals. The slurry is allowed to dry, the organic or nitrate binders decomposed, and the wire strings are then sintered under prescribed conditions.

Rod extrusion employs a dough of different binder and higher oxide solids content than that used with bead thermistor slurries. The stiff dough is deaired and placed in an auger or plunger press.

Long rods of thermistor dough are extruded through a die orifice and supported on grooved boards. After slow solvent removal by evaporation, rods of desired size are cut from the extrusion stock and prepared for sintering.

Disk and washer shapes are formed by uniaxial compaction in hardened steel or carbide dies using high-speed compaction presses. Sensor device diameters of from 0.1 to 3.0 cm and thicknesses of less than 0.1 mm have been produced. Compaction uniformity is greatly enhanced by the use of spray-dried powder. Spray-drying produces spherical powders from slurry suspensions by flash evaporation of the suspension vehicle in a hot circulating gas environment. Ash-free organic binders are generally used to control compaction and lubricate die walls.

Tape cast suspensions have the same consistency as bead slurries. These are generally prepared using techniques described for alumina substrate manufacture [32]. A well-dispersed slurry is deaired and distributed over a flat surface using a doctor blade. The tape thickness needs to be controlled to very tight tolerances. Upon drying, the flexible sheet is blanked into desired substrate sizes and stored for sintering. Some NTC producers have laminated tape sheets together for thickness variety and accuracy control. Individual NTC devices are cut from sintered NTC substrate sheets using dicing or ultrasonic drilling equipment.

Laminated tape cast thicknesses of NTC formulations can be controlled commercially over a wide thickness range, 0.10–2.5 mm. Good control of as-cast tape thickness (0.1–1.00 mm) is easily achieved.

NTC films can be produced on various substrates by the thick-film screening of viscous thermistor material suspensions through masked metal mesh patterns. Film elements when dried and sintered can be adjusted to desired resistance values by abrasion. The incorporation of such NTC sensor elements into system circuitry using thick-film conductor interconnects has been demonstrated [33,34]. The properties and stability of such NTC system circuits is not widely known. The concept is, however, quite attractive from a systems design and manufacturing viewpoint and some development efforts are being pursued for volume applications.

Other film sensors, based on vacuum deposition [35] and thin flake casting techniques [36], have been described. Technical disclosure of NTC films produced by chemical vapor deposition (CVD) and organometallic film decomposition is meager but appears to be feasible. This may be a fruitful area for exploratory research.

c. Sintering and Chemistry/Microstructure Development. The importance of the sintering cycle on device property prediction is probably well established but not well documented for NTC sensor manufacture.

1. The sintering of multivalent oxide spinels so as to obtain well-defined electrical properties is a complex and multifaceted process (see the comments in Section II.A.3). Two widely different thermal treatment approaches can produce essentially the same product capabilities.
2. The need to protect proprietary information has limited disclosure of technical accomplishments. Most manufacturers attempt to guard their processing "tricks," equipment designs, and product development intentions. In this environment, NTC technology and sensor advancements have to some degree been inhibited.

A review of the technical and patent literature since the early 1940s, however, allows one some insight into the principles of NTC sintering, oxidation-reduction effects, and properties prediction. Careful examination of related ferrite R&D and other technical ceramics activities also provides additional relevant information. The reader is referred to standard ceramic reference texts [37,38] and review articles for sintering fundamentals. A short review is offered below.

The sintering process is the last major stage in the development of sensor properties. The sensor unit prior to sintering is comprised of a tight array of fine particles in intimate contact. Control of particle-particle coordination, particle size, and defects (cracks, macropores, contamination, etc.) is important for quality manufacture. The composite structure is subjected to a thermal excursion that matures or fuses the body into its final shape. The resultant electrical, chemical, and mechanical properties are fairly well defined after the completion of the sintering cycle.

The sintering process causes the compact to densify to a volume as small as half the original compact. Linear shrinkages of 12–25% are quite common. NTC producers attempt to hold shrinkage in the range 15 to 18 linear percent for accurate dimensional control.

The sintering process also determines the microstructural features of the NTC body—typically, a single-phase polycrystalline ceramic. After elimination of organic binder constituents, the device is heated to 1100–1300°C, where densification, final oxygen equilibration, and grain growth to sizes in the range 4–30 μm are allowed to proceed. Careful temperature control is maintained to assure product uniformity and desired beta/resistivity value. Product yields for tight-tolerance devices can be affected by minor time-temperature-atmosphere variations.

The attainment of high-density, closed-pore microstructures is desirable to achieve high-quality, stable NTC products. Sintered sensor devices then need to be cooled in a uniform, predictable manner. Care to prevent thermal shock cracking and/or surface reoxidation is necessary to assure sensor reliability.

d. Material Properties Definition/Quality Assurance. The formulation, powder synthesis, forming, and sintering processes are coordinated and controlled to produce desired NTC sensors with well-defined thermal, physical, and electrical responses. Electrical characterization of a representative sample is done to predict yields and be sure that the NTC sensors meet specifications requested by the user. Statistical data on resistance, beta, and stability are generally accumulated for high-reliability products. Use of specialized precision baths capable of being controlled to $<0.01°C$ accuracy are not uncommon at high-quality thermistor facilities. Standard reference sensors traceable to NBS temperature standards are often required for precision calibrations. Thermistor manufacturers often use their own sensor products as secondary standards and probes to control precision baths. Equally important are precision resistance bridges, digital multimeters, and other critical measurement instruments.

Qualification of some NTC sensor products requires time constant, frequency, and/or powder dissipation characterization to assure conformance. A well-equipped quality assurance laboratory and competent staff capable of setting up and measuring such parameters are required.

Table 7 lists general thermophysical properties of various materials used in ceramic sensor manufacture. Accurate values for specific NTC formulations are not, however, well publicized. Independent measurements on actual device materials are recommended for critical applications.

e. Finishing Operations

(1) Electrode application. The attachment of electrical contacts to NTC sensors is generally accomplished with precious metal electrode materials. Bead thermistors have embedded platinum alloy wires cofired into intimate contact with the sensor material. The similar thermal expansion behavior of bead and electrode materials, small structure, and reasonable shrinkage during sintering enable the device to have good mechanical stability and reliability up to the 300°C thermal limit of bead sensors.

Most disk and chip sensor manufacturers use specialty silver electrode materials for applications below 150°C. A silver-glass powder mixture (suspended in an organic binder and solvent vehicle) is applied to opposing electrode surfaces. Such electrode application is accomplished by spraying, dipping, and screen printing, or transfer/roll printing.

After solvent and binder elimination, the silver-glass mixture is matured and intimately bonded to the NTC body by belt furnace heating to 500–800°C for a cycle time of 10–30 min. The exact maturing cycle is dependent on the glass frit-silver particle mixture, particle size, NTC surface quality, and glass melting/wetting behavior. Experience has shown that silver thickness, thermal cycle, and glass–silver particle packing controls are required to achieve dense, low-contact-resistance electrodes. Many NTC producers have relied on well-established, time-tested silver electrode formulations offered by thick-film conductor paste manufacturers. Performance tests done on standard NTC test devices are used to qualify new lots or alternative silver paste sources for future usage.

Novel metal contacts and application techniques have been utilized for special sensor applications where reliability and performance justify the expense. Gold (to eliminate silver migration under voltage stress) and vacuum-deposited or sputtered contacts are such examples.

It is important to evaluate electrode integrity periodically to assure product quality and reliability. NTC devices with soldered leads are typically used as sensor devices. Pull testing of wires soldered onto small devices is an ideal method for mechanically evaluating silver and silver bond integrity.

Poorly sintered leads tend to be porous, weak, and absorb fluxing agents. Overfired silver contacts tend to have high contact resistance and weak contact/sensor interfacial strength. In addition, the possibility of glass migration and chemical reaction at the NTC material surface is a distinct possibility. Appropriate control of the electrode application and heat treatment assures the NTC producer of high-quality production. Pull testing of soldered leads, microstructure examination of sectioned devices (thickness and densification checks), and general monitoring of process variables are particularly helpful.

(2) Electrical testing. The batch qualification testing described in Section IV.A.6.d was used to justify full manufacture of the NTC sensor order required by a customer. The same basic measurement concerns (bath and bridge accuracies, time required for sensor stabilization, special measurement requirements, etc.) are important for cost-efficient sensor manufacture. Production quantities, however, may require more automated testing equipment and/or statistical procedures to assure product acceptability.

Broad tolerance requirements (±10% of R_0) often pass statistical sampling testing or comparative bridge sorting. Tighter-tolerance units might be individually sorted at one or more temperatures. Precision sensor devices have been made by matching two separate units [39] or by resistance adjustment (abrasion) [40] of the sensor. The cost of manufacture and eventual user's cost will be determined by device yields and labor expended to measure and control the resistance and beta requirements.

f. Protection-Encapsulation-Probe Assembly. NTC devices perform an admirable sensing function when placed in an acceptable chemical and mechanical environment. The real world, however, calls for the device to be encapsulated to prevent the detrimental effects of humidity, chemical attack, contact corrosion, and so on. In most cases, the active lead contacts must be isolated to achieve proper electrical measurement.

The variety of NTC sensor configurations, wide temperature range, environmental conditions, and time-response requirements makes for an interesting design task for the systems and materials specialist. Much of the groundwork for design considerations has been discussed earlier. The following discussion and reference to devices shown in Figures 8 to 13 will be used to demonstrate a number of pertinent points.

(1) Thermal response. The mechanical integrity of NTC devices is usually enhanced by glass (beads) and polymeric coatings. The brittle nature of ceramic material is generally known and conservative mechanical designs are generally used to protect them from harsh, stressful environments. As discussed earlier, such coatings will severely affect the thermal response of the sensor. Innovative approaches to achieve thin coatings or to isolate the sensor from stress potential are considered for fast time response. NTC probe catalogs show that metal housing assemblies are a popular approach. The increase in thermal mass can still affect the time response.

An alternative approach is to surface mount the sensor to a low-mass, high-conductivity substrate. This provides a good thermal link to the measurement environment. High-thermal-conductance coatings can provide electrical and chemical protection and if thin enough, will not have too great an impact on time constant or measurement accuracy.

For sensor assemblies that undergo large thermal excursions, the thermal expansion behavior of assembly components can be critical. Careful design pays dividends in long sensor life and high reliability.

A variety of polymeric, metal, and inorganic coatings or housings have been utilized for protection of sensors from harsh chemical environments. The thermal limitation of the coating must be taken into consideration when the assembly is subjected to processing and use excursions. Safety margins are recommended for long-term elevated-temperature use. During encapsulant selection, the compatibility of both sensor and electrode materials must be assured.

(2) High-temperature stability. The high-temperature stability of most thermistor compositions can be substantially improved by hermetic sealing of the device. This technique is accomplished by melting a thin glass sleeve over bead thermistors at temperatures in the region 700–900°C.

Sachse [21] has pointed out that most thermistor compositions (possibly Cu-bearing systems might be excluded) can withstand short thermal excursions to 800°C without serious oxygen exchange with the gas environment. The glass encapsulation procedure was developed to ensure sensor chemical integrity and postseal hermetic protection.

Care in selecting sealing glass compositions is required to ensure mechanical integrity (thermal expansion match) and minimal chemical reaction. The use of low-alkali glasses allows sealed elements to operate continuously at temperatures up to 300°C without serious degradation.

One barrier in the development of high-temperature NTC sensors capable of measurement and control of systems to greater than 500°C is the diffusion of glass components (alkalis) into the sensor body. Stability assessment generally appears initially satisfactory for a number of glass encapsulant materials, but major resistance instability often begins after an initial incubation period, as shown in Figure 25.

7. NTC Sensor Applications

Previously, we described the evolution of various NTC devices since their introduction in the mid-1930s. A general breakdown of present NTC applications is as follows:

- Zero-power thermal sensing
  - Temperature measurement (general and precision)
  - Heat flow
  - Radiation sensing
  - Temperature compensation and control
- Self-heating sensing
  - Voltage regulation
  - Liquid-level sensing
  - Vacuum and pressure guages
  - Gas analysis and chromatography
- Indirectly heated sensing
  - Dc-ac conversion
  - Phase shifting
  - Amplitude modulation

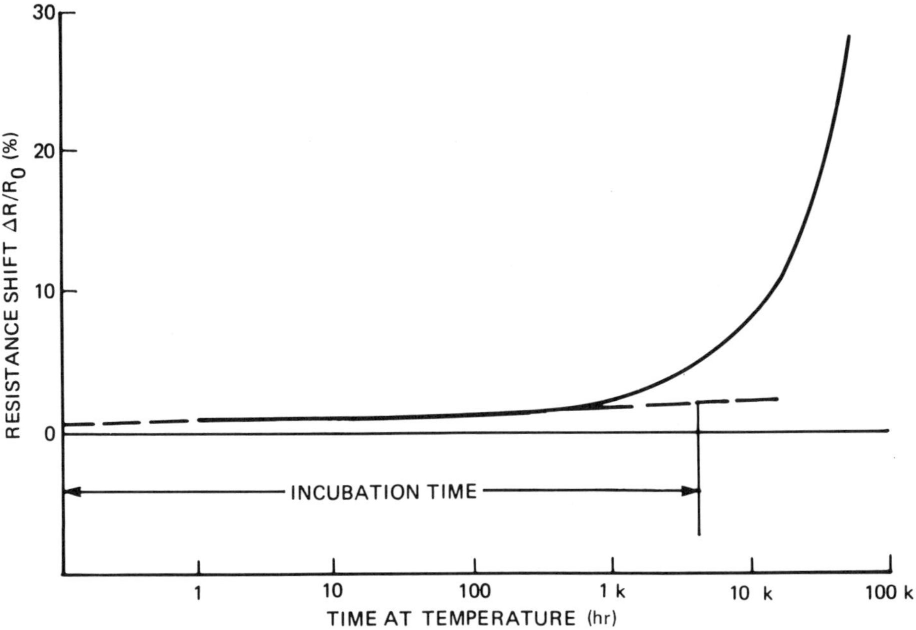

Figure 25 Resistance instability of a glass-coated NTC thermistor at elevated temperature (>350°C).

A more definitive treatment of NTC device design and applications can be found in the texts of Sachse [21] and Macklen [24]. A fairly comprehensive treatment for the design and use of an NTC sensor for medical temperature monitoring is given below for illustrative purposes.

a. NTC Thermistor for Medical Temperature Monitoring

(1) Application. It is estimated that temperature monitoring of patients in hospital and clinical environments is occurring some 2 billion times annually. Electronic monitoring systems are being introduced which more accurately measure patient temperature in shorter time intervals. In addition, patient monitoring using disposable sensing units have become more widely used in intensive care, operating room, and other clinical situations where speed, accuracy, and staff convenience is important.

Electronic systems replace mercury expansion thermometers for obvious reasons. In addition, they offer opportunities for computer interfacing and record maintenance not previously available from alternative approaches. The hospital supply industry appears to be focusing on thermistor sensors as the cost-effective system.

(2) Sensor performance goals. Patient monitoring in the range of 94–110°F for body surface, internal, and oral temperatures (variable probe lengths) are required. The electronics systems design should produce an analog or digital output signal that interfaces with a display, printer, or computer. The probe assembly should contain the NTC sensor having fast time response (low mass, good thermal coupling) in a housing that is familiar and not threatening to the patient. The system accuracy (<0.2F) requires the thermistor device to sense ±0.1 F full range during the shelf life of the probe assembly. Exposure to normal hospital chemicals, possibly harsh sterilization, and thermal excursions may also be required. High-volume manufacture of the sensor probe assembly at modest cost is a reasonable goal.

(3) Design solutions

(a) Sensor. The fast thermal response requires a small (<0.0003 cm^3) sensor with high α (>4%/°C). Interchangeable designs require very accurate beta control and tight resistance match at or near the nominal temperature range. The device stability after resistance establishment must withstand processing variables associated with probe assembly and life storage—<0.2% resistance shift and predictable.

(b) Probe assembly. The thermal impedence must be low so that the patient temperature can be monitored within a short time interval (10–30 sec, depending on the monitoring technique). Probe design requires a thermal isolation of the sensor from the main body of the probe assembly. The probe covering must be electrically insulating and must not adversely influence sensor performance due to chemical and/or thermal exposure.

(4) NTC sensor manufacture. A process flow and control monitoring diagram suitable for producing NTC sensors for medical applications is shown in Figure 24. The thermistor formulation is prepared as a slurry and cast into thin sheets about 10–14 mils thick. After drying and cutting, the sheets are sintered to closely controlled resistivity and beta tolerances. Tight control allows the utilization of all material in the subsequent resistance adjustment operation.

After careful electrical characterization of the sintered product, manufacturing yields are computed and the sintered lot is ready for electroding, lead attachment, and resistance adjustment. Samples of the product are monitored for resistance shifts and stability compared to previous samples. Quality control charts of electrode thickness, resistance shift, lot yields, and other pertinent parameters are reviewed to assure manufacturing control and predictability.

The resistance adjustment operation is the critical step in sensor production. Abrasion techniques have traditionally been used to accurately adjust resistances to less than 0.1% of a nominal value. The units are then cleaned, aged, and statistical samples are monitored for resistance shifts.

(5) Probe assembly. Standard resistance NTC sensors may be used for different probe assembly designs. The leaded devices are placed in fixtures that control how the sensor will be thermally coupled to the probe measurement head. The automation equipment for probe assembly manufacture depends on probe design, production volumes, and cost goals. Emphasis is generally given to sensor accuracy and probe reliability. A statistical sample is taken for production monitoring, stability, and performance testing. Single-point temperature testing on a 100% basis is usually done to assure sensor performance. After cleaning, packaging, and sterilization, the sensor probe is ready for use.

(6) Quality assurance. Monitoring of the product during each manufacturing operation has been emphasized in the processing description above. Such control monitoring can be costly but is necessary to assure production yields and critical equipment utilization. In addition, the ability to predict and control the quality and reliability of this tight-tolerance sensor device will determine the success of the product.

B. PTC Thermistors

1. History

The pioneering work of Vervey et al. [3] on "valence-compensated semiconduction" led to the discovery and patenting [41] of the PTC effect found in doped-$BaTiO_3$-based ceramics. This unique phenomenon—an orders-of-magnitude shift from a low base resistivity to a high resistivity over a narrow temperature range—was universally recognized as a potential solid-state switching device. By the late-1950s intense R&D activities were being conducted worldwide in an attempt to commercially exploit this phenomenon for electrical and electronic applications. Significant progress followed with respect to understanding the phenomenom, advancing materials capabilities, and developing devices for sensor and switching applications.

Initial applications utilized PTC ceramic components as sensors for motor protection, self-controlled heaters for crystal ovens, and special timed current limiters for electronic circuit applications. The biggest initial application occurred with the development of an inexpensive television degaussing component that replaced a complex electronic circuit for the fledgling color television industry.

Several unique applications were introduced in Japan in the late 1960s and early 1970s: specialty heaters for food processing (rice steamers) and room air heating [42] (harmonica and honeycomb type). In addition, the temperature utilization was extended as high as 300°C with the development of $(Pb,Ba)TiO_3$ formulations.

Important automotive applications were developed at Texas Instruments, Inc., in the 1970s. PTC elements for fast heating of the bimetallic coil and actuation of the carburator choke butterfly valve were introduced. Fast actuation of the choke substantially reduces exhaust hydrocarbon pollutants during the cold start phase of automobile usage. A low-voltage honeycomb heater employed to assure complete evaporation of fuel droplets in cold carburetors was introduced in 1979. This component, when properly matched with an even-faster PTC heated choke system, has resulted in excellent four-cylinder cold-start engine performance and lower hydrocarbon (HC) and nitrous oxide (NOx) emissions.

A significant effort to develop sensing systems utilizing small PTC elements as resistance switches was undertaken in Europe. The introduction of system designs that use these elements in liquid-level detection, thermal protection, and other control applications by Philips [24] is particularly noteworthy.

Currently, development efforts are focusing on the high-performance PTC elements. The short-term dissipation of >1000 W of energy in small elements and the low-cost manufacture of high-performance devices should have a significant impact on motor controls, industrial heater, and consumer markets.

Future PTC advances and innovations are likely to come in the following areas:

- Computer-assisted PTC production
  - Sintering and properties control
  - Powder synthesis: closed-cycle process control
- High-power applications
  - Timing devices: motor-start assist
  - Cost-effective heaters
  - High-temperature materials
- Systems approach to PTC utilization
  - Close scrutiny of customer requirements
  - Longer development engagement intervals
  - Closer attention to product reliability/safety

2. PTC Device Construction

Typical PTC devices are shown in Figure 26. The device resistance, time constant, and dissipation behavior—like those of its NTC counterpart—are determined by Eq. (42) and parameters outlined in Sections IV.A.3 to IV.A.5 and the Appendix.

3. PTC Electrical Behavior: Zero Power

To fully appreciate the capabilities of PTC devices, it is helpful to understand the resistivity-temperature behavior of the material. Examination of Figure 27 illustrates the salient features of PTC behavior and some characteristics of importance in the design and utilization of such sensor devices. This figure includes device resistivities obtained under low-voltage as well as near zero power—short pulse voltage application. Zero-power measurements are a quick, useful method of assessing PTC device capabilities. Actual device performance, however, is not always predicted by zero-power testing, and characterization under use conditions as well as life-test evaluations are more desirable.

Figure 27 depicts an element with a room temperature resistivity of 50–60 Ω-cm, a switch temperature (knee or T_c) at 120–130°C, a PTC magnitude [$\log(\rho_{max}/\rho_{min})$] of about 5 at zero power – <1 V applied, and a voltage sensitivity (the log ratio of a pulsed zero power resistivity measurement at 100 and ∿1 V/mm, respectively). The term "end point" refers to the ability of the device to withstand a pulsed voltage with some safety margin without approaching the condition of thermal runaway if placed in a powered condition.

The resistivity-temperature slope ($d\rho/\rho_0 dT$) above the Curie temperature T_c is another important parameter determining the operating or switch temperature for the device. Extremely high slopes have been observed for optimized processing of $BaTiO_3$ ceramic elements—of the order of 150%/°C. The substitution of most solid solution elements (i.e., Sr, Ca, Pb for Ba or Sn, Zr or Ti) results in a significant reduction in slope, often to less than 25–30%/°C.

PTC devices are generally used in dc or low-frequency applications where resistivity-frequency dependence is not important. The reduction in apparent resistivity with applied voltage at high frequency can be significant [43]. This is observed with other semiconducting materials possessing insulating grain boundaries, as discussed in Section II.

4. PTC Electrical Behavior: V-I Curves

The thermal response of PTC devices subjected to continuous voltage-current conditions can best be represented by the V-I curve

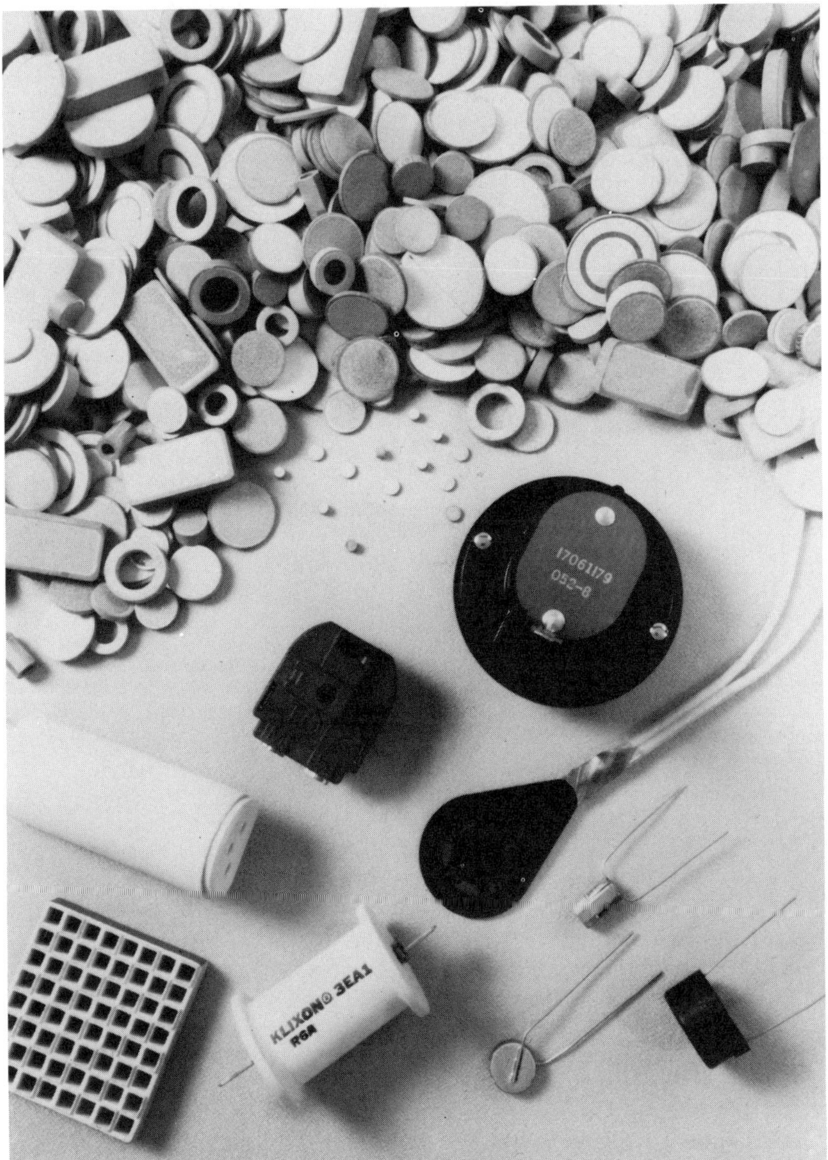

Figure 26 PTC elements and device assemblies for heater and component applications. (Photo courtesy of Texas Instruments, Inc.)

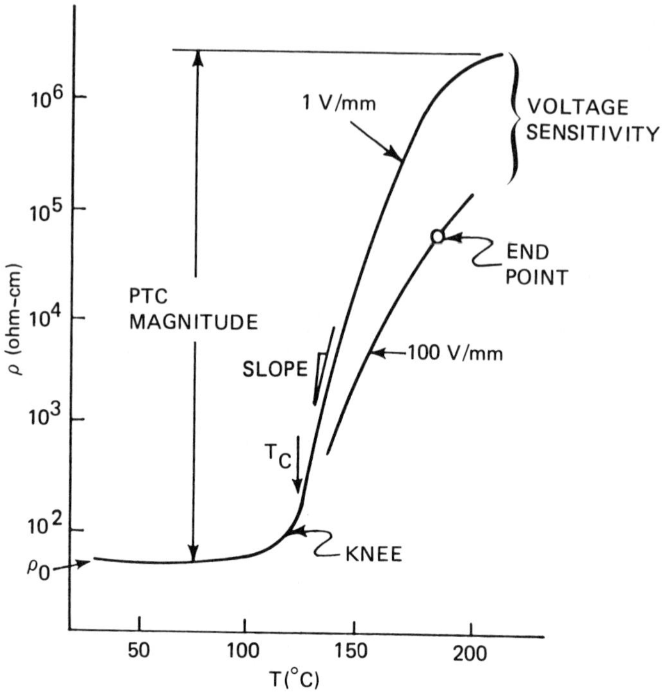

Figure 27 Resistivity-temperature characteristics of $BaTiO_3$ PTC ceramics. PTC characteristics: R–T curve (ρ–T); ρ = RA/l (Ω-cm); slope = (1/ρ) (dρ/dT); PTC = $\log(\rho_{max}/\rho_{min})$; voltage sensitivity = $\log(\rho_{100}/\rho_1)$.

shown in Figure 28, curve A. The salient feature of this curve is the self-limiting power condition that occurs when the device is heated above its Curie temperature. Further voltage increase to V_{max} can give rise to PTC heating to a point where the "end point" is exceeded. A thermal runaway condition will result when the dρ/dT derivative (Figure 26) becomes $\leqslant 0$.

It is interesting to compare the V–I curves of PTC and NTC devices (Figure 28, curve B). One can see that the potential of thermal runaway is always prevalent for NTC devices if heat dissipation does not equal or exceed the power generated in the system. As pointed out earlier, powered devices should have an adequate safety margin built into the device design so as to assure the performance and useful life of the component. A safe voltage V_s is depicted in Figure 28 (curve A) for the PTC device in question.

The timing and dissipation behavior of PTC devices are alluded to in the Appendix. A summary of different PTC elements is offered in Table 9 for device size and heat dissipation mode comparisons. The overriding feature governing the heat dissipation and time constant behavior appears to be the low thermal conductivity of the PTC material.

5. Device Chemistry

The chemistry of the perovskite-structured PTC ceramic sensors, like that of most grain boundary—controlled ceramic components, is rather complex. One initially starts with high-purity (99.9+) raw materials. Usually, the titanium and barium sources are the more critical ingredients for PTC manufacture. Ideally, acceptor dopant impurity concentrations (alkalis, Mg, Al) must be low (<100 ppm). The purity level of the PTC formulation must then be controlled throughout the device processing to assure that the desired PTC performance and product reliability is attained.

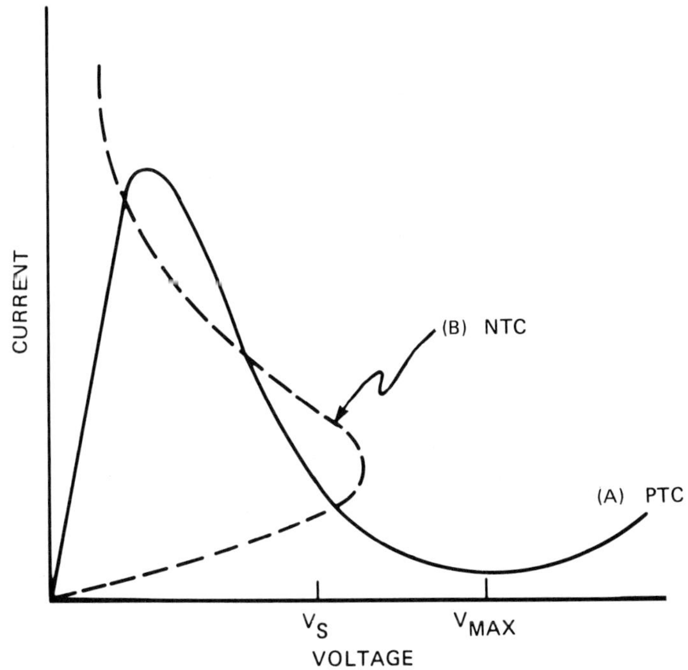

Figure 28 Voltage-current behavior of (A) PTC and (B) NTC thermistors.

Table 9 PTC Device Thermal Comparisons

Device	Typical application	Device size (g)	Power dissipation (W)		Time to control mode (sec)	PTC switch temperature (°C)
			Maximum	Control mode		
Sensor	Motor protect	0.02	Nil	Nil	<3	60–200
Current timer	TV degausser	1–2	1100	0.1	<1	60–80
Current timer	Motor assist	3–20	2000	2.0	1–3	100–130
Heater	Auto choke	1–1.5	150	0.1	10–30	90–180
Honeycomb	Air heater	20–30	400	400	<1	180–220
Honeycomb	Auto EFE	20–30	300	300	<1	150–180

Dopant additions are made to the "host" titanate composition to achieve PTC property enhancement. Most practical PTC formulations comprise $BaTiO_3$ as the major ingredient, and this will be considered the host composition for the remainder of this discussion.

Table 10 lists the general categories of chemical additions for producing PTC devices from a $BaTiO_3$ formulation. The purpose of such additives is described briefly in the remainder of this section.

a. Isovalent Additives. Semiconducting $BaTiO_3$ undergoes a tetragonal-to-cubic phase transformation at ∿130°C> and a lower-temperature orthorhombic-to-tetragonal transition at ∿10°C. The substitution of structurally compatible divalent cations (Sr, Ca, Pb) for barium in the perovskite lattice causes these phase transformations [(A) tetragonal to cubic and (B) orthorhombic to tetragonal] to shift significantly in temperature, as shown in Figure 29. It is important to note that all additions reduce the low-temperature transition (B) to below 0°C. Calcium substitution is unique in that it does not greatly influence the high-temperature transition (A).

PTC ceramic semiconducting elements with desired resistance switch (T_c or PTC knee) temperatures (Figure 3) can therefore be produced by control of the solid-solution formulations. In addition, small calcium substitutions might be introduced to improve the low-temperature-resistance aging behavior associated with domain reorientation [44] and proximity of the rhombohedral-to-tetragonal phase transition. In some cases the Ca substitution also improves the grain size and microstructure of the PTC ceramic [45].

b. Altervalent Dopants for Conduction Control. It was shown earlier (Section II.B.2) that lanthanum La^{3+} dopant, when substituted for Ba^{2+} in the perovskite lattice, will produce n-type semiconductivity. From a crystal chemistry viewpoint, any trivalent substitution of proper size on the barium site results in the one-for-one creation of a Ti^{3+} ion at octohedral Ti^{4+} positions in order to preserve charge neutrality. The general formula is

$$Ba^{2+}_{1-x} M^{3+}_{x} Ti^{4+}_{1-x} Ti^{3+}_{x} O_3 \qquad (45)$$

where x is the level of substitution of the trivalent M^{3+} ion. It is possible to substitute any of the rare earth elements and produce semiconducting behavior in the perovskite lattice. PTC manufacturers, however, have focused on Y, La, and Gd as conduction dopants for production purposes.

Table 10 Chemistry Considerations for PTC Sensor Manufacture: Additions Impacting Barium Titanate Semiconductor Properties

Addition type	Addition function	Typical substitution level for Ba or Ti	Addition elements used	Impact on PTC properties
Isovalent	Solid-solution substitution for Ba or Ti in the titanate	0–0.20	Sr, Pb, Ca, Sn	Adjusts PTC switch or Curie temperature
Altervalent	Semiconduction or resistivity control	0–0.008	Y, La, Sb, Ce, Nb	Resistivity adjustment, grain size refinement
Boundary	Barrier layer control	0–0.002	Mn, Fe, V, Cu	Voltage sensitivity, high PTC magnitude
Sintering	Liquid-phase sintering control	0.005–0.020	Si, Ti, Ge	Lower firing temperature, improve PTC quality
Poisons	Impurities that degrade PTC quality	0.0002–0.0050	Na, K, Al, P, Mg, transition metals	PTC quality impaired if levels not controlled

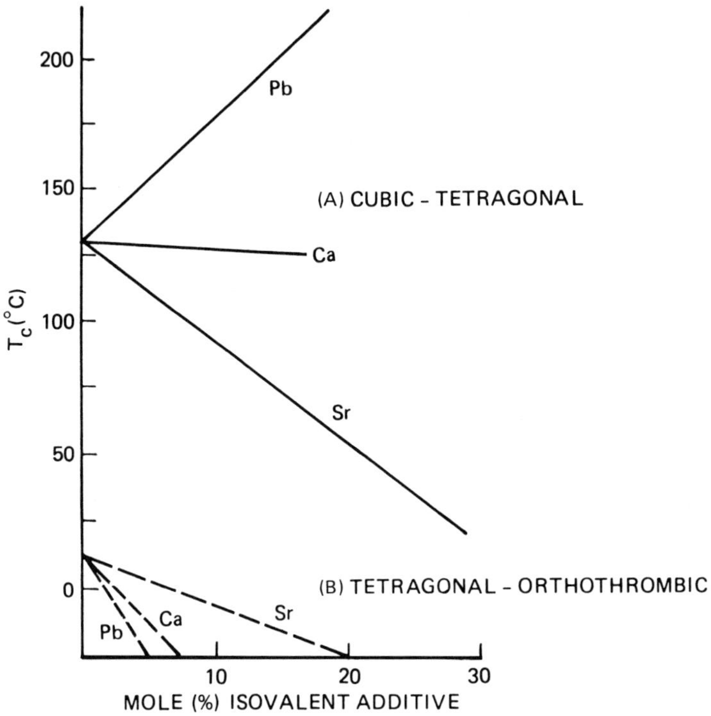

Figure 29 Curie temperature control of PTC ceramics with isovalent additives: (A) cubic-tetragonal phase transformation; (B) tetragonal-orthorhombic phase transformation.

Pentavalent dopants (Sb, Nb, Ta) can also produce semiconduction by substitution at Ti^{4+} lattice sites. These dopants (the exception being Sb), however, are not generally used for production.

The ability of trivalent rare earth substitution to optimize semiconduction and affect the microstructural development is shown in Figure 30. A minimum in resistivity is achieved between 0.2 and 0.3 mol % rare earth element substitution. In addition, a significant reduction in grain size occurs after the resistivity minimum is reached. The implication of grain size, size control, and PTC performance is discussed subsequently.

c. *Barrier Layer Modifiers*. Certain transition metal ions (Mn, Fe, V) have been shown to enhance the PTC effect (Section II.B.2). For example, a PTC resistivity increase of magnitude greater than

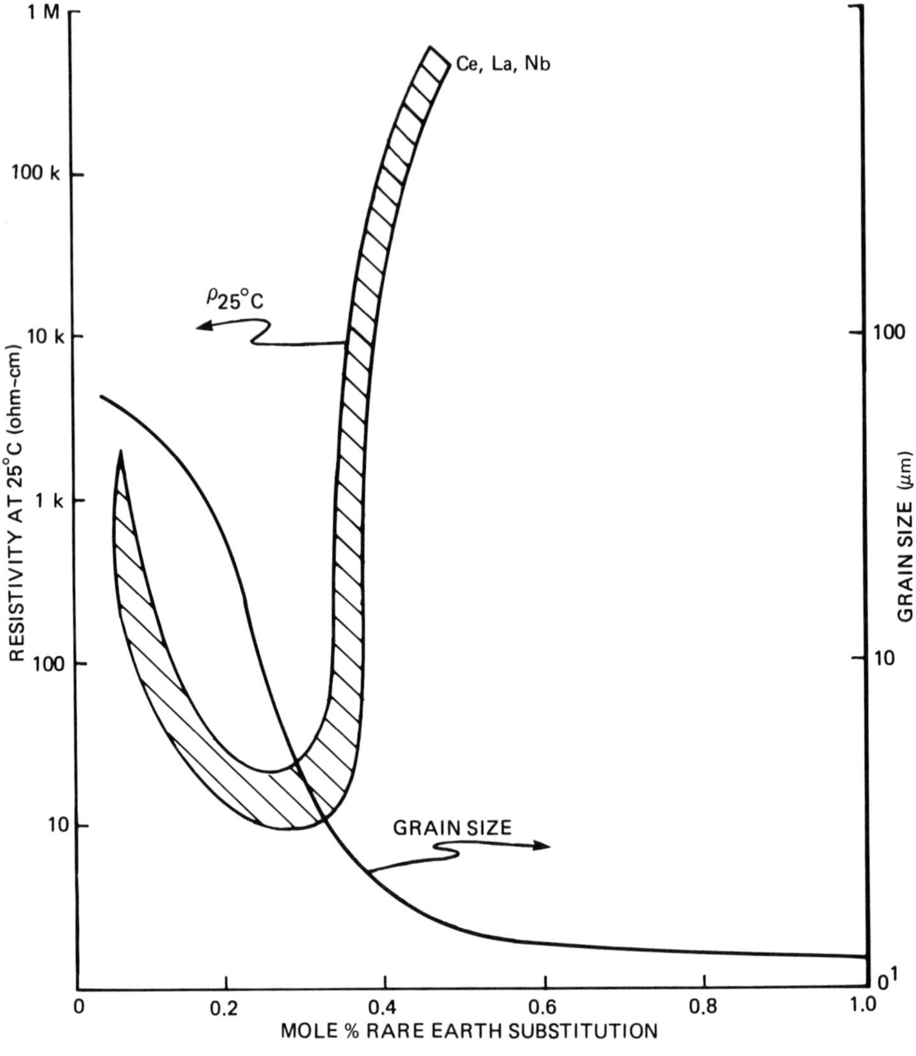

Figure 30 Impact of altervalent dopants on semiconduction and microstructure of $BaTiO_3$ PTC ceramics.

seven orders has been reported and some selected data are shown in Figure 31. The optimized formulations shown were developed in the laboratory during an experimental matrix study. The dopant, sintering aids, and modifier concentrations as well as processing conditions were controlled using a high-purity $BaTiO_3$ raw materials source. Transition from the laboratory to manufacture, however, resulted in a PTC magnitude enhancement approaching 5.5 to 6 orders.

Most active modifiers are capable of some degree of substitution into the perovskite lattice during the sintering process. The impact of the modifier addition, however, appears to affect predominately the region immediately adjacent to the grain boundary interface. The enhancement of the PTC effect has been shown to occur during the cooling of PTC elements in the range 1250–1000°C. The modifier addition, in particular Mn, is believed to influence the manner in which grain boundaries impede electron flow by virtue of enhancing oxidation and/or interface state creation. Optimum modifier concentrations promote resistivity maxima, yet present only a slight barrier resistance increase over unmodified or unoxidized boundary layers at $T < T_c$.

The complexity of the barrier layer and its optimum control is still the subject of intense research. It is sufficient to state that chemistry, microstructure, and processing all play an important role in achieving enhanced PTC behavior, as pointed out by Kulwicki [23] and Daniels et al. [14].

d. PTC Poison Impurities. A number of elements, some of which are found in the lower-cost titanate raw materials sources, can diminish or "poison" the PTC behavior in semiconducting $BaTiO_3$. These include:

Alkali metal ions: Na, K, Li
Transition metals: Fe, Ni, Co, Cu, etc.
Acceptor dopants: Mg, Al, B
Anions: Cl, S, SO_4

Close scrutiny of raw materials is thus required to ensure low voltage sensitivity, uniform grain size, and high PTC magnitude.

e. Sintering Aids: Grain Boundary Phases. Prior to the discovery of the PTC effect, titanate ceramics were made from raw materials of the type in which Al, Si, and other impurities did not unduly influence the nonlinear dielectric functions for which they were intended. As PTC developments progressed, the impact of certain impurities at <100 ppm on PTC performance became known. The

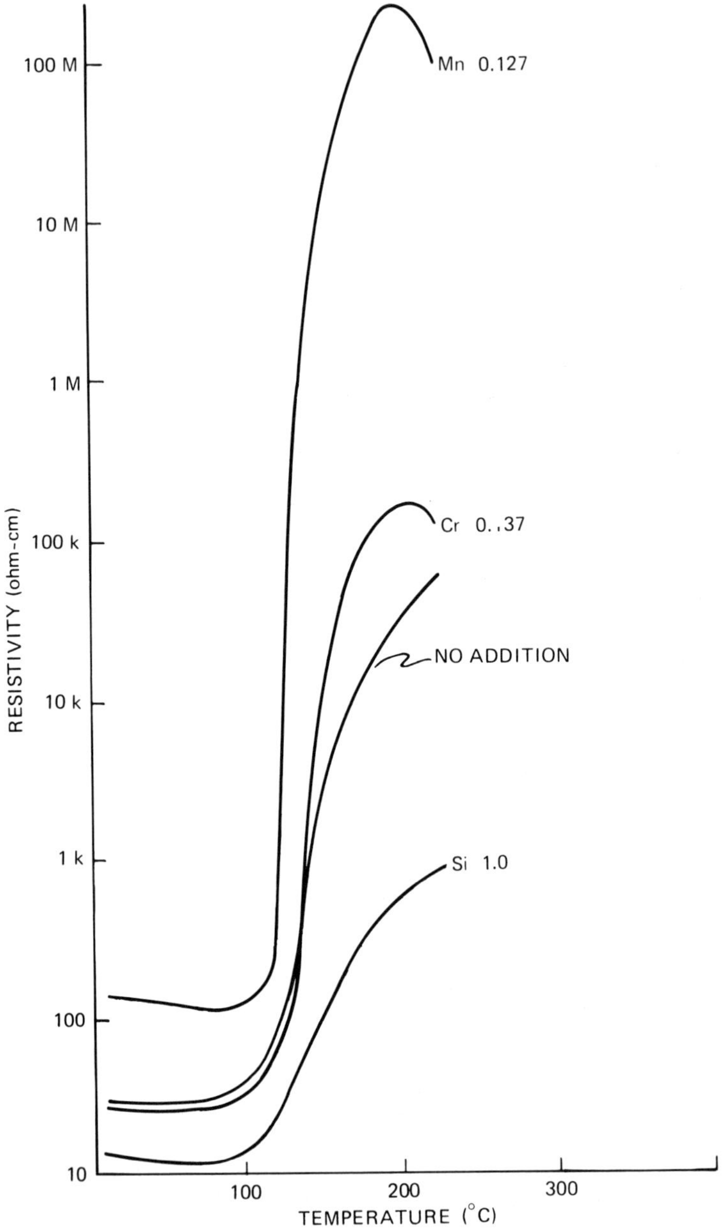

Figure 31 Impact of modifier and sintering additives on resistivity-temperature behavior of $BaTiO_3$ PTC ceramics. Numbers next to additive element indicate the mole percent addition. (After Ref. 51.)

natural inclination for researchers was to use the highest-purity raw materials sources and "dope" the base composition with the desired conduction generators, modifiers, and/or sintering aids. Investigators often doped with undesirable impurities as well, to determine tolerable levels and interactions with other impurities.

Two sintering aid additions—excess Ti and Si—have evolved from this work and are used by most PTC manufacturers to promote densification, reduce sintering temperatures, and aid in microstructural control. Examination of the $BaO-TiO_2$ phase diagram (Refs. 12, 104, Chapter 2) shows that a eutectic temperature on the TiO_2 excess side of $BaTiO_3$ occurs at 1317°C. The liquid phase created at temperatures above the eutectic is quite fluid, readily wets the titanate grains, and induces rapid densification of the powder compacts. The addition of small amounts of SiO_2 (0.5–3.0 mol %) further reduces the eutectic temperature [46] allowing sintering and grain growth to proceed just above 1250°C. Microchemical analysis of grains and second phases at grain boundaries indicate that a substantial portion of the Si and excess Ti additions have segregated to the boundary phase.

The boundary second phase is believed to serve an additional function besides aiding in sintering (i.e., the collection of other impurities that could adversely affect grain resistivity and depletion of carriers immediately adjacent to the boundary interface). The (BaTi) silicate liquid phase can serve as a host for alkali, Al, Mg, P, and other potential "poisons" that influence PTC performance. The solution of such impurity elements can occur as the fine $BaTiO_3$ crystallites (0.3 μm size) undergo grain growth during sintering. As the boundary surfaces sweep across the crystallites, impurities are segregated to the liquid phase and remain isolated from the recrystallized semiconducting perovskite grains.

6. PTC Powder Synthesis

There are several high-purity raw materials sources that have been used for PTC powder synthesis. The best and most cost-effective process choice, as always, is dictated by an assessment of manufacturing yields and materials costs. Four process methods are outlined in Table 11 and represent alternative approaches to making formulations suitable for PTC manufacture. Some general comments concerning each process—advantages and shortcomings—are discussed below.

a. Organic Titanate. Suitable titanium sources derived by reaction of $TiCl_4$ and organic compounds are tetraisopropyltitanate (TPT) and similar organic titanates, as produced by DuPont [47]. These

Table 11 Powder Synthesis Processing for PTC Powder Manufacture

Process	Raw material source: Ti	Raw material source: (Ba,Sr,Pb)	Synthesis	Particle reduction	Relative cost	Ref.
Organic	TPT	Acetate	Flame calcine	Milling	10	47
Oxalate	Cl_4	Aqueous	Calcine	Milling	8	48
Precipitate	TiO_2	Aqueous	Calcine	Milling	5	49
Slurry	TiO_2	Carbonate	Calcine	Milling	3	50

materials have good purity and have been used as titanium sources for many high-quality titanate electronic ceramics. The organic process options outlined in the patent literature require the dissolution of the organic titanium compound with the other formulation ingredients. Both aqueous and organic solutions have been used. The advantage of the organic titanate process relates to the mixing of the cation species on the atomic scale. With subsequent drying, the precursor formulation is converted to the perovskite phase at relatively low temperatures (750–950°C).

The organic titanate process has some disadvantages. The titanium source and other required soluble salts, although pure and plentiful, are more expensive than other raw materials sources of similar purity. In addition, the calcination of organic precursors can proceed with strong exothermic reactions, making the resultant titanate powder difficult to control from the viewpoint of agglomerate reduction and predictable powder reactivity. The method is presently used for specialty products where PTC performance requirements outweigh raw materials costs and processing difficulties.

b. *Oxalate Process*. Chemical precipitation of barium titanyl oxalate has been utilized to produce high-purity PTC formulations [48]. This method has been extended to a production process by TAM Ceramics (United States) and Fuji Titanium Co. (Japan), which are presently marketing the product for high-purity applications. Although the relative cost is high, the powder consistency and purity meet PTC quality standards.

The advantage of this preparation method relates to the control of chemical uniformity, purity, and ease of reduction of the calcined agglomerates to a desired compaction and powder reactivity state.

The precipitation process, however, requires careful control of the solution chemistry. Each chemical formulation requires process optimization to achieve the stoichiometry control required for PTC manufacture.

c. *Precipitation Process.* Another chemical process [49] for the preparation of PTC titanates is the dissolution of soluble salt constituents (except Ti) in deionized water. The proper amount of TiO_2 is dispersed in this solution and the resulting slurry mixed with $(NH_4)_2CO_3$ to precipitate the salts as carbonates or hydroxides. The resulting slurry is filtered, dried, and calcined to produce the desired PTC titanate. The resultant calcined agglomerate can easily be reduced to the desired sinterable powder by ball milling. Powder homogeneity and sintering reactivity are achieved by control of the precipitation and calcining conditions.

This process offers distinct cost advantages over the chemical methods described earlier, in that cheaper TiO_2 and soluble salts can be selected, purified, and introduced into the process. Establishing a consistent, dependable TiO_2 source suitable for the PTC quality required is the biggest hurdle for this process. In addition, the precipitation of the soluble salts in intimate mixture with the dispersed TiO_2 must be consistently maintained. A well-defined chemical process could be instrumented and microprocesser controlled for continuous production.

d. *Slurry Mixing Process.* The cheapest and probably the most widely used process for PTC titanate powder synthesis is the TiO_2-carbonate slurry method. The raw materials sources are carefully selected for chemical purity and dispersed in deionized water. This mixing is generally done in a ball mill using noncontaminating media to assure intimate mixing of all constituents. After milling the slurry is consolidated (filtered and dried), then calcined at 900–1150°C for sufficient time to achieve the desired titanate crystallite agglomerates. The agglomerates are reduced to the desired powder particle size and sintering reactivity, again using a noncontaminating comminution method (i.e., ball milling).

A process flow outline of the TiO_2-carbonate slurry process is shown in Figure 32. The process is simple, conducive to scale-up, and derived from long experiences with volume processing of titanates for capacitor manufacture. Slurry process powders might be considered inferior from a chemical homogeneity, particle size, and purity viewpoint. However, good-quality PTC devices (six orders PTC magnitude) have been produced with proper raw materials selection and the use of interactive process controls during device manufacture.

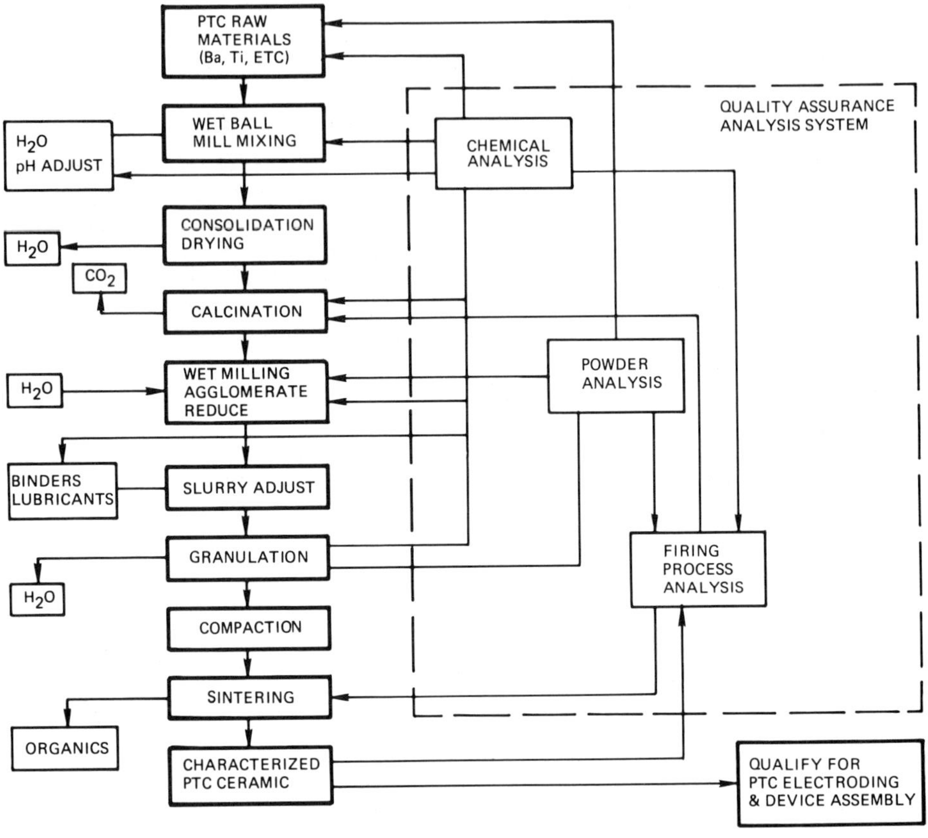

Figure 32 Typical materials inputs, unit operations, process flow, and quality assurance interrelationships for PTC element manufacture.

7. Forming

The majority of PTC devices are formed using uniaxial compaction presses standard to the electronic ceramics industry. A brief description of the preparation of powders suitable for uniformly filling the die cavity and compacting the shape is in order.

A temporary organic binder is usually required to achieve uniform density of the device shape when compaction forces are applied during the pressing cycle. It is important to select a binder system that achieves the compaction uniformity and device shrinkage control desired but does not introduce impurities or contamination that will affect PTC performance.

Quite often in large-scale manufacture, the binder addition is the last step in the agglomerate reduction operation for powder synthesis. The powder-binder slurry is then consolidated into free-flowing granules with either flash evaporation (spray drying) or drying-granulation equipment. The pressing powder granual size is controlled to achieve uniform die cavity fill.

The device compaction cycle is carefully controlled to achieve uniform density throughout the device shape. Density variations within the shape could result in differential shrinkage and undue stresses during the temporary binder removal and sintering of the device. In addition, poor compaction controls could introduce pressing defects such as voids and cracks. The reliability of such production lots would be suspect if such defects were allowed to go undetected through the remaining manufacturing operations.

Other forming methods for producing PTC devices have been used. Honeycomb PTC elements for air heaters (see Figure 33) have been made by extruding thixotropic powder-binder suspensions into long tubes and slicing individual elements from the dried honeycomb extrusion. Flat, thin elements can be produced from tape-cast sheets. Complex shapes such as special honeycomb EFE heaters (Figure 34) have been made successfully by injection molding.

Figure 33 PTC honeycomb air heater and hair dryer. (Photo courtesy of NGK.)

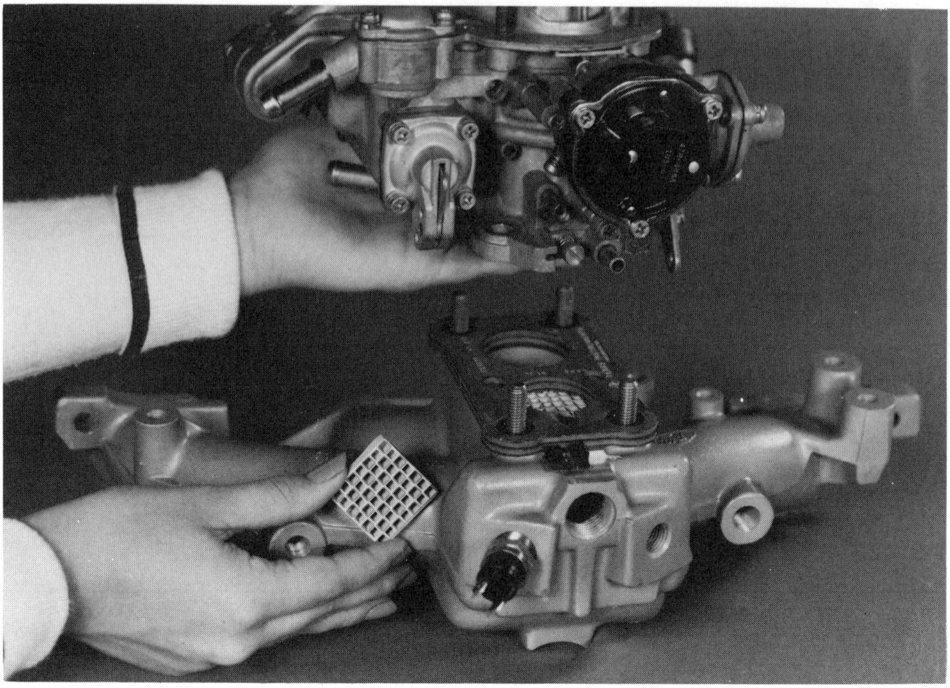

Figure 34 Low-voltage PTC honeycomb heater for automotive cold-start fuel evaporation systems. (Photo courtesy of Texas Instruments, Inc.)

PTC film devices (either thick or thin film) would be an attractive development breakthrough for device utilization. This possibility, however, has not been forthcoming because substrate contamination occurs at temperaures where the enhanced PTC properties are achieved.

8. Sintering and Microstructure Development

a. Binder Removal. The initial stage of heat treatment concerns the removal of temporary binders from the formed device. Rapid heating of the element may be all that is necessary, but some binder systems require slow heating to prevent warping, cracking, bloating, or particle rearrangements from occurring.

b. Liquid-Phase Formation. The furnace load containing the devices may then be heated up to the temperature range where the liquid phase is formed ∿1250°C [51]. It is important to establish a uniform heat balance over the furnace load, as the liquid phase is formed to assure uniform grain size formation within the device and over the whole production lot. Furnace loading and heating rate should be adjusted and controlled to assure microstructure development uniformity.

At the onset of liquid-phase formation, the densification of the device will proceed at a rapid rate. The chemistry of the system, in particular the conduction dopant level and sintering aids additions, will dictate the grain growth and densification behavior. Microstructural uniformity will be determined by the control of the sintering cycle and to an equally important degree, the chemical uniformity of the powders, even down to the individual crystallite size level.

With the formation of the liquid phase, grain growth and incorporation of the conduction dopants in the perovskite lattice occur. Continued heating to the peak temperature, ∿1300–1400°C, results in the establishment of the bulk semiconducting properties of the device. The bulk resistivity of the semiconducting grains are estimated at 0.1 ohm-cm when cooled to room temperature. The load is held at peak temperature long enough to reach a reasonable equilibrium condition where product uniformity—electrical and microstructural—is established.

c. *Grain Boundary Free Carrier Depletion or Oxidation During Cooling*. Daniels et al. [14] and others [13,15] have reported on mechanisms whereby the grain boundary region immediately adjacent to the grain boundary surface becomes depleted of carriers. This change is associated with a metastable equilibrium being established upon cooling the PTC composition from the sintering temperature through the oxidation region between 1250 and 1000°C. It is obvious that oxidation can proceed quickly along grain boundaries to establish a barrier region of very thin dimensions adjacent to the boundary interface. The net result electrically is a boundary layer–controlled device with resistivity, voltage, and frequency characteristics falling somewhere between those of a ZnO varistor and a $SrTiO_3$ boundary layer capacitor.

The free carrier depletion in the boundary layer region is dependent on the microstructure or grain size of the PTC element and the ratio between the grain boundary and bulk diffusion rates of the species that cause free carrier depletion. As pointed out earlier, the addition of modifiers allows the materials technologist to control the "oxidation" of the grain boundary region and thus optimize PTC properties.

Experimental evidence indicates that boundary layer "oxidation" is nearly complete below 1000°C. Further cooling does not affect electrical properties [except for the exponential increase in resistivity (NTC effect)] until the Curie temperature is approached. As stated in Section II.B.2, the spontaneous polarization in the domains partially compensates for the space charge at the grain boundary surfaces. A dramatic drop in apparent resistivity of the PTC ceramic is observed upon cooling through the Curie temperature T_C.

Base resistivities of commercial PTC ceramics can be varied over the range 30–10,000 Ω-cm. The maximum resistivity values (ρ_{max}) of such materials can vary depending on impurity levels, dopant/modifier concentrations, microstructure, and sintering profiles used. The general curves shown in Figure 3 give some indication of electrical characteristics being offered by PTC manufacturers.

d. Microstructure Development. Various microstructures for PTC ceramic devices are contained in Figures 35 to 37. Figure 35 illustrates the as-fired surface microstructure of an early PTC device with base resistivity in the range 400 Ω-cm. The material is quite suitable for low-voltage switch temperature-sensing applications, where voltage sensitivity and ρ_{max} requirements are not critical. The large grain size (15 μm average), low liquid phase, and large grain size variation were typical of materials used up to 1970.

A significant improvement in the microstructure for high-power devices is shown in Figure 36. The average grain size has been reduced to the 5-μm range, while the base resistivity has been

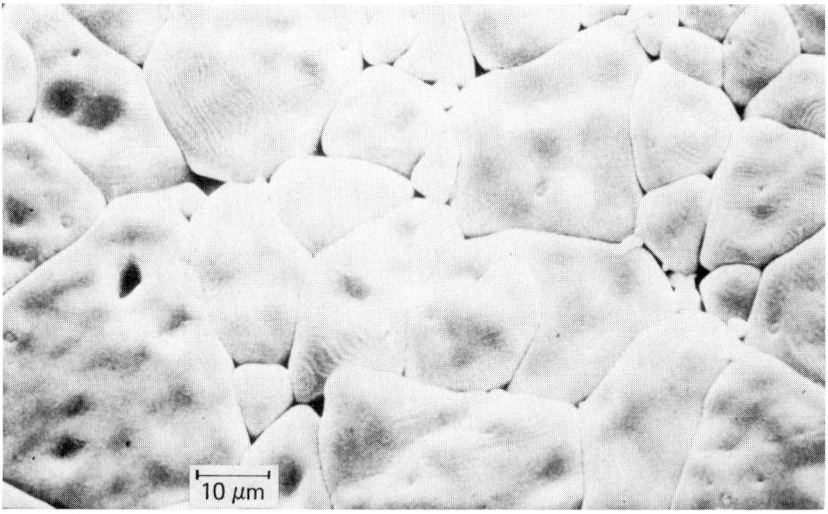

Figure 35 As-fired surface microstructure of a large-grained PTC ceramic. Large grain size would be indicative of low voltage sensitivity and modest end-point capability.

Figure 36 As-fired surface microstructure of a fine-grained PTC ceramic. Fine and uniform grain size would be indicative of high-performance PTC capability.

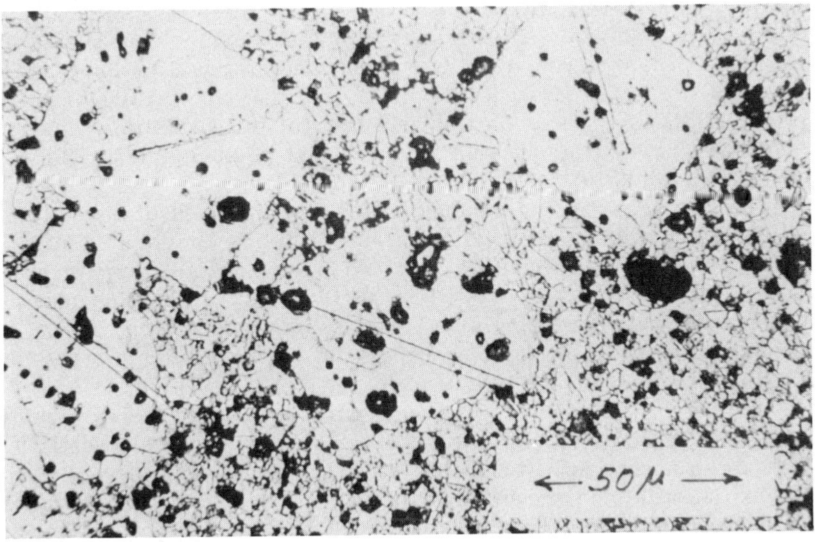

Figure 37 Polished and etched bimodal microstructure of a PTC ceramic. Duplex structure would be indicative of chemical inhomogeneities and/or density gradients during firing.

improved to below 80 Ω-cm. Improved voltage blocking capability, voltage sensitivity, and high current inrush capability were achieved with careful control of the dopants, modifiers, sintering aids, and microstructural control during sintering.

A bimodal grain size microstructure is demonstrated in Figure 37. This is indicative of chemical inhomogeneities or density gradients being present in powder compacts during sintering. Such flaws can create micropockets of resistivity variation within the PTC ceramic that could result in thermal-mechanical failures during high-power electrical excursions. High-performance PTC devices require consistently uniform microstructures and low resistivity variation [52] within the device to assure reliable performance over the expected life. Therefore, the stringent interactive process controls required to maintain PTC chemistry, purity, and microstructural control are essential for reliable high-quality device manufacture.

9. Electrical Contacts

Kulwicki [52] has ably summarized the requirements for achieving "ohmic" contacts on n-type semiconductors such as PTC ceramics. In effect, a metal with a low work function (Al, Zn) will produce an ohmic contact. High-work-function metals (Au, Pt, Pd, Ni, Cu, Ag) tend to produce rectifying or Schottky barrier contacts.

Oxide semiconductors can acquire adsorbed layers and surface states that impede the flow of electrons across the semiconductor/metal contact interface. To achieve "ohmic" behavior, this surface layer must be neutralized by either mechanical removal or chemical disruption. For n-type semiconductors, a metal with strong oxygen affinity (B, Ga, P, Ti) is particularly helpful in eliminating the adsorbed layers on PTC ceramic surfaces and achieving low resistance to electron flow.

Ohmic contacts can be achieved using combinations of metals with oxygen affinity and only partially acceptable work function. For example, ohmic Ni and Ag electrodes can be made by doping with a more active metal (i.e., Ni:B, Ni:P, Ag:Ga or Ag:Zn). The necessity for high oxygen affinity of at least a part of the electrode metal contact appears to be at least as important as the need for a low work function.

A partial list of various electrical contacts used for PTC devices is contained in Table 12. No one contact system appears to satisfy all PTC device requirements. High current capacity, solderable leads, oxidation and corrosion resistance, and low contact resistance are a few of the concerns in producing reliable, long-life contacts.

Table 12 Typical Electrode Systems for PTC Devices

Electrode system	Method of application	Typical device usage	Electrode material purpose: Ohmic	Electrode material purpose: Conduction[a]
Ni:P/Ag	Electroless Ni and electrolytic Ag plate	TV degaussers Motor assist	Ni:P	Ag
Al/Cu	Plasma or flame spray	Auto choke and degaussers	Al	Cu
Ga/In	Ultrasonic solder	Diode ovens	Ga	In
Zn:Ag/Ag	Thick-film screen Double firing	TV degaussers Current limiters	Zn:Ag	Ag
Al	Plasma or flare spray	Honeycomb heaters	Al	Al

[a]Electrode metals shown in column are in addition to ohmic electrode. These provide reliable system performance depending on connection method (i.e., soldering, pressure contacts).

10. Reliability

The ability to withstand hostile environments, mechanical and/or electrical stress, and chemical degradation is essential for proper PTC operation. To this end, life testing under realistic conditions is required to attain confidence in the reliability and life expectancy of the product. This is particularly necessary for devices that are being subjected to the stressful condition cited below.

a. Voltage/Cycling Behavior. PTC devices used for starting 110-V refrigerator motors, for example, must operate reliably at high voltage gradients and sudden thermal changes (short, high-current surge) for a 10-year (500,000-cycle) life expectancy. The presence of cracks, other defects, resistivity anomalies, and microstructure variations could have an adverse effect on reliability if the defect induces a high-enough stress level to cause premature failure. Guidelines have been developed by PTC manufacturers on what constitutes a critical flaw and the margin required to assure safe operation. The impact of other reliability considerations and their interdependence on thermal and/or electrical stressing must also be assessed.

b. Chemical Degradation. PTC devices have been known to degrade when exposed to chemical environments that can disrupt the free carrier depletion layers. The presence of reducing agents contained in plastics heated in vacuum [53] or reducing environment exposure for long periods at temperatures above 150°C range could result in a resistance decrease, significant ρ_{max} degradation, and thermal runaway under power voltage conditions. Evaluation of degradation kinetics in accelerated aging conditions is a useful method to analyze this potential hazard and obtain product lifetime prediction at the early design stages.

Other forms of resistivity instability have been observed and at present are not well understood. These include low-temperature resistance shifts caused by domain reorientation (discussed in Section IV.B.5) and diffusion of impurities.

The device performance and stability assessment of a PTC ceramic requires detailed knowledge of the thermal, mechanical, electrical, and environmental behavior that the device will encounter. The representation shown for NTC sensors in Figure 21 is equally applicable for PTC device reliability considerations.

11. PTC Applications

As discussed above, utilization of PTC ceramic materials began in the late 1950s with temperature-sensing applications and has progressed to dynamic heater applications where high energy dissipation (300 W per element) is a reality. A representation of the PTC applications tree is shown in Figure 38. PTC products are broken down into the following applications branches:

- Low-power temperature sensing
- Static heat
  - Constant-temperature heaters
  - Current/voltage control
  - Heat dissipation change
- Dynamic heat-timing designs
  - Circuit timers
  - Current timers (TV degaussing)
  - Motor assistance/starting
- Dynamic heat—automotive <15 V
  - Choke timers
  - Auto fuel evaporation
- Dynamic heat—high heat dissipation
  - Honeycomb air heater

A comprehensive treatment of device design and applications can be found elsewhere [21,24]. The following is offered as an example of how a PTC thermal heater is incorporated into a complex system design to serve a thermal sensing transducer function not easily satisfied by other thermal heating devices.

a. Application: Heater for Automotive Actuator (Carburetor Choke). Fast-acting automotive chokes require the supply of an immediate self-controlled heat source to actuate a bimetal coil. The coil actuator controls the butterfly valve of the carburation system, allowing choke control crossover from a rich to a lean fuel mixture in an internal combustion engine. The system thermal environment (outside/engine temperature), the carburetor/engine design, and heater/actuator time response are critical factors affecting cold-start engine performance and exhaust pollution control.

The PTC element can be classified in the three-dimensional grid notation as [electrical thermal, thermal]. The total automotive choke transducer system, however, is more difficult to classify but might be considered as [electrical, thermal, mechanical] for choke actuation or as [electrical, thermal, chemical] for pollution control.

Several alternative systems have been developed based on thermal, electrical, and mechanical considerations. An exhaust gas system utilizes waste engine heat by diverting exhaust gases to the carburetor choke control system. The heat transfer to the bimetal coil actuates the valve. The system is much slower than the PTC heater in that the exhaust gases take a longer time to transfer sufficient energy to actuate the bimetallic elements. A thick-film

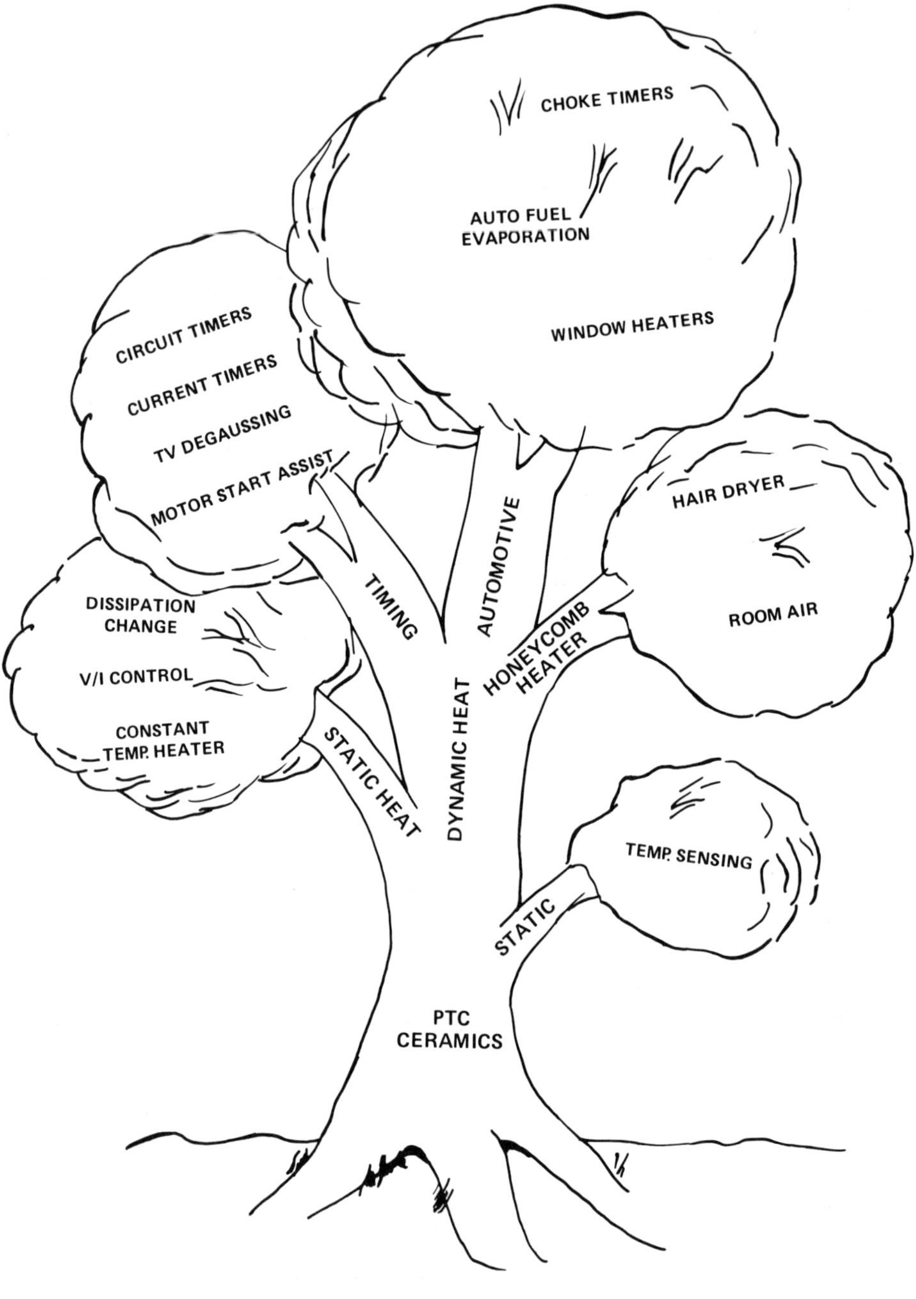

Figure 38 PTC applications tree.

heater has also been utilized in choke systems. The thermal response for these designs is an improvement over earlier designs, but the low-temperature coefficient resistors require thermal protection and still do not perform as well as PTC heater designs. Since its volume production introduction in 1972, the PTC heater/actuator system has dominated choke designs for internal combustion engines.

b. Device Performance Goals. It is desirable to heat the choke actuator system in a controlled and predictable manner. A PTC heater disk when energized by the automotive power source can generate as much as 180 W of power. The heat capacity and thermal coupling of the heater disk to the rest of the choke system and engine will dictate the duration of the high power surge.

As the PTC device is heated through the phase transformation, the sharp resistance increase results in a self-limiting power condition where the heat generated matches the heat dissipation to the rest of the system. By properly choosing the initial resistance, Curie temperature, and heat capacity of the PTC heater, a safe, predictable self-controlled heat flow to the bimetal coil can be established. A controlled thermal coupling of the heater to the actuator coil must then be designed to produce the desired automotive cold-engine-start characteristics.

Complex engine performance requirements, such as exhaust emission requirements, drivability, outside temperature, and earlier engine usage, will affect such a system design. Furthermore, the harsh environment found under the hood of a car places severe constraints on materials choices for long-life capability.

A typical automotive electric choke system utilizing PTC heaters for fast butterfly valve actuator response is shown in Figure 39. The PTC heater disk is mounted on a metal plate using a thermally and electrically conducting silver-filled polymer adhesive. The metal plate holds the bimetallic coil in the choke housing. The spring tension of the coil is adjusted by rotating the choke housing. This will achieve proper positioning of the butterfly valve to achieve the desired engine performance. For cold starting, the butterfly valve is initially closed as the PTC element is energized. Heat generated by the element flows to the bimetallic coil actuating the valve to an open condition. The air-fuel mixture is modified from rich to lean, effectively reducing hydrocarbon emissions during the critical early stages of driving. If the engine should shut down before full warm-up, the choke system is designed to match carburetor performance, engine mass, and outside temperature conditions so that smooth engine performance is assured.

Figure 39 Low-voltage PTC heater for automotive electric choke actuator. (Courtesy of Texas Instruments, Inc.)

c. PTC Device Design Solutions. The device parameters that can be adjusted to achieve optimum system performance are:

Magnitude of PTC effect
Initial resistivity of the PTC material
Contact area to the metal heat transfer plate
Curie temperature and device operating power
Time constant to reach the operating condition

These materials and dimensional considerations can be controlled to easily satisfy the self-controlled heater requirements for automotive choke control systems presently used in internal combustion engines.

d. PTC Device Manufacture. Manufacturing goals are to produce a medium tolerance (±20% R_0) PTC disk with Curie temperatures from 100 to 180°C in volume quantities and capable of automated handling in choke assembly equipment. A suitable manufacturing process for making good-quality PTC elements was outlined previously. Special attention should be given to automated sample testing and interactive process controls to meet device performance requirements. This can be met by using AQL sampling procedures since resistance tolerances and other pertinent performance parameters can be controlled.

C. Gas Sensors

Perhaps one of the most desirable but until recently least developed sensors has been the gas or chemical sensor. Potential applications in monitoring and controlling chemical reactions and combustion cycles are myriad in the chemicals, food, and energy conversion industries. Recent economic and political forces have, however, served to encourage the development of auto exhaust sensors, which presently represent the first large-scale deployment of gas sensors. In this section we examine the operating characteristics that such devices must satisfy and two of the more successful ceramic sensor designs which fulfill the stringent requirements of this application.

1. Auto Exhaust Sensors

Increased concern with pollutants over the past decade has resulted in legislation that limits allowable emissions of hydrocarbons (HC), carbon monoxide (CO), and nitrous oxides (NOx) from automobile exhausts. Aside from total engine redesign, the only practical alternative available to automobile manufacturers for satisfying emission requirements was the use of exhaust catalysts to assist in the complete combustion of the fuel and the reduction of NOx to nitrogen. As Figure 40 illustrates, a single so-called "three-way" catalyst can affect the simultaneous removal of these pollutants, but only if the engine is operated within a very narrow range of air-to-fuel (A/F) ratios centered at ∿14.5—the stoichiometric ratio. Since no conventional carburetor or fuel injection system could possibly maintain the required A/F ratio over the range of driving conditions normally encountered, efforts were directed toward developing a sensor in which the output could be utilized in a feedback mode to continuously adjust the injection system. In this manner the exhaust oxygen level could be maintained within the narrow window required by the three-way catalyst [54].

Fortuitously, the oxygen partial pressure increases nearly 15 orders of magnitude in passing through the stoichiometric A/F ratio, as illustrated in Figure 41. Consequently, in search for a practical auto exhaust gas sensor, one needs to look for some property of a material or device which exhibits a marked sensitivity to P_{O_2} excursions in this range of conditions.

In considering the types of alternative approaches possible, it is useful to reexamine the three-dimensional format illustrated in Figure 6. One could conceive of a self-generating sensor of the type [chem, y, O] or a modulating transducer of the type [x, y, chem], in which x is the input energy, y the output energy, and z the modulator. Of all the possible forms of output energy, electrical is clearly the most compatible with available electronic control systems. Although

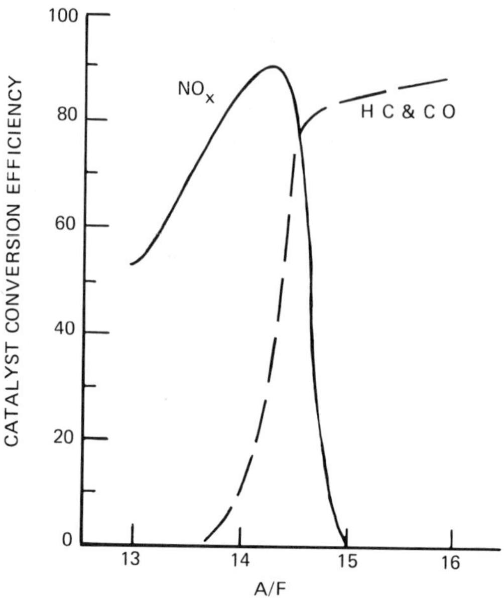

Figure 40 Three-way catalyst conversion efficiency dependence on air fuel ratio. (From Ref. 19.)

other inputs in a modulating transducer could be visualized (e.g., thermal), electrical is again the most practical given our ability to generate and control such signals readily. This therefore leaves us with only two likely alternatives: (1) Self-generating sensors [chem, electr, O], and (2) Modulating sensors [electr, elect, chem]. Each is examined in some detail in the following sections.

2. Self-Generating Galvanic-Type Oxygen Sensor

Solid ionic conductors (i.e., solid electrolytes), if maintained between appropriate gases of different activities (e.g., oxygen), will form galvanic cells in the same way that "batteries" can be formed with liquid electrolytes. In particular, if an oxygen ion conductor is used, a self-generated voltage given by the well-known Nernst equation (see Section II.C.2)

$$\Delta V = \frac{RT}{4F} \ln \frac{P'_{O_2}}{P''_{O_2}} \tag{46}$$

is induced by imposing a P_{O_2} gradient across the electrolyte. If we choose P''_{O_2} to be a fixed and known reference gas (e.g., air) and P'_{O_2} a the unknown exhaust gas, then the measured electromotive force ΔV can be used to determine the P_{O_2} of the exhaust gas at any temperature T.

To satisfy Eq. (46), the solid electrolyte must satisfy a number of requirements, including high ionic conductivity and high electronic resistivity over an extensive range of oxygen activities and temperatures which the devices will probably be exposed to during operation. To date, only stabilized zirconia (ZrO_2) ceramics have found widespread use in auto exhaust gas sensors as the oxygen electrolyte.

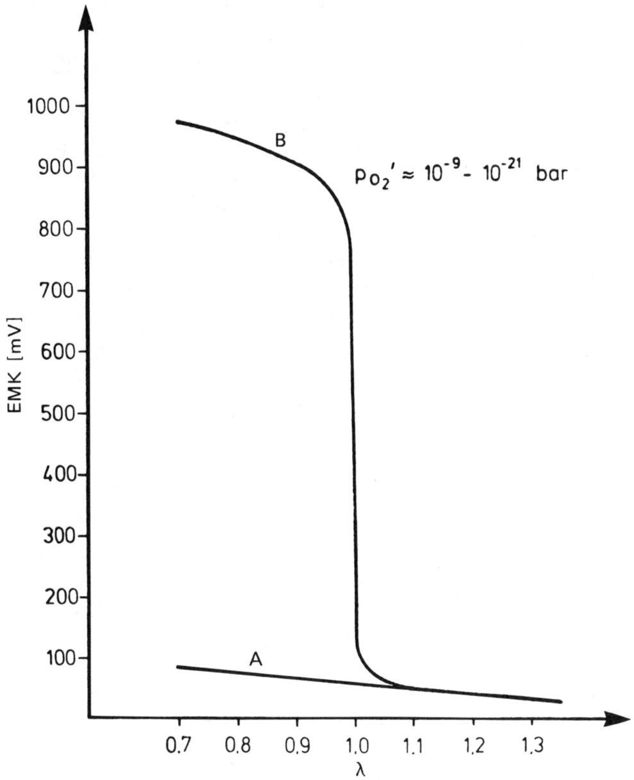

Figure 41 Theoretical dependence of electromotive force of oxygen concentration cell with and without catalytically active electrode as a function of air-to-fuel ratio: (A) catalytically inactive exhaust electrode; (B) catalytically active exhaust electrode. $\lambda = 1.0$ corresponds to stoichiometry. (From Ref. 55.)

The cubic fluorite phase is stabilized by the addition of, for example, a minimum of 7 mol % Y_2O_3 or 15 mol % CaO, which form solid solutions with ZrO_2. The lower-valent additives (i.e., Y^{3+}, Ca^{2+}) enter the lattice substitutionally on Zr^{4+} sites and are compensated for by doubly ionized oxygen vacancies, V_O'' to maintain charge neutrality. Given the high solubilities of the lower-valent cations, this results in the generation of $>10^{21}$ cm^{-3} mobile ionic species. This, combined with the relatively open oxygen sublattice, results in exceptionally high oxygen ion conductivities in these materials.

Adequate and even higher levels of oxygen ion conduction are found in other cubic fluorite oxides (e.g., CeO_2, ThO_2) as well as oxides with other crystal structures (e.g., Bi_2O_3, Ta_2O_5), as illustrated in Figure 42. However, all of these other oxides tend to become either n- or p-type semiconductors under sufficiently reducing or oxidizing conditions. In effect, this serves to partially short circuit the Nernst emf, rendering the device unreliable. Only zirconia remains predominantly ionic over the conditions of interest—hence its universal use.

For acceptable oxygen sensor manufacture, the composition and fabrication process must be chosen to optimize a number of properties. Oxygen ion conductivity, density (to ensure gas imperviousness), microstructure, grain boundary phases, impurities, mechanical strength, and thermal shock resistance are a few of the important considerations. Although Y_2O_3 is more costly than CaO, it is nearly always chosen as the stabilizer due to the notably higher ionic conductivity (about one order of magnitude) it induces in the zirconia relative to the CaO stabilized ceramic. This becomes particularly important at the lower operating temperatures of the device ($\sim$350°C) [55].

Small percentages of other additives, such as silicates, are often added to accelerate the degree of densification at temperatures below that required for purer materials. Dueker et al. [55] report that the liquid phase thus formed during sintering results in a denser and stronger structure. Furthermore, they found that the ionic conductivity initially increased and reached a maximum at 0.5% silicate flux addition. Above that value the conductivity dropped, presumably due to the increased proportion of poorly conducting, secondary phases at the grain boundaries. Beeckmans and Heyne [56] further suggested that the presence of such secondary phases could also markedly degrade the response time of such devices. Clearly trade-offs must be made in selecting additives so as to maintain a balance between optimum electrical conductivity, mechanical strength, response time, cost, and reliability.

A schematic view of a practical zirconia oxygen sensor is illustrated in Figure 43. The ceramic is formed into an impervious tube which serves to separate the air reference gas in the inside of the

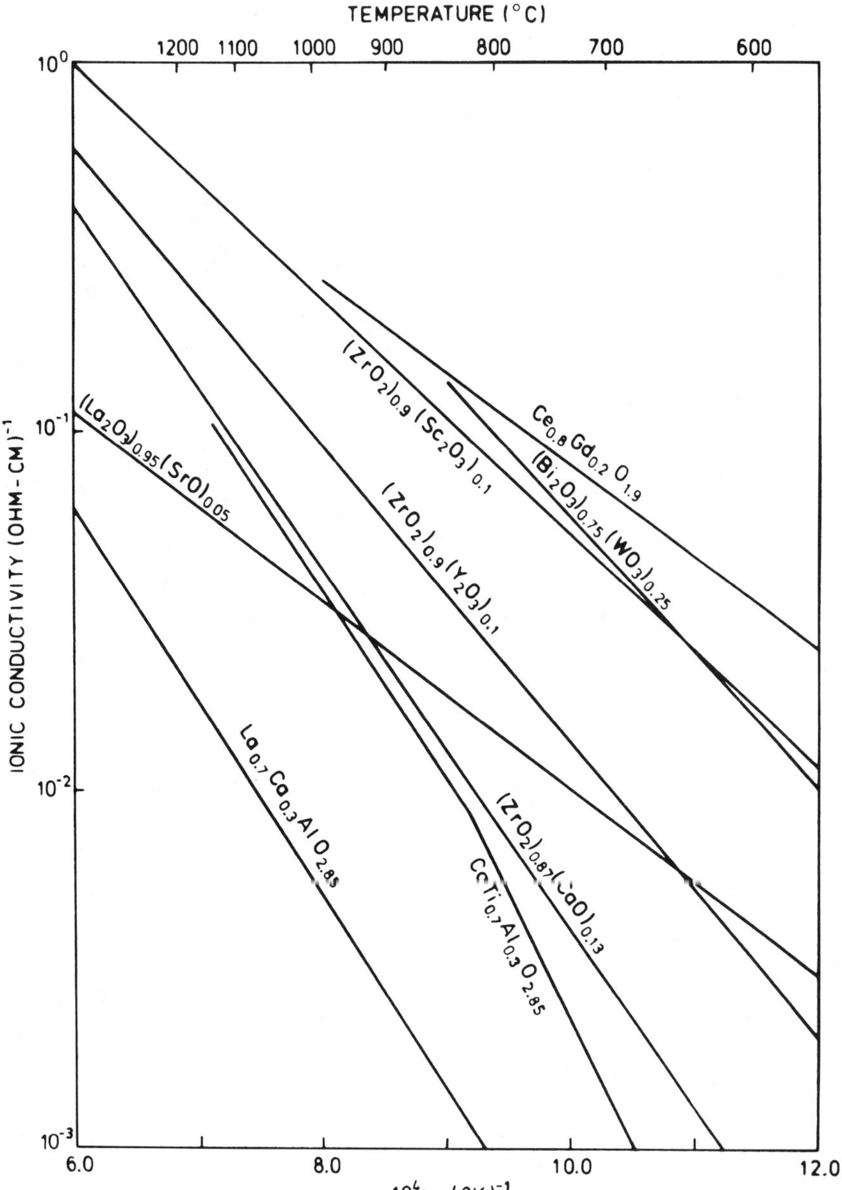

Figure 42 Temperature dependence of oxygen ion conductivity for a number of oxides considered for use in oxygen sensors. (From Ref. 60.)

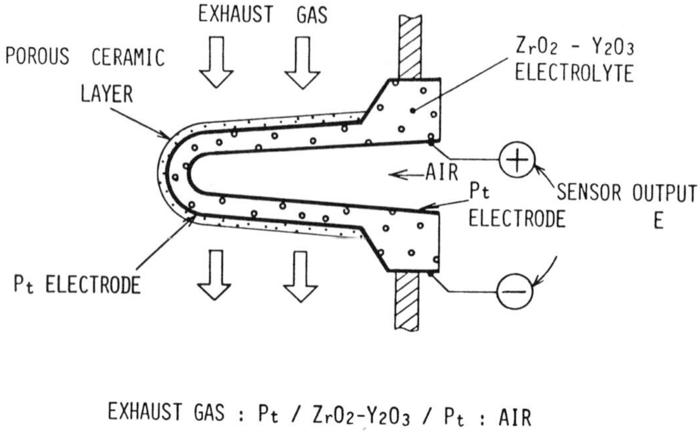

Figure 43 Schematic view of a practical zirconia auto exhaust sensor. (Courtesy of Toyota Motors.)

tube from the auto exhaust gas on the outside. Both inner and outer surfaces of the tube are coated with porous layers of platinum which serve dual purposes. First, the metallic conducting layer enables measurement of the induced Nernst emf, which is the result of reactions of the form

$$\tfrac{1}{2}O_2 + 2e' \leftrightarrow O^{2-} \tag{47}$$

at the three phase boundaries (i.e., oxide: platinum: gas). Second, it is essential that the electrodes aid in catalyzing the reactions illustrated in Figure 43. If residual unreacted gases (HC, CO, and oxygen) come in contact with the active three-phase boundary at the electrolyte, no sharp change in sensor output is observed when traversing the stoichiometric A/F ratio, as illustrated in Figure 41, curve A. This illustrates the importance of the device-electrode interface, not only in the operation of this type of sensor, but also in other types of gas sensors.

One last interesting feature of such devices is the use of a porous layer of an inert oxide such as spinel to cover the outer platinum electrode. Such protective layers reduce the penetration of exhaust gas constituents such as lead, halogen, sulfur, and phosphorus compounds, which can poison the platinum catalyst [58]. It has also been suggested that by slowing the rate at which the exhaust gases reach the active interface, they have more time to reach the state of equilibrium required for proper operation of the device.

3. Modulating Resistive-Type Oxygen Sensor

Previously, we demonstrated that the electronic carrier density can be a relatively strong function of the oxygen partial pressure due to variations in the stoichiometry of the compound. In general, the carrier density and thus the electrical conductivity will exhibit a P_{O_2} dependence given by

$$\sigma = A(T)P_{O_2}^{\pm 1/x} \tag{48}$$

in which the coefficient $A(T) = A_0 \exp(-E/kT)$ includes defect formation and ionization terms and the factor x typically ranges between 4 and 6, depending on the nature of the ionic defects and their ionization state. A positive exponent, corresponding to p-type semiconductivity, is obtained for oxides that tend toward oxygen excess, while a negative exponent, corresponding to n-type semiconductivity, is found for oxides that tend toward oxygen deficiency. Given the approximately 15 orders of magnitude change in P_{O_2} expected at the stoichiometric A/F ratio, an easily detected change in resistance of nearly three to four orders of magnitude can be expected from a resistive type of device. The two oxide systems that have been most extensively investigated for such applications have included $Co_{1-x}Mg_xO$ ($x < 0.85$) [57], a p-type conductor, and TiO_2 [19], an n-type conductor.

As in the galvanic-type sensor a number of materials requirements must be satisfied for proper operation. These requirements are summarized in Table 13. A strong P_{O_2} dependence is self-explanatory,

Table 13 Materials Requirements for Resistive Exhaust Gas Oxygen Sensors

1. High electronic conductivity strongly dependent on oxygen partial pressure
2. Small temperature coefficient of resistance
3. High mobility of oxygen-related lattice defects (e.g., oxygen vacancies)
4. High porosity and small particle size (for ceramics)
5. High catalytic activity and good oxygen-transfer characteristics
6. Wide oxygen partial pressure and temperature range of stability

Source: Ref. 19.

while a high electronic conductivity generally minimizes polarization effects and simplifies measuring circuitry.

One of the major drawbacks of the resistive type of device is its relatively strong temperature dependence [see Eq. (19)]. Thus even if the P_{O_2} is held constant, a resistance reading will change with variation in exhaust temperature. For TiO_2, with an activation energy of ∿1.6 eV, the TCR at 400°C amounts to ∿4% per °C. Given the often large temperature excursions encountered in exhaust manifolds, this could easily render the device useless.

Two approaches have been proposed to overcome this difficulty. The first is to self-heat the device, with integrated resistive elements, to a temperature well above the normal excursions encountered in the exhaust. Aside from somewhat complicating the design of the device, it also requires the expenditure of electrical energy and more careful device design to maintain the device at a desired elevated temperature.

A second, more innovative approach, has been proposed by Ford Motor Company [19]. In this design, two elements are utilized: an oxygen-sensing porous TiO_2 ceramic and a densified TiO_2 ceramic insensitive to P_{O_2} variations but used as a matching thermistor for temperature compensation. By using these elements in a voltage-divider circuit, any resistance shifts induced by temperature changes are canceled out by near equivalent shifts of resistance in the other.

The ability to prepare two such different elements from the same material is related to items 3 and 4 in Table 13. We earlier emphasized that the P_{O_2} dependence of the electronic carrier density is tied to changes in the bulk composition of the material. Consequently, the achievement of a new equilibrium value for the resistivity after a redox reaction relies on the diffusion of ions from the surface of the element to its interior, or vice versa. For the response time of the sensor to be within reasonable limits, high ion mobilities are required (item 3, Table 13). An acceptable response time can be achieved by shortening the diffusion length (i.e., the use of small grain size and high-porosity ceramics). What the work at Ford did was to ensure that the sensing element satisfied these requirements but not the compensating thermistor, which was prepared in a highly densified form. The microstructures of these TiO_2 sensing elements are compared in Figure 44. The successful application of these principles are illustrated in Figure 45, which shows nearly the same output change at the stoichiometric A/F ratio at 350 and 850°C. Items 5 and 6 in Table 13 apply to galvanic sensors as well and have already been discussed in that section.

In comparing the two types of sensors (i.e., galvanic versus resistive), one finds advantages with each approach. Because the output of the galvanic sensor is self-generated, no auxiliary

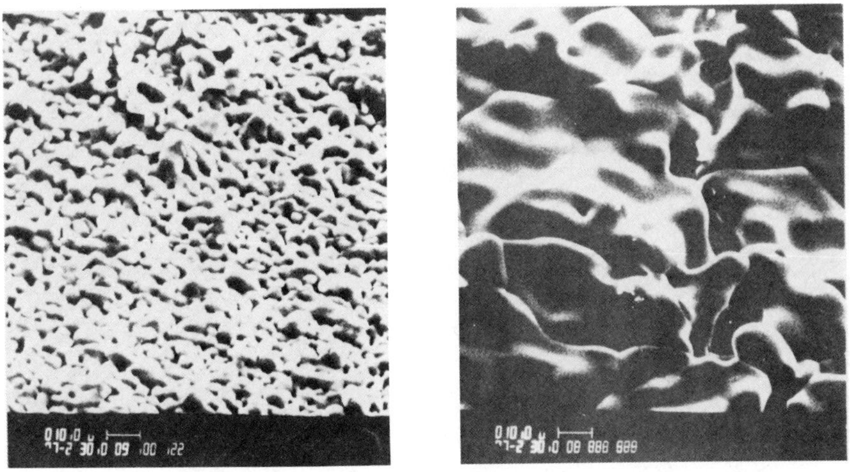

(a) (b)

Figure 44 Scanning electron microscope photograph of TiO_2 sensor element (a) and the compensating thermistor (b) (Magnifications × 700.) Note the considerably lower density and grain size of the oxygen sensor compared to the thermistor. (From Ref. 19.)

circuitry is required for its operation as is the case for the resistive type. Furthermore, since the output is thermodynamically determined (assuming equilibrium), no periodic calibration is necessary. Because the resistance of a ceramic element is affected not only by bulk properties but also by microstructure, cracks, and so on, a drift in resistance with time for given temperature and P_{O_2} is possible and periodic recalibration might be required. Finally, since the galvanic output is linearly rather than exponentially dependent on temperature, no special temperature-compensation techniques are generally necessary.

The main advantages of the resistive device are related to its simplicity. Because no reference gas is necessary, one may dispense with the relatively large dense tube design of the zirconia cell and its associated high-temperature sealing features. Instead, a small bead of material with embedded leads is all that is necessary. Esper et al. [58] also report that precious-metal impregnation of the porous ceramic provides adequate abrasion resistance without need for a protective ceramic overcoat.

In summary, several working designs have been developed for large-scale application of gas sensors in the demanding environment of an automobile exhaust manifold. Employing the three-dimensional

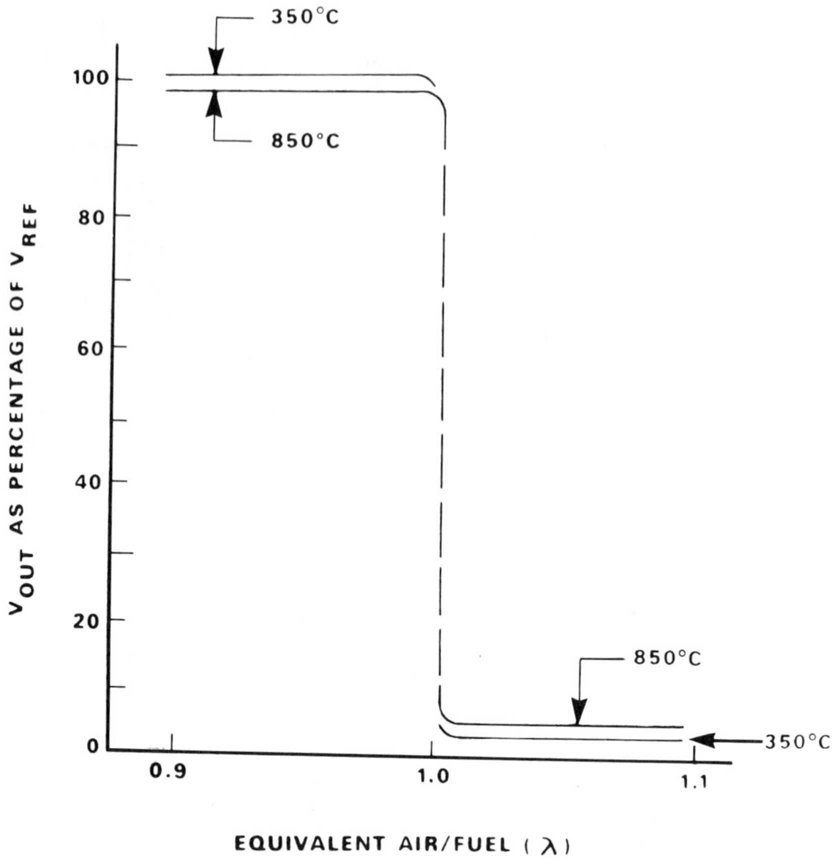

Figure 45 Dependence of TiO_2 sensor voltage on air-to-fuel ratio. Note weak dependence on operating temperature. (From Ref. 19.)

approach illustrated in Figure 6, devices working on different principles could be identified. Application of fundamental chemical and physical principles together with innovative materials development and fabrication were shown to lead to several workable alternatives, each with its own special features. Rapid progress in other sensor technologies is likely to proceed along similar lines in the future.

V. FUTURE PROSPECTS

The march toward the development and use of more complex and efficient processes and devices in the near future is inevitable.

Demands for feedback control systems with increasing levels of sophistication must therefore grow in parallel. Given the rapid progress experienced in the development of inexpensive signal processors over the last decade, attention must now focus on the development of complementary low-cost, compact, and sophisticated input and output transducers. Given their marked chemical and mechanical stability and low-cost potential, ceramic sensors are found to play a major role in future developments.

In this chapter we have attempted to provide the reader with both a basis for understanding and predicting the operation of ceramic sensors as well as an appreciation of the complexities that go into the fabrication process and subsequent device analysis. This follows the belief that rapid future sensor device development will proceed only if a satisfactory synthesis of device theory and processing is achieved.

Aside from their relative stability and low cost, ceramics exhibit an exceptionally wide range of physical properties as exemplified in other chapters of this book. One can therefore expect that as the need for different types of transducers expands (see three-dimensional matrix in Figure 6), ceramics, often with compositions or structures unlike those present in today's sensors, will provide the most cost-effective solutions.

APPENDIX: THERMOPHYSICAL PROPERTIES

The physical size of a sensor, its thermophysical properties, and the coupling to its surrounding environment determines how energy stimuli interact with the sensor body. This sensor/energy interaction integrated over a finite time span results in electronic, optical, and/or mechanical responses. The sensor's chemical constitution, its crystallinity, and microstructure will dictate the magnitude of the response.

Heat Flux Sensors Under Low Electrical Fields

Consider a cylindrically shaped sensor of diameter ϕ and height ℓ, with a low electrical potential maintained across its output or measurement contacts. The sensor surfaces at temperature T_1 respond to a heat flux q from an infinite heat sink environment at temperature T_2. Assume initially that the sensor has a very high thermal conductivity, k_s, so that the heat flux determined by the interfacial heat transfer coefficient, h, between the heat sink and the sensor interface will dictate the time-temperature response of the sensor.

Equation (49) is a standard heat transfer differential used for determining time (t)—temperature response. The sensor parameters include volume V, exposed surface area A, density ρ, and specific heat C_p.

$$dq = Ah(T_2 - T)\ dt = V\rho C_p dT \tag{49}$$

Upon integration for the boundary conditions above, this reduces to

$$t = \frac{V}{A}\frac{\rho C_p}{h}\ln\frac{T_2 - T_1}{T_2 - T} \tag{50}$$

where t is the time required to heat the sensor from its initial temperature to the instaneous temperature T, after introducing the sensor into a chamber at temperature T_2.

A special condition arises when $\ln[(T_2 - T_1)/(T_2 - T)]$ equals unity. This time-response value, known as the time constant, τ_{TC}, is defined as the time required for a body to reach 63.2% of the final temperature change after the step change in environmental conditions. For cylindrical sensors this is expressed by

$$\tau_{TC} = \frac{\phi \ell}{2\ (\phi + 2\ell)}\ \frac{\rho C_p}{h} \tag{51}$$

Equation (51) has configurational ($\phi \ell$), heat transfer (h), and materials, (ρ, C_p) aspects to the expression. This allows one to systematically evaluate how dimensional variations will affect the sensor time response. For example, the effect of sensor diameter on τ_{TC} is shown in Figure 46.

The plot and notations shown in Figure 47 illustrate how time response is influenced by the diameter-to-height ratio, x. For thin planar devices ($5 < x > 2.0$), τ_{TC} is considerably reduced below that for rods or spherically shaped devices. It is also interesting to note that for the boundary conditions imposed so far, the diameter-to-height ratio time response curve is dependent of cylindrical mass.

Equation (51) is helpful in assessing device behavior for constant x in order to optimize sensor dimensions for desired τ_{TC}. Sensor and probe assembly designs can also be optimized with respect to thermal capacity and heat transfer coefficient to achieve time-response goals (see Section III.B.3).

Practical sensor devices and probes do not necessary approximate the boundary conditions assumed under Eqs. (49) to (51). Ceramic

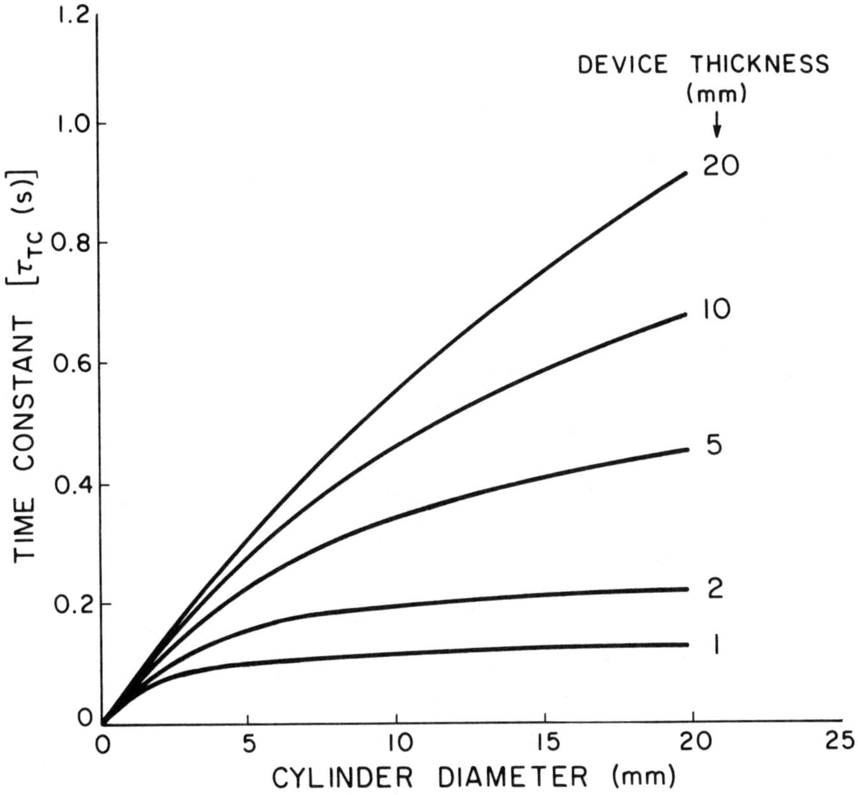

Figure 46 Effect of cylindrical device dimensions on sensor thermal time response.

sensor materials usually possess low thermal conductivity and provide considerable thermal resistance, R_s, to heat flux. The equations above require modification to predict actual sensor probe thermal response.

Consider a cylindrical plate sensor configuration $(2 < \times > 4)$ that is approaching thermal equilibrium at the chamber temperature T_2. A one-dimensional analysis, assuming near-steady-state conditions, approximates the expression

$$\frac{q}{A} = \frac{dT}{\Sigma R} = \frac{dT}{R_i + R_s} = \frac{dT}{1/h + \ell/2k_s} \qquad (52)$$

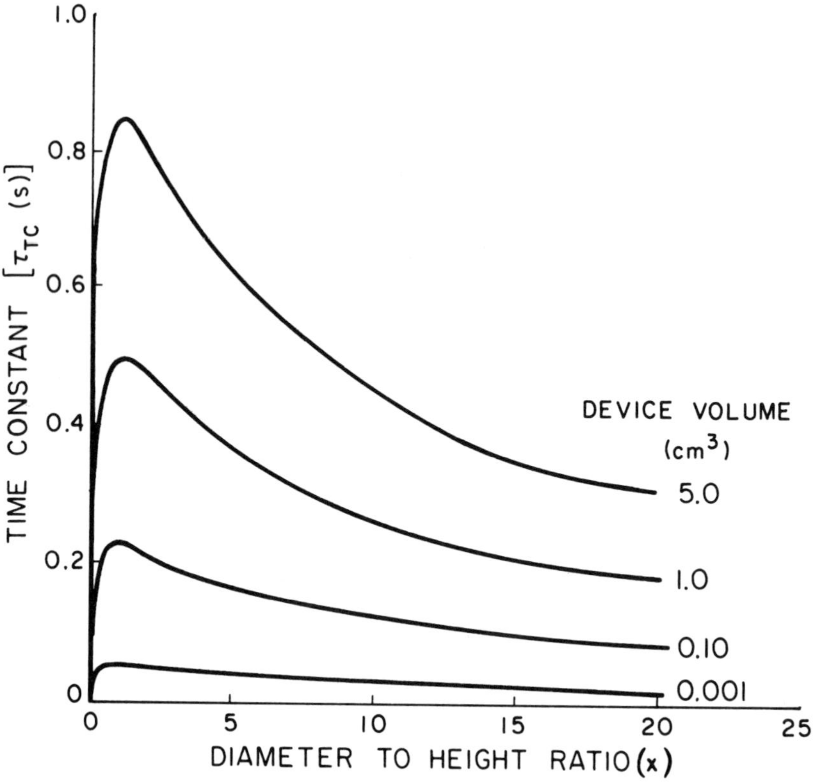

Figure 47 Effect of diameter-to-height ratio on sensor time response for different device volumes.

where the overall thermal resistance to the heat flux q is the sum of the interfacial barrier (R_i = 1/h) and the sensor resistance (R_s = $\ell/2k_s$) to the plate centerline. The term k_s is the thermal conductivity of the sensor material.

Consider how q/A varies with changes in thermal resistance of the sensor material. For very large k_s and/or small ℓ, the term 1/h will dominate the thermal resistance term, ΣR, and govern the time response of the system. The real world, however, indicates that sensor thermal resistance can have an appreciable and often predominant role in time response. Figure 48 illustrates how sensor thickness influences the heat flux in bare plates immersed in different thermal environments. The curves have been normalized with respect to h to be roughly equivalent to a still air environment transmitting 0.008 cal/sec-°C to the sensor interface. A material thermal conductivity typical of ceramic sensors (0.015 cal/sec-cm-°C) was used.

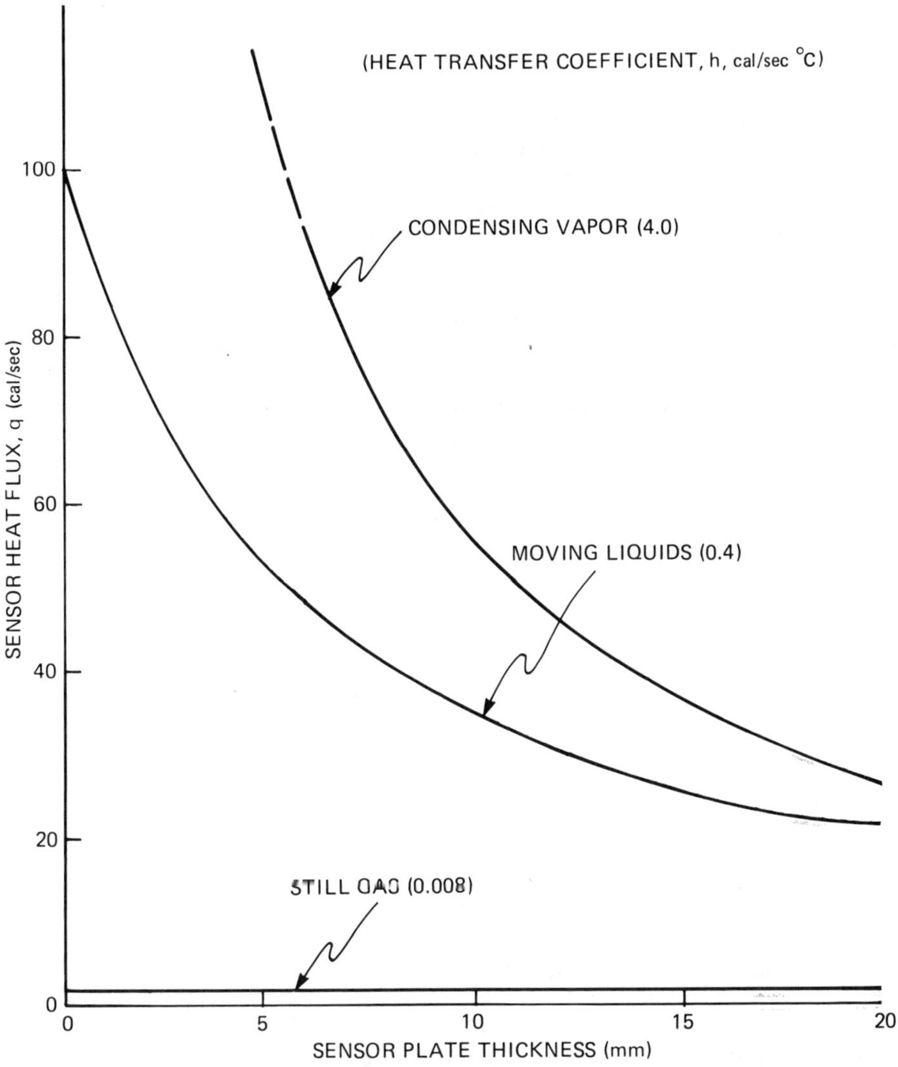

Figure 48 Influence of sensor thickness on heat flux for sensors in different thermal heat transfer environments. Note that a still air environment is considered unity. Values in parentheses are estimates of the heat transfer coefficient h for the heat transfer environment listed.

The curves derived from Eq. (52) demonstrate that sensor thicknesses greater than 0.10 cm inhibit heat transfer and govern t_{TC}, particularly for high heat flux conditions.

The near-steady-state heat flux behavior discussed above does not adequately describe variable dq/dt conditions outlined in Eq. (49). Advanced heat transfer studies, in particular finite element analysis, have become popular tools for sensor design and optimization. This method, properly executed, can result in significant development cost/time savings and improved designs.

Sensors Subjected to Joule Heating

Sensors in an isolated environment when subjected to direct internal heating exhibit a two-step thermal response as depicted in Figure 48. Initial warmup is comprised of uniform heating of the cylindrical device, where

$$dq = \frac{V^2}{R} = \frac{\pi\phi^2 \ell C_p \Delta T}{Ah} + \text{heat dissipation} \tag{53}$$

When thermal dissipation matches power generated, Eq. (54) is operative:

$$\text{Power} = \frac{V^2}{R} = K\ (T_s - T_2) \tag{54}$$

where V, I, and R are device electrical parameters, K is the dissipation constant (W/°C), and T_s is the operating temperature of the powered sensor element.

The heat transfer between the environment and the energized device is governed by similar factors expressed in earlier equations. Energized devices, however, will self-heat and dissipate that energy to the environment. Equilibrium is established when energy generated equals the energy dissipated. Time response to equilibrium for zero-power sensors could be considerably different compared to equivalent self-heated devices. Temperature differential, power level, thermal dissipation resistance, and sensor thermal mass are a few of the factors governing powered time response.

Sensor Modifications

Consider the impact of coupling the sensor device, S, by surrounding it with a coating material, C, of thickness ℓ_c. The thermal inertia (ΣR) of this structure would respond differently when exposed to the infinite heat sink described earlier. Composite materials properties (heat capacity, thermal conductivity) and configurational conditions making up the sensor probe would require expansion of the heat transfer equation (52) and different dissipation to the infinite heat sink. The physical construction of the sensor and coating (see Figure 49) and their relative masses complete the time-temperature response picture. A variety of materials and configurational options are feasible and can result in sensor designs whose time responses can be varied over wide limits. Examples of different sensor probe assembly designs were shown earlier.

Heat Transfer

Heat transfer across interfaces to the environment can play an important role in sensor device performance. Table 14 illustrates typical heat transfer coefficient values across boundaries from heat sink environments.

Careful consideration of interfaces between different states of matter are therefore required when analyzing thermal transient phenomena.

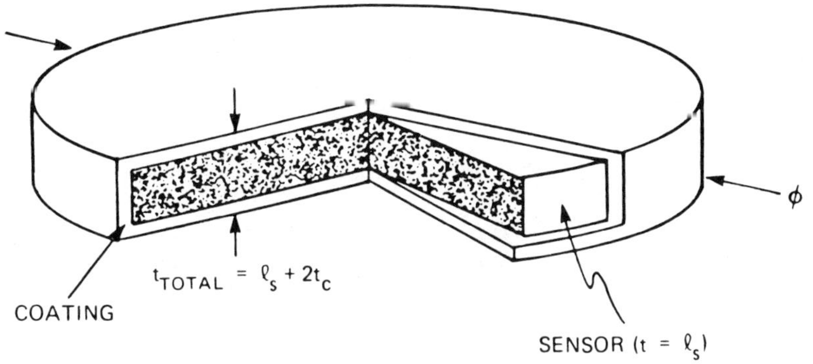

Figure 49 Cylindrical sensor with thin coating for thermal resistance modeling. For $\phi/\ell_s > 2.5$, $t_c < 5\ \ell_s$; $\Sigma R \approx 1/h + t_c/k_c + \ell_s/k_s$.

Table 14 Heat Transfer Across Different Interfacial Boundaries

Heat sink condition	Heat transfer coefficient (h) Range (cal/sec-°C)
Still gases	0.004–0.021
Moving gases	0.008–0.21
Moving liquids	0.125–4.10
Boiling fluids	0.800–200
Vapor condensation	2.0–20
Polymer/glass film[a]	0.14–0.50
Metals interface[a]	2000–20,000

[a]Configuration is a 0.01-cm-thick material interface.

REFERENCES

1. F. A. Kröger, *The Chemistry of Imperfect Crystals*, 2nd ed., North–Holland, Amsterdam, 1974.
2. R. H. Bube, *Electronic Properties of Crystalline Solids*, Academic Press, New York, 1974.
3. E. J. Vervey, P. W. Haaijman, F. C. Romeijn, and G. W. Oosterhant, *Philips Res. Rep.*, *5*, 173 (1950).
4. H. L. Tuller, Chapter 8 in this volume.
5. M. H. Sukkar and H. L. Tuller, *Adv. Ceram.*, *7*, 281 (1984).
6. R. Wernicke, *Phys. Status Solidi (a)*, *47*, 139–144 (1975).
7. S. M. Sze, *Physics of Semiconductor Devices*, Wiley-Interscience, New York, 1969.
8. J. Appel, *Solid State Phys.*, *21*, 193 (1968).
9. H. L. Tuller and A. S. Nowick, *J. Phys. Chem. Solids*, *38*, 859 (1977).
10. J. S. Kasper, in *Solid Electrolytes*, P. Hagenmuller and W. Van Gool, eds., Academic Press, New York, 1978, p. 217.
11. U.S. Patent 2,091,259, Aug. 31, 1937.
12. G. E. Pike and C. H. Seager, *J. Appl. Phys.*, *50*, 3414 (1979).
13. W. Heywang, *J. Am. Ceram. Soc.*, *47*, 484 (1964).
14. J. Daniels, K. H. Hardtl, D. Hennings, and R. Wernicke, *Philips Res. Rept.*, *31*, 487 (1976).
15. H. Ihrig and W. Puschert, *J. Appl. Phys.*, *48*, 3081 (1977).
16. H. L. Tuller, in *Nonstoichiometric Oxides*, O. T. Sorensen, ed., Academic Press, New York, 1981, p. 271.

17. H. Windischmann and P. Mark, *J. Electrochem. Soc., 126*, 627 (1979).
18. M. H. Sukkar, H. L. Tuller, and K. H. Johnson, in *Grain Boundaries in Semiconductors*, H. J. Leamy, G. E. Pike, and C. H. Seager, eds., North–Holland, Amsterdam, 1982, p. 1410.
19. E. M. Logothetis, *Ceram. Eng. Sci. Proc., 1*, 281 (1980).
20. S. Middelhoek and D. J. W. Noorlag, *Sensors Actuators, 2*, 29 (1981/82).
21. H. B. Sachse, *Semiconducting Temperature Sensors and Their Applications*, Wiley, New York, 1975.
22. H. Futaki, *Jap. J. Appl. Phys., 4*, 28 (1965).
23. B. M. Kulwicki, *Adv. Ceram., 1*, 138 (1981).
24. E. D. Macklen, *Thermistors*, Electrochemical Publications, Ayr, Scotland, 1979.
25. H. Futaki, *Hitachi Hyoron, Extra No. 30*, Sept. 1962.
26. I. T. Sheftel, E. V. Kurlina and G. N. Tekster–Proskuryakova, *Solid State Phys. (USSR), 3*, 2726 (1961). (See the References for additional articles.)
27. U.S. Patent 2,258,646, Oct. 14, 1941.
28. J. A. Becker, C. B. Green, and G. L. Pearson, *Trans. Am. Inst. Electr. Eng., 65*, 711 (1946).
29. H. W. Trolander and J. J. Sterling, *IRE Trans. Bio-Med. Electron., BME-9*, 142 (1962).
30. H. W. Trolander, D. A. Case, and R. W. Harruff, *Temp. Meas. Control Sci. Ind., 4*, 997 (1972).
31. G. G. Larson, R. J. Arnott, and D. G. Wickham, *J. Phys. Chem. Solids, 23*, 1771 (1962).
32. D. J. Shanefield and R. E. Mistler, *West. Electr. Eng, 15*, 26 (1971).
33. A. Ikegami, H. Arima, H. Tosaki, Y. Matsuoka, M. Ai, H. Minorikawa, and Y. Asahino, *Proc. IEEE Comp. Conf., 80CH 1568-5*, 1980, p. 98.
34. H. Arima, A. Ikegami, K. Abe, S. Iwanaga, and T. Isoga, *Proc. 1980 Int. Microelectr. Symp.*, 1980, p. 272.
35. Technical Literature, Victory Engineering Inc., Somerville, N.J.
36. U.S. Patent 2,633,521, Mar. 31, 1953.
37. W. D. Kingery, H. K. Bowen, and D. R. Uhlmann, *Introduction to Ceramics*, 2nd ed., Wiley, New York, 1976.
38. K. Okazaki, *Ceramic Engineering for Dielectrics*, Gakken–sha, Tokyo, 1978.
39. Technical Literature, Fenwal Electronics, Inc., Framingham, Mass.
40. U.S. Patent 3,316,765, May 2, 1967.
41. Netherlands Patent 631,321, May 23, 1951.
42. U.S. Patent 3,927,300, Dec. 12, 1975.
43. E. Andrich, *Electr. Appl., 26*, 123 (1965–66).
44. R. C. Bradt and G. S. Ansell, *J. Am. Ceram. Soc., 52*, 192 (1969).

45. W. Y. Howng and C. McCutcheon, *Am. Ceram. Soc. Bull.*, *62*, 231 (1983).
46. T. Fukami and H. Tsuchiya, *Jpn. J. Appl. Phys.*, *18*, 735 (1979).
47. Technical literature, E. I. DuPont, Wilmington, Del.
48. W. S. Clabaugh, E. M. Swiggard, and R. Gilchrist, *J. Res. Natl. Bur. Std.*, *56*, 289 (1956).
49. U.S. Patent 3,983,077, Sept. 28, 1976.
50. O. Saburi and W. Wakino, *IEEE Trans. Component Parts*, *CP-10*, 53 (1963).
51. H. Ueoka and M. Yodogawa, *IEEE Trans. Manuf. Technol.*, *MFT-3*, 77 (1974).
52. B. M. Kulwicki, *J. Phys. Chem. Solids*, 45, 1015, 1984.
53. K. Okazaki, *Adv. Ceram.*, *1*, 23 (1979).
54. P. McGeehin, *J. Brt. Ceram. Soc.*, *80*, 37 (1981).
55. H. Dueker, K. F. Fiese, and W. D. Hoecker, Paper 750, *SAE Congr.*, Detroit, Mich., Feb. 1975, p. 223.
56. N. M. Beeckmans and L. Heyne, *Electrochim. Acta*, *21*, 303 (1976).
57. K. Park and E. M. Logothetis, *J. Electrochem. Soc.*, *124*. 1443 (1977).
58. E. J. Esper, E. M. Logothetis, and J. C. Chu, Paper 790, *SAE Congr.*, Detroit, Mich., Feb. 1979, p. 140.
59. L. Heyne, in *Solid Electrolytes*, S. Geller, ed., Springer–Verlag, Berlin, 1977, p. 169.
60. R. M. Dell and A. Hooper, in *Solid Electrolytes—General Principles, Characterization, Materials, Applications*, P. Hagenmuller and W. Van Gool, eds., Academic Press, New York, 1978, p. 291.
61. J. Daniels and R. Wernicke, *Philips Res. Rep.*, *31*, 544 (1976).

6

Application and Characterization of ZnO Varistors

LIONEL M. LEVINSON and HERBERT R. PHILIPP
General Electric Company, Schenectady, New York

I. INTRODUCTION

Metal oxide varistors are novel ZnO-based ceramic semiconductor devices with highly nonlinear current-voltage characteristics similar to back-to-back zener diodes, but with much greater current- and energy-handling capabilities. The study and use of these materials represent a rapidly developing field in the technology of electronic oxide ceramics.

The varistors are produced by a ceramic sintering process which gives rise to a structure comprised of conductive ZnO grains surrounded by thin insulating barriers. In this chapter the physical properties of the phases constituting the resulting three-dimensional series-parallel junction network are discussed, and the novel electrical properties interpreted in terms of the material microstructure and grain-to-grain conduction mechanism.

The applications of ZnO varistors are predominantly in the field of circuit overvoltage protection, although some uses of the material as an active circuit element have been described [1,2]. The current overvoltage protection application derives from the fact that circuits and systems containing low-cost, high-reliability solid-state components are now frequently designed for use on 120-V ac power lines. It is generally accepted, however, that proper design practices must take into account the fact that solid-state components cannot in general withstand the amount of overvoltage imposed on the circuitry by typical power system transients. Thus a problem that was of little consequence in the heyday of vacuum-tube technology is becoming increasingly acute.

The magnitude of the problem faced is graphically illustrated in Table 1, which gives surge data from a survey on the nature of surge voltage transients in residential power circuits. The data, collected from a variety of homes in urban and rural locations in New York, Florida, and South Carolina, indicate that surges in the 1000-V range and above are prevalent with a frequency of the order of several surges per day. Such transients usually result from switching surges or lightning storms and can play havoc with unprotected solid-state components connected to a power line. Product reliability therefore requires that the solid-state component be provided with adequate protection from these transient overvoltages.

In this chapter we assess the current status of a relatively new and generally superior [4] transient protection device, the ZnO varistor. The ZnO varistor, often called a "metal oxide varistor," was initially developed in Japan [5] and has been available in the United States since 1972 under the trade name GE-MOV® varistor. Most of the data presented here will relate to measurements on GE-MOV® varistors, but we wish to emphasize at the outset that by and large, all ZnO varistors will exhibit analogous behavior.

We shall proceed as follows. After a description of the electrical characteristics of the varistor, we discuss the microstructure and fabrication of the devices. Varistors have an interesting and rich behavior as circuit elements and we describe briefly the effect of frequency and voltage on the varistor impedance. Finally, we discuss our current understanding of the mechanism of the highly nonlinear current-voltage characteristic of ZnO ceramic varistors.

II. VARISTOR ELECTRICAL CHARACTERISTICS

ZnO varistors are ZnO-based ceramic semiconductor devices with highly nonlinear current-voltage characteristics similar to back-to-back zener diodes but with much greater voltage-, current-, and

Table 1 Surge Voltages Observed in a Variety of Urban and Rural Homes in New York, Florida, and South Carolina

Location number	Most frequent surge Crest (V)	Most severe surge Duration (µsec or cycles)	Average surge per hour	Crest (V)	Duration (µsec or cycles)
1	300	10 µsec	0.07	700	10 µsec
2	500	20 µsec	0.14	750	20 µsec
3	300	1 cycle	0.05	600	1 cycle
4	300	2 cycles	0.2	400	2 cycles
5	250	1 cycle	0.01	400	1 cycle
6	800	1 cycle	0.03	1800	1 cycle
7	300	4 cycles	0.1	1200	10 µsec
8	1500	1 cycle	0.2	1500	1 cycle
9	2000	1 cycle	0.4	2500	1 cycle
10	1500	1 cycle	0.15	1500	1 cycle
11	1400	1 cycle	0.06	1700	1 cycle
12	600	3 cycles	0.05	800	3 cycles
13	200	30 µsec	0.1	400	15 µsec
14	1000	1 cycle	0.1	5600	4 cycles

Source: Ref. 3.

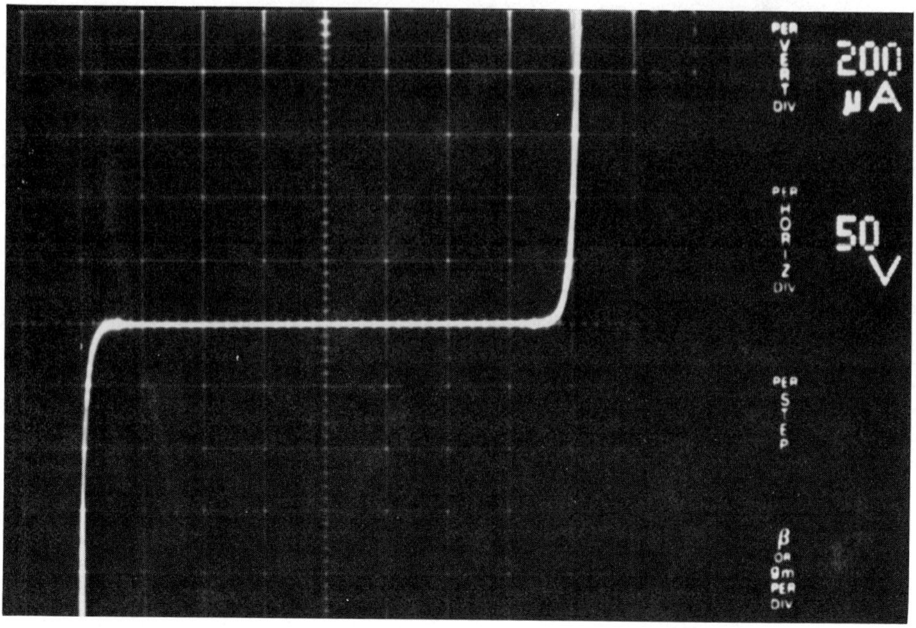

Figure 1 Current voltage trace (actual photo) of a typical ZnO varistor. (From Ref. 22.)

energy-handling capabilities. Figure 1 shows an oscillograph of the symmetrical current-voltage characteristic of this material. The device can be regarded as "insulating" up to a certain field, the breakdown field F_{BR}, above which it becomes conducting.

A more complete examination of the device behavior requires the display of current and voltage on a logarithmic scale. Figure 2 gives current density j versus applied field F for a typical varistor. Note that there is an enormous variation in current (a factor of 10^{11}), while the applied field varies about a factor of 3. From Figure 2 it is also evident that there is wide range of current density (10^{-6}–$\approx 10^2$ A/cm^2) over which the log current–voltage relation is approximately linear, that is, we can write

$$\frac{j_1}{j_2} = \left(\frac{F_1}{F_2}\right)^{\alpha} \tag{1}$$

Equation (1) is the so-called empirical varistor power law equation. If $\alpha = 1$, we would have an ohmic device. As $\alpha \longrightarrow \infty$ we have

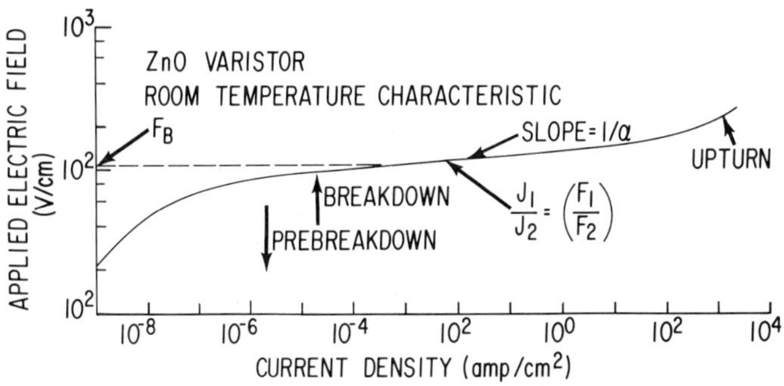

Figure 2 Log-log plot of current density versus applied field for a ZnO varistor. The varistor "breakdown field" F_{BR} is indicated. The exponent α equals the inverse slope of the curve and is a measure of device nonlinearity. (From Ref. 22.)

a perfect varistor (i.e., the current varies infinitely for small changes in the applied field). The exponent α is 25–50 or more in a typical ZnO varistor. Actual values of α, breakdown field, and off-state resistance in typical devices are available elsewhere [6] and will not be specified here.

At very low and very high currents, Eq. (1) is no longer a good representation of the varistor characteristic. The low-current region has been denoted "prebreakdown" and corresponds physically to the transport of very low level currents through the varistor at voltages below the breakdown region. The very high current deviations from the power law Eq. (1) are believed to result from an ohmic series resistance derived from the ZnO grains in the varistor. This region has been labeled "upturn" in Figure 2.

The use of ZnO varistors to protect sensitive equipment from transient overvoltages is in principle extremely simple; the varistor is directly connected across the power line in parallel with the load to be protected. This is indicated schematically in Figure 3. Care is taken to choose a varistor having a breakdown voltage slightly greater than the maximum design voltage applied to the system to be protected. In normal operation the varistor is insulating (i.e., it operates in the prebreakdown region of Figure 2). If a transient is incident so that the total voltage applied to the load rises above the varistor breakdown voltage, the varistor current rapidly increases along its characteristic current-voltage curve (Figure 2), becoming a conducting shunt path for the incident transient pulse. No "delay time" has been detected for varistor operation. Figure 4 gives the measured response of some varistor material to a subnanosecond rise time 500-V amplitude square pulse.

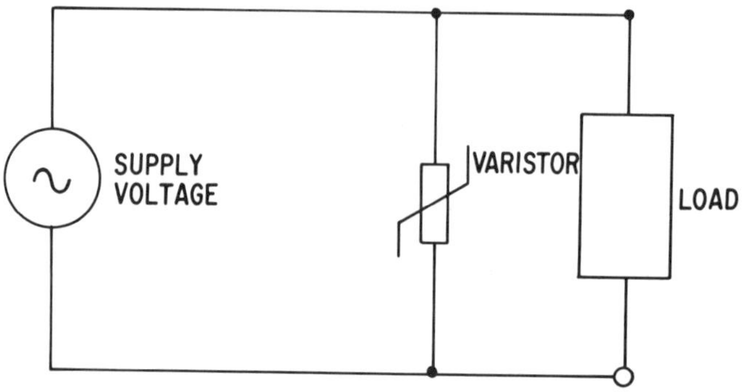

Figure 3 Typical application of ZnO varistor as a transient protective element. (From Ref. 22.)

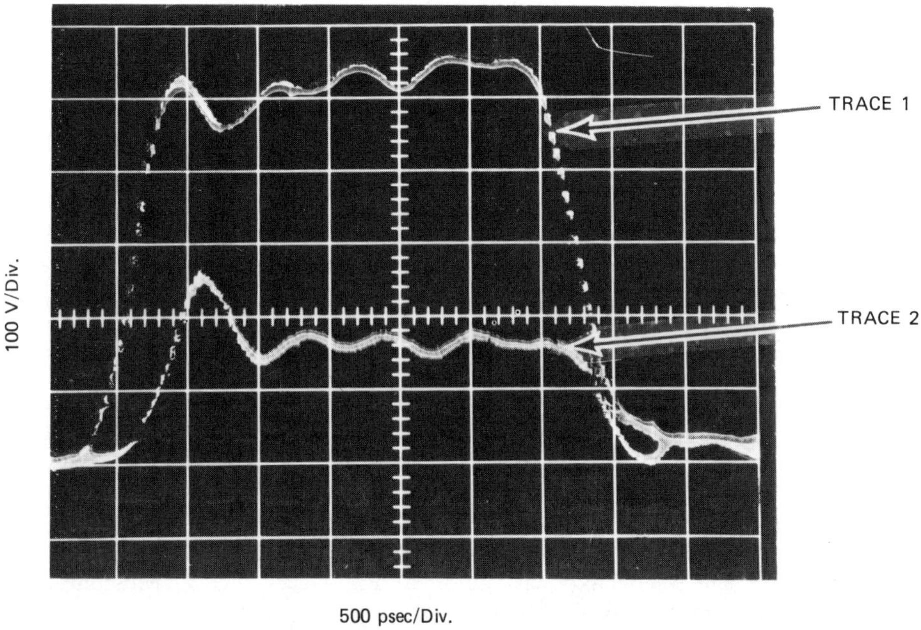

Figure 4 Response of a ZnO varistor to a fast-rise-time (500-psec) pulse. Trace 1 gives the voltage across the load in the absence of the varistor. Trace 2 corresponds to the voltage across the load when clamped to a lower value by the varistor. No varistor "turn-on" delay is evident. (From Ref. 22.)

The traces of Figure 4 were obtained using varistor material in a very low inductance configuration placed across a 50-Ω load. Voltage measurements were obtained using a sampling oscilloscope with an effective 75-psec rise time. Trace 1 is the voltage applied to the load in the absence of the varistor. Trace 2 is the clamped voltage (determined by the circuit impedance and varistor current-voltage characteristic) when the varistor is placed in parallel with the load. From Figure 4 it is evident that the applied pulse is clamped by the varistor in times less than the pulse rise time (500 psec). No varistor delay or turn-on time is thus observed.

The depiction of a ZnO varistor for transient overvoltage protection as given in Figure 3 is schematic. In any real application other device parameters such as energy absorption capability, maximum surge current, ambient temperature, wattage, and so on, are all relevant to the case under consideration. Detailed application notes are available in Reference 6.

III. VARISTOR MICROSTRUCTURE AND FABRICATION

ZnO varistors are highly complex multicomponent oxide ceramics whose electrical behavior depends both on the ceramic microstructure of the device and on detailed processes occurring at the ZnO grain boundaries. The primary constituent of the varistor is ZnO, typically 80 mol % or more. As a result of the varistor processing, the ZnO in ZnO varistors is semiconducting, with $\rho_{ZnO} \leqslant 1$ Ω-cm at room temperature. In addition to the ZnO, the varistor contains smaller amounts of a number of other metal oxide constituents (hence the name "metal oxide varistor"). A typical composition given by Matsuoka [5] contains 97 mol % ZnO, 1 mol % Sb_2O_3 and 1/2 mol % each of Bi_2O_3, CoO, MnO, and Cr_2O_3.

Fabrication of ZnO varistors follows standard ceramic techniques. A simplified flow diagram is given in Figure 5. The ZnO and other constituents are mixed, for example, by milling in a ball mill. The mixed powder is dried and pressed to the desired shape. The resulting pellets are sintered at high temperature (typically, 1100–1400°C). The sintered devices are then electroded, often (but not invariably) with a fired silver contact. Leads are attached by solder and the finished device may be potted in epoxy and tested to meet required specifications.

The microstructure of the resulting device is depicted schematically in Figure 6. The varistor is comprised of conducting grains, size d, surrounded by an extremely thin (∿10s of angstroms) adsorbed layer [7,8]. Current flows between the electrodes as indicated. Typical grain sizes are $d \sim 10$ μm and the grain resistivity is $\rho_{ZnO} \lesssim 1$ Ω-cm. As will be discussed later, an electrically

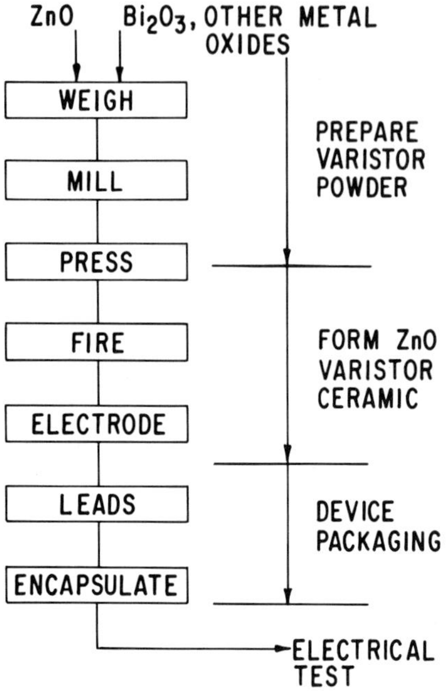

Figure 5 Simplified flow diagram for the fabrication of ZnO varistors. (From Ref. 22.)

insulating boundary region, thickness t ∿ 1000 Å, is also formed at each ZnO grain boundary. This boundary or "depletion" region is wholly within the ZnO grain.

The actual microstructure of a ZnO varistor is considerably more complex than the idealized depiction of Figure 6. In Figure 7 we give a photomicrograph of a polished and etched section of a typical varistor. Three phases—(1) grains, (2) intergranular phase, and (3) particles—are evident.

The grains are the predominant phase in Figure 7 and consist of small (∿20 μm) conducting ZnO microcrystals. The straight lines evident in some of the ZnO grains correspond to twin boundaries delineating different ZnO crystal planes having different etch properties. The whitish areas lying largely between the ZnO phase at triple points have been labeled "intergranular material" and are bismuth-rich. Actually, the true integranular barrier relevant to varistor action is not visible in Figure 7. Varistor action is controlled by depletion layers lying within the ZnO grains at grain-grain interfaces.

In addition to the ZnO grains and intergranular material, a third phase, labeled "particles," is visible in Figure 7. X-ray studies [9]

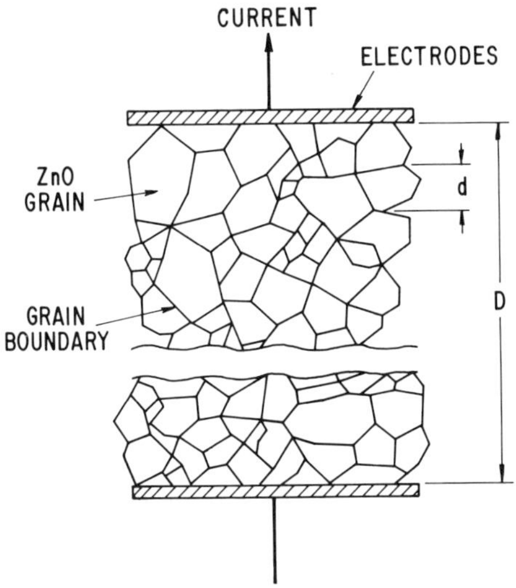

Figure 6 Schematic depiction of the microstructure of a ZnO varistor. Grains of conducting ZnO, average size d, are completely surrounded by adsorbed thin (few atomic layers) oxide. Electrodes are attached and current flows as indicated.

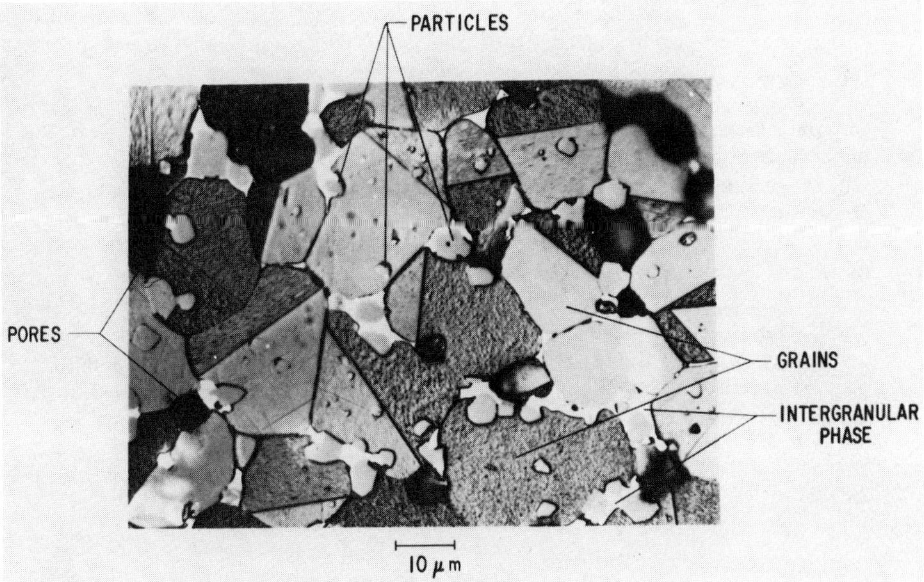

Figure 7 Optical photomicrograph of a polished and etched section of a typical varistor. The various features are discussed in the text. (From Ref. 22.)

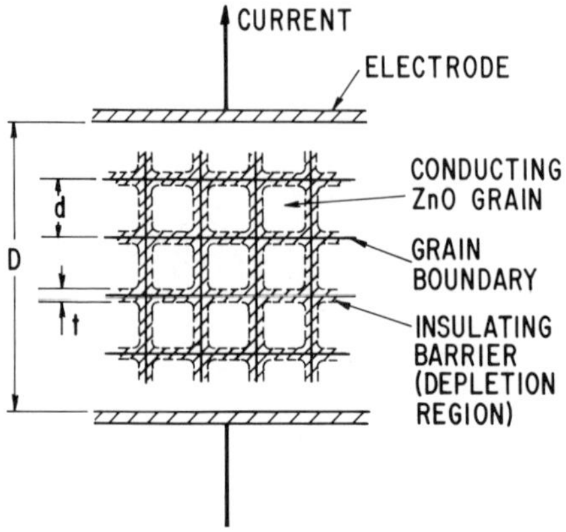

Figure 8 "Block model" of a ZnO varistor having grain size d (∿20 μm) and intergranular depletion barrier thickness t 1000 Å). D is the electrode separation.

have demonstrated this phase to have a spinel-type structure with the approximate formula $Zn_7Sb_2O_{12}$. These spinel particles are insulating and play only a secondary role in determining device properties. Their presence is a result of the particular composition and processing of the varistor of Figure 7. Other varistor formulations have other (or none) such secondary phases.

To analyze varistor behavior it is useful to represent the varistor microstructure by the block model shown in Figure 8. This model presumes the device to be assembled of conducting ZnO cubes of size d separated from each other by an insulating barrier region of thickness t. It should be emphasized that the insulating barrier is not a separate phase but is largely a representation of the back-to-back depletion layers at the ZnO grain boundaries.

Referring to Figure 7, we note that in this sample the varistor grain size d is about 20 μm, and from Figure 2 the macroscopic average breakdown field F_{BR} specified at a current density = 1 mA/cm^2 is around 130 V/mm. It follows that the macroscopic breakdown voltage per intergranular barrier v_g is

$$
\begin{aligned}
v_g &= F_{BR} d \\
&= 2.6 \text{ V per barrier}
\end{aligned}
\tag{2}
$$

The foregoing value of breakdown voltage per grain barrier is characteristic of ZnO varistors when the grain breakdown voltage is evaluated by measurements upon bulk samples. We actually expect this value to be lower than the true breakdown voltage per grain. This follows since the current always seeks the easiest path (i.e., the path with fewest barriers between the electrodes). Hence the number of grains for the current path is lower than the average number of grains between the electrodes, or equivalently, we should increase the value of d used to estimate v_g. Indeed, microscopic data on varistor behavior of single grain junctions (to be presented below) gives $v_g \approx 3.2$ V per grain for varistor materials .

It should also be emphasized that an important and significant feature of the behavior of ZnO varistors is that macroscopic breakdown voltages of 2 or 3 V per barrier are observed for a wide variety of ZnO varistor materials. Substantial variations in device processing and composition have relatively minor effects on v_g.

From the schematic of Figure 8 it is clear that the electrical characteristics of ZnO varistors are related to the bulk of material (i.e., the device is inherently multijunction with varistor action shared between the various ZnO grain boundaries). This then implies that tailoring the device breakdown voltage V_{BR} is simply a matter of fabricating a varistor with the appropriate number of grains, n, in series between the electrodes. Thus to achieve a given breakdown voltage, one can change the varistor thickness (for fixed grain size) or alternatively, one can vary the grain size to increase the number of barriers n, keeping the device thickness constant. In either case

$$V_{BR} = nv_g = \frac{Dv_g}{d} \tag{3}$$

where D is the electrode spacing and d is the ZnO grain size. Typical varistor values for protection of equipment on 100 V ac power lines are $V_{BR} = 200$ V, d = 20 μm, D = 1.6 mm and n = 80.

The block model (Figure 8) approach to understanding the behavior of ZnO varistors is equally useful in developing a reasonable equivalent circuit representation of this material. How this comes about is outlined in the following section.

IV. VARISTOR EQUIVALENT CIRCUIT

Measurement of the dielectric constant ε of ZnO varistors typically gives values $\varepsilon \sim 1000$. This high value, while perhaps initially surprising, is quite easily understood in terms of the varistor model of Figure 8 and the equivalent circuit of Figure 9. Since $d >> t$ (Figure 8), it is clear that the volume between the varistor electrodes is largely occupied by conducting ZnO grain material. In

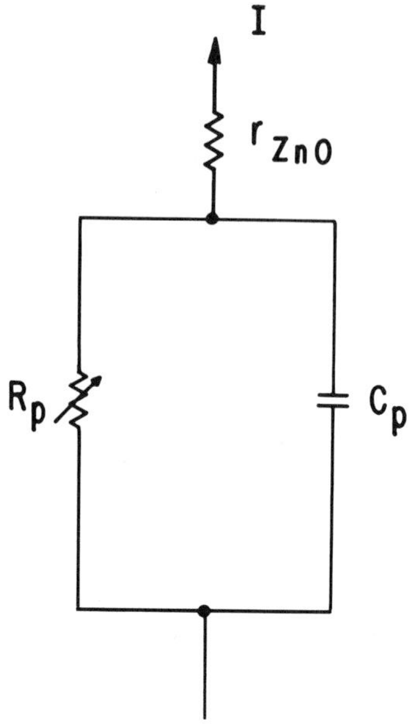

Figure 9 Simple equivalent circuit representing a metal oxide varistor as a capacitance in parallel with a voltage-dependent resistor. C_p and R_p are the capacitance and resistance of the intergranular layer, respectively. r_{ZnO} is the ZnO grain resistance. For low applied voltages R_p behaves as an ohmic loss. (From Ref. 22.)

fact, the thickness of insulating dielectric lying between the sample electrodes is not D but Dt/d. Hence, if ε_g is the dielectric constant of the ZnO in the depletion layers, we expect the varistor capacitance C to be given by

$$C = \varepsilon_g \varepsilon_0 \frac{A}{Dt/d}$$

$$= \frac{d}{t} \varepsilon_g \varepsilon_0 \frac{A}{D} \tag{4}$$

that is, the effective dielectric constant is increased by a factor of d/t.

Measured values of the effective dielectric constant are $\sim$1000 [10], implying d/t $\sim$ 100 since $\varepsilon_g \sim 10$ for ZnO. Taking d $\sim$ 10 μm [11], we find t $\sim$ 1000 Å.

This value represents the thickness of the depletion layer at the ZnO grain boundary [12–14]. The depletion layer largely controls the varistor low-voltage capacitance. The thickness of non-ZnO material between the grains is about two orders of magnitude less [7].

Returning to Figures 8 and 9, it is reasonable to describe the varistor intergranular barriers by a parallel capacitor (C_p) resistor (R_p) pair. R_p is clearly voltage dependent, and in addition both R_p and C_p vary with frequency and temperature (see below). In series with R_p and C_p is the small resistance, r_{ZnO} of the ZnO grains. We can usually ignore r_{ZnO} except for very high varistor currents or at very high varistor frequencies [15].

In Figures 10 to 12 we plot the measured dielectric constant ε, dissipation factor $D = \tan\delta$, and parallel resistivity ρ_p of a GE-MOV® varistor. The high dielectric constant evident in Figure 10 is consistent with our previous discussion and reflects the two-phase nature of ZnO varistors. ε decreases somewhat, from 30 to 10^5 Hz, with a sharper dispersive drop evident in the range $10^5 - 10^7$.

In Figure 11 the dissipation factor D is peaked in the vicinity of 300 kHz. This peak is reminiscent of a (broadened) Debye resonance, and some discussion as to the origin of the resonance peak can be found in Reference 10. It is interesting that in the range $30 - 10^8$ Hz the dissipation factor varies only by a factor of 10. This has immediate consequences for the frequency dependence of the parallel resistivity since

$$\rho_p = \frac{1}{\varepsilon\varepsilon_0 \omega D} \tag{5}$$

Thus noting from Figures 10 and 11 that to a first approximation D and ε are constant, we expect $\rho_p \sim 1/\omega$ (i.e., the varistor resistivity should drop as the inverse frequency).

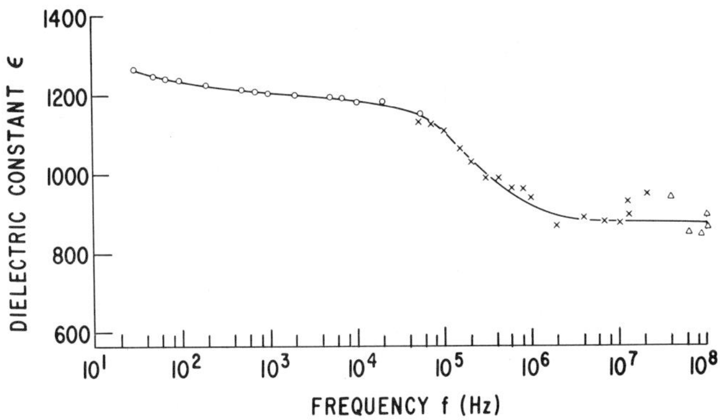

Figure 10 Variation with frequency of the room-temperature dielectric constant of a ZnO varistor. (From Ref. 22.)

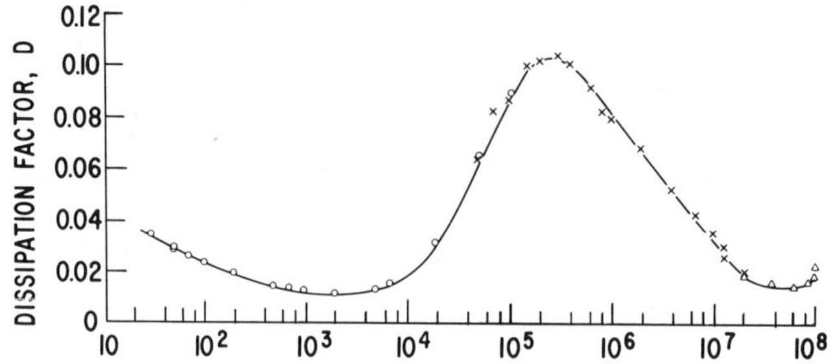

Figure 11 Dissipation factor D = tan δ versus frequency at room temperature. (From Ref. 22.)

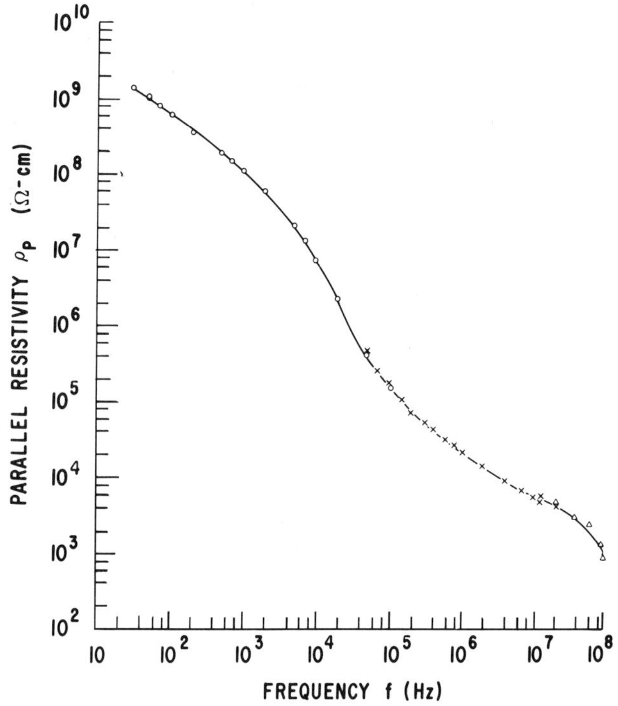

Figure 12 Parallel resistivity ρ_p of a ZnO varistor at room temperature. (From Ref. 22.)

This behavior is illustrated in Figure 12 where we plot the parallel resistivity ρ_p versus frequency. From the figure we have $\rho_p \approx 10^9$ Ω-cm at 30 Hz and $\rho_p \approx 10^3$ Ω-cm at f = 10^8 Hz, implying that the varistor resistivity drops 6 decades when the frequency increases by 6.5 decades. Clearly, R_p of Figure 9 cannot be regarded as a simple frequency-independent leakage path. Superimposed on the general drop in R_p with frequency, we observe a dispersive anomaly in the region 10^5–10^7 Hz, as expected from Figures 10, 11 and Eq. (5).

In addition to the frequency dependence of C_p, D = tan δ, and R_p, measurement indicates that these quantities also vary with temperature. Relevant data and an analysis of the results are given in References 10 and 15.

Finally, it is important to emphasize that the data of Figures 10 to 12 were obtained from measurements at voltages very low compared to the varistor breakdown voltage V_{BR}. In this voltage region the varistor is an essentially linear circuit element. As the applied voltage approaches V_{BR}, more complex phenomena can be anticipated, and an example of such behavior is given by the voltage dependence of the capacitance described in the following section.

V. MECHANISM OF VARISTOR BEHAVIOR

Any complete theory of the electrical behavior of ZnO varistors must account for features that depend on the varistor ceramic microstructure and also for the detailed behavior of a single ZnO grain-intergranular barrier heterojunction. By and large the simplified block model of Figure 8 has proved adequate to understand the implications of varying varistor microstructures, although the idealized representation of the ceramic of Figure 7 should certainly be used with caution. Our current understanding of the phenomena associated with a single ZnO grain-intergranular barrier junction is less complete, and perhaps this imperfect understanding is not unexpected, since it is inordinately difficult to obtain well-controlled electrical or chemical data on an irregular grain interface embedded between two grains in a ceramic.

An account of our present understanding of varistor theory is given in Reference 16 and summarized here. We also outline the largely agreed upon experimental facts relating to ZnO varistor behavior, and present in general terms the approach and results of the theory derived in Reference 16.

As described in Section II, the current-voltage relation of ZnO varistors can be described empirically by the power law relation

$$I = kV^{\alpha} \tag{6}$$

where the parameter α [$=d(\ln I)/d(\ln V)$] is a measure of the device nonlinearity. The parameter α varies with voltage. In the breakdown region, α can attain values well in excess of 50 and can exceed 100 under special conditions.

In addition to the explanation of the high observed values of α, a consistent theory of varistor conduction should explain the following features observed in these devices:

1. The breakdown mechanism is substantially independent of the temperature with a small negative temperature coefficient of breakdown voltage V_{BR} at fixed current [17]:

$$10^{-4} \lesssim \frac{1}{V_{BR}} \left(\frac{dV_{BR}}{dt} \right)_I \lesssim 10^{-3}$$

2. The varistor breakdown voltage is independent of carrier concentration in the ZnO grains [18].
3. Both increased temperature and increased ZnO carrier concentration increase the varistor leakage currents [11,17–19]. The leakage current (prebreakdown) is exponentially activated with temperature.
4. The true microscopic grain-to-grain varistor breakdown voltage is about 3–3.5 V per grain barrier [20] (see also below) and is quite insensitive to the nature and amount of the various additives, the details of the firing procedure, and the processing steps used in the device fabrication. Indeed, ZnO varistors without Bi_2O_3 have recently been reported [21] and we have verified that these varistors (and other ZnO varistor systems without Bi_2O_3) have breakdown voltages of about 3.5 V per grain boundary.
5. The varistor capacitance decreases somewhat with applied dc bias [12,22] and then rises sharply as the varistor enters the breakdown region.

The theory outlined in Reference 16 associates the highly nonlinear varistor conduction with electron tunneling through ZnO grain depletion layers abruptly thinned by hole creation when the conduction band in the grain interior drops below the valence band at the grain surface.

Figure 13 shows schematically the band diagram and current flows for a single ZnO grain junction biased to voltage $V = V_R - V_L$.

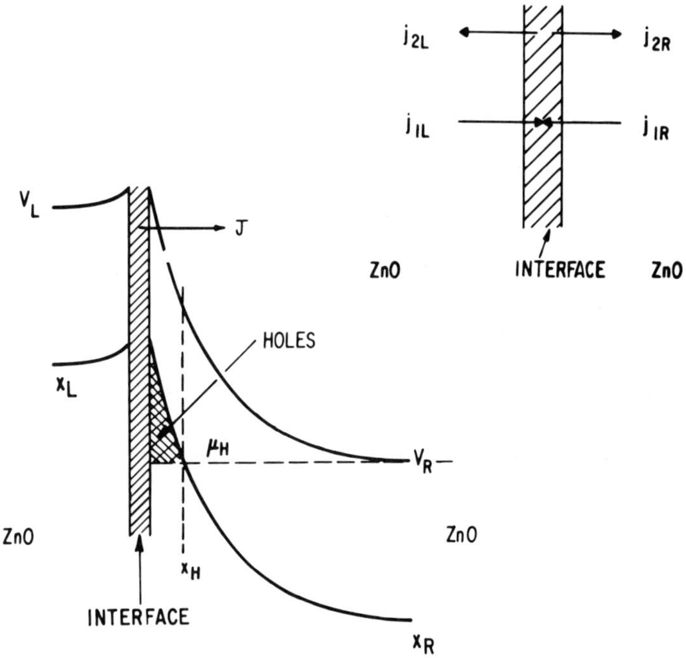

Figure 13 Current flows and band diagram for a biased ZnO grain junction. Long-lived holes are injected on the right-hand side of the junction wherever the hole chemical potential is below the valence-band edge. Electron tunneling from the interface oxide is indicated by the current J. (From Ref. 16.)

At V = 0 (not shown), $V_R = V_L = V_B$, where V_B is the zero-voltage barrier height. The grains are presumed separated by a very thin adsorbed oxide layer whose primary function is to store an amount of negative charge equal in magnitude to the positive charge in the ZnO depletion layers. The ZnO depletion layers constitute the barrier to electron flow.

Transport through the grain junction is viewed as a two-step process, whereby electrons go from the ZnO grain to the oxide interface in one step (j_{1L} and j_{1R}), and then on through the interface to the next ZnO grain in the second step (j_{2L} and j_{2R}). The interface oxide is expected to have a high trap density ($>>10^{19}$ cm^{-3}), and we therefore assume that the barrier height for thermal excitation of electrons out of the interface (= V_B) is independent of V.

In general, the current transport represented by j_{1L}, j_{2L}, j_{1R}, and j_{2R} will be a weighted average of thermal excitation and tunnel currents. Our study has shown that j_{1L}, j_{2L}, and j_{1R} (but not j_{2R}) are dominated by thermal processes, so

$$j_{1L} = j_0 \exp(-\beta V_L) \tag{7}$$

$$j_{1R} = j_0 \exp(-\beta V_R) \tag{8}$$

$$j_{2L} = j_0 S \exp(-\beta V_B) \tag{9}$$

where $\beta = 1/kT$.

In Eqs. (7) to (9), j_0 represents the electron flux in the ZnO grains incident on the depletion layers. We have taken the current out of the interface oxide to be proportional to the normalized interface oxide charge density $S = \sigma/\sigma_0$. Charge conservation requires that the interface charge be the negative of the charge in the depletion region plus the hole charge, so that S is found self-consistently.

Electron flow from the interface through the right-hand depletion layer must also take account of the tunneling process. Thus

$$j_{2R} = j_0 S \exp(-\beta V_B)\ \Lambda(V_R) \tag{10}$$

where

$$\Lambda(V_R) = (U)^{-1} \int_0^{V_B} dE \exp(\beta E) \exp[-W(E, V_R)] \tag{11}$$

and

$$U = \int_0^{V_B} dE \exp(\beta E) \exp[-W(E, V_B)] \tag{12}$$

The integral in $\Lambda(V_R)$ is over the interface energy from the top of the barrier (E = 0) down to the Fermi level ($E = V_B \approx 0.8$ eV) [11,22] in the interface. The factor exp(−W) is the WKBJ tunneling probability through the barrier, and $\exp[-\beta(V_B - E)]$ is the probability of a particle being thermally excited to this energy. The factor U is a normalization integral.

At low voltages, tunneling is improbable and transport is by thermal excitation. As V increases, the depletion layer thins and tunneling becomes more likely. At a critical value V_c of V_R (not to be confused with the so-called varistor "breakdown" voltage V_{BR}, which is defined as the varistor voltage at a current density of 1 mA/cm^2), we obtain

$$V_c = E_{gap} - V_B \approx 3.2 - 0.8 = 2.4 \text{ eV} \tag{13}$$

and holes first appear in the valence band of the right-hand ZnO grain. The pronounced current nonlinearity corresponding to the initiation of varistor breakdown is due to the dramatic change in the conduction-band shape near the interface caused by the presence of the valence-band holes. The dominant current flow is produced by electrons tunneling from the interface near its energetic top, as shown by the arrow in Figure 13. This tunneling is greatly enhanced since the tunneling barrier is now much thinner.

Using the Thomas–Fermi approximation for the potential, Eqs. (7) to (12) have been evaluated numerically [16]. Figure 14 gives calculated values of the current-voltage curves for the parameters indicated. More extensive plots of the solutions to Eqs. (7) to (12) are to be found in Reference 16. Here we shall simply note that the leakage current in the prebreakdown region is exponentially activated with temperature, but the breakdown current is substantially temperature independent. In the breakdown region we find that

$$\frac{1}{V_{BR}}\left(\frac{dV_{BR}}{dt}\right)_I$$

is negative and of the right order of magnitude, as required.

It is interesting to compare the predictions of the model outlined above with measured data, bearing in mind that the model is derived for a single grain-grain junction, whereas many measurements of ZnO varistors evaluate average properties, that is, the experimental results are a complex average of numerous parallel-series paths traversing many junctions in the device under study.

The model described contains only one unknown parameter, namely the depletion layer barrier height V_B. This parameter is easily evaluated from the activation energy of the very low voltage linear varistor leakage characteristic. All other constants used are either known ZnO material parameters or universal physical constants. Nevertheless, we *predict* the varistor breakdown voltage V_{BR} (by definition the voltage at a current of 1 mA/cm^2) to be 3.2 V per junction at T $\sim$ 300 K, $n_0 \sim 10^{17}$ electrons/cm^{-3} (see Figure 14). In addition, we predict [16] that wide variations in temperature T or ZnO donor density n_0 produce little change in V_{BR}, as verified experimentally.

To compare the theoretical predictions with experiment requires measurements on a single ZnO varistor grain junction. Such data were obtained by carefully evaporating a pair of opposing Al electrodes having a 10-μm separation on the surface of a varistor disk.

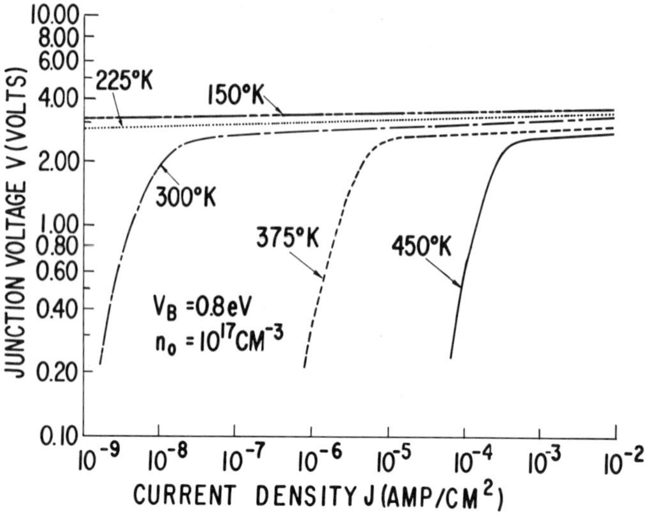

(a)

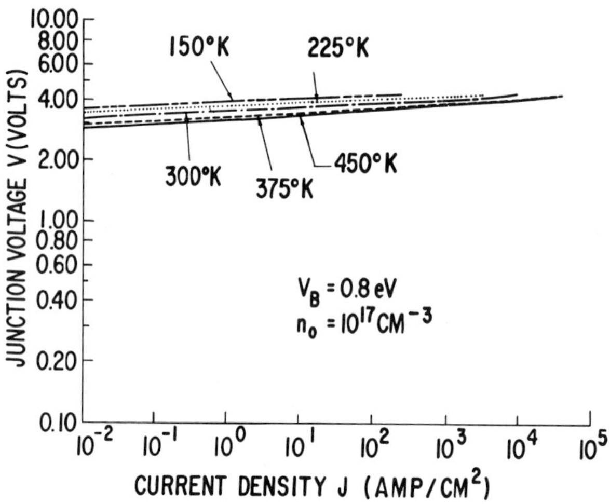

(b)

Figure 14 Log-log plot of the predicted variation of current density J with applied voltage V for a single varistor grain junction: (a) low current densities; (b) high current densities. The curves are calculated for the temperatures indicated. We have set the ZnO grain donor density $n_0 = 10^{17}$ cm^{-3} and the junction barrier height $V_B = 0.8$ eV. (From Ref. 16.)

The varistor grain size was about 25 μm and the electrodes lay on either side of a single ZnO grain junction. A photomicrograph of the experimental configuration is given in Figure 15.

In Figure 16 we plot the room temperature current-voltage characteristic measured experimentally on a single ZnO varistor junction of the type shown in Figure 15. Assuming that the area of grain-grain contact is $\approx$10 μm × 10 μm, a current of 10^{-9} A corresponds to

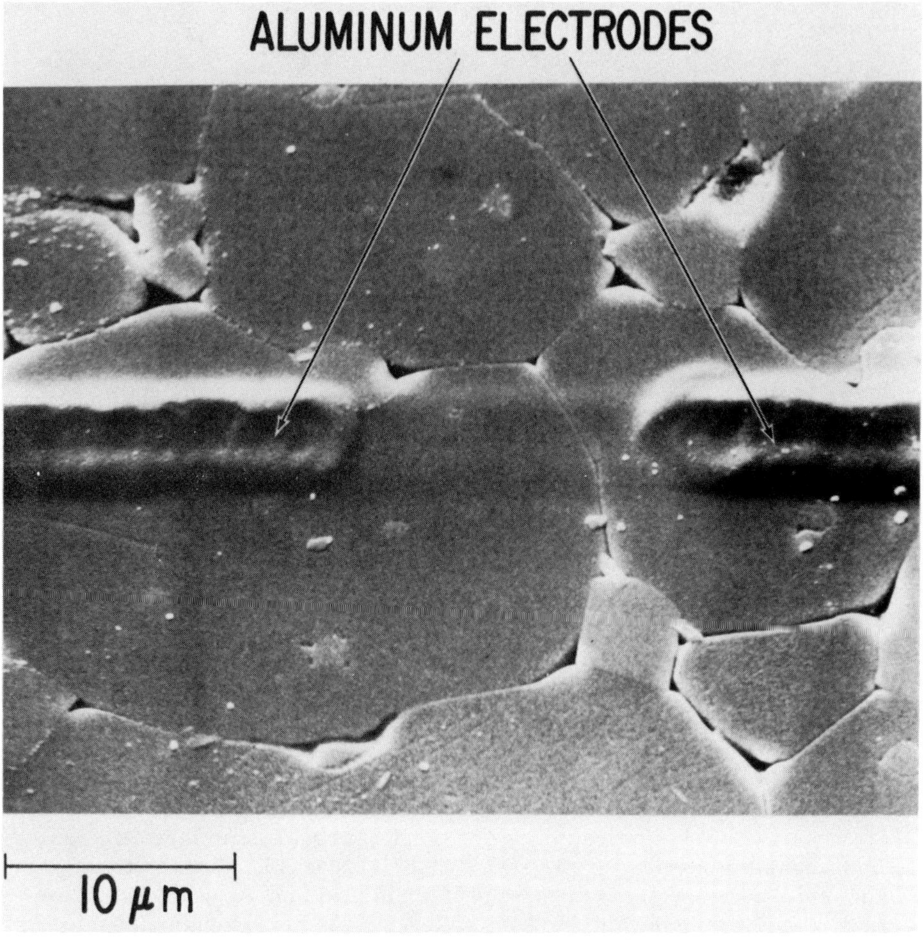

Figure 15 Photomicrograph of evaporated Al electrode configuration on the surface of a ZnO varistor. The electrodes have ohmic contact to two ZnO grains separated by a single grain barrier. The measured current-voltage curve is given in Figure 16. (From Ref. 16.)

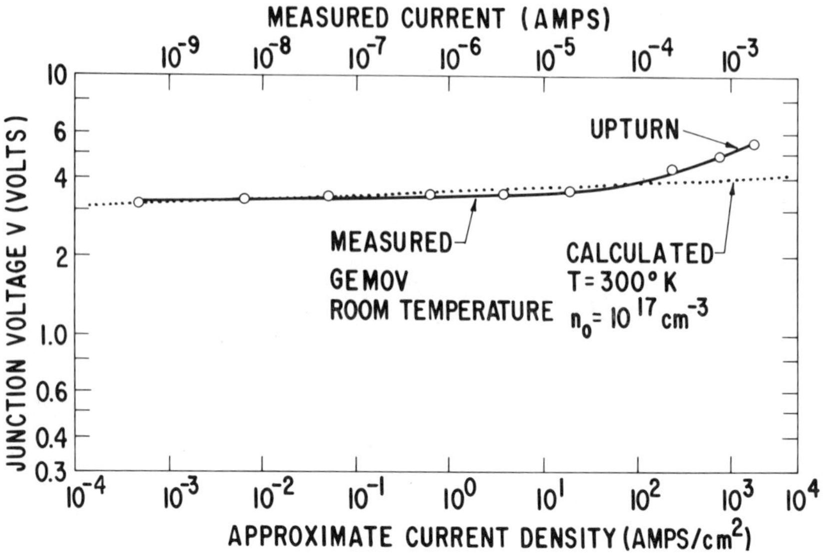

Figure 16 Experimental and calculated current-voltage characteristic of a single ZnO varistor grain junction. The observed decrease in non-linearity at high current densities is due to the series resistance of the ZnO grains [15]. This has not been included in our model. (From Ref. 16.)

a current density of 10^{-3} A/cm^2. Thus from Figure 16, we obtain the experimental value of $V_b \approx 3.3$ V per grain junction. For the sample of Figure 16, $n_0 = 10^{17}$ electrons/cm^3 [16], and we have plotted the appropriate calculated curve. The agreement between our model and the experimental single-junction measurements is remarkably good except at very high current densities, where the experimental voltage rises above the theoretical curve due to the series resistance of the ZnO grains [15]. This effect has not been included in our model.

We have also repeated the single grain measurements on a variety of other varistor types and compositions (some without bismuth) and find breakdown voltages close to that of Figure 16. Van Kemenade and Eijnthoven [20] have made similar single-junction studies. They find that $V_b \approx 3.5$ V at 10^{-6} A, which we presume corresponds to a current density of about 1 A/cm^2. From Figure 14 we predict that $V_b = 3.6$ V at $j = 1$ A/cm^2. Again the agreement is remarkably good and both the model and the experimental data exhibit similar values of V_b for wide variations in the varistor composition and processing.

We may also compare the predicted and experimentally measured temperature dependence of the varistor breakdown voltage, $1/V_{BR}$ $(dV_{BR}/dT)_I$. This quentity is usually measured at a current density of 10^{-3} A/cm^2. From Figure 14 the predicted value is $1/V_{BR}$ $(dV_{BR}/dT)_I = -8 \times 10^{-4}$ K^{-1}, between 150 and 300 K, in good agreement with experiment [11,18,23].

In Figure 17 we give the predicted values of the nonlinear exponent α corresponding to the curves of Figure 14. Values of α well in excess of 50 are found with the higher values of α at lower temperatures. This is confirmed experimentally. Note also that the peak in α moves to lower currents at lower temperatures. This is also corroborated experimentally [17].

In Figure 18 we present plots of current density J versus applied junction voltage for the indicated values of donor density in the ZnO grains. The point to be emphasized is that increased ZnO carrier concentration n_0 leads to increased varistor leakage but does not appreciably affect the varistor breakdown. This is an important feature of the model.

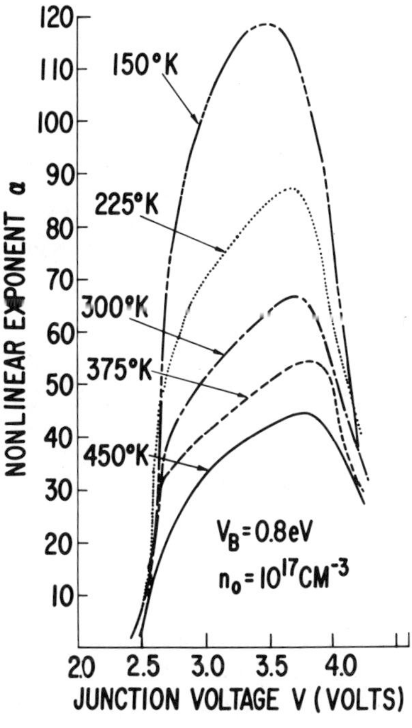

Figure 17 Variation of the varistor coefficient of nonlinearity $\alpha = d \ln I/d \ln V$ for the curves of Figure 14. The abscissa is the voltage V applied to the junction. (From Ref. 16.)

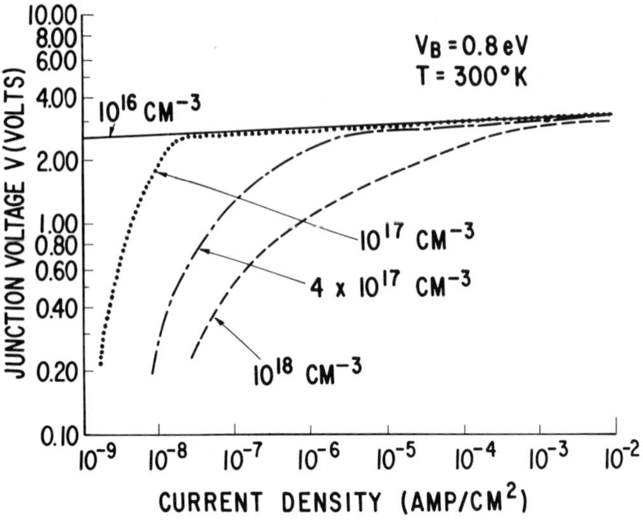

(a)

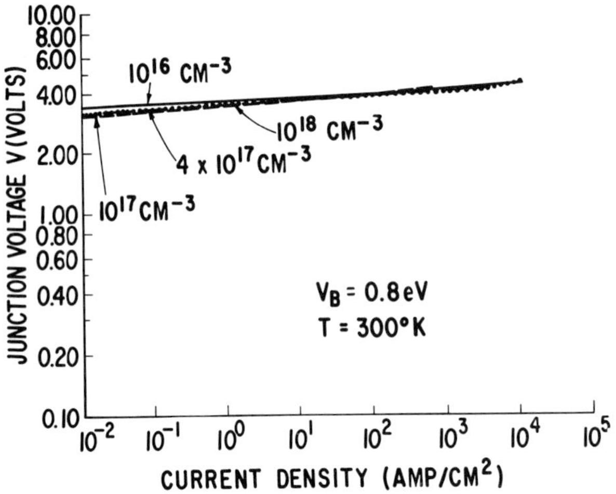

(b)

Figure 18 Log-log plot of the predicted variation of current density j with voltage V for a single varistor grain junction: (a) low current; (b) high current. The curves are calculated for the ZnO grain donor densities indicated. We have set T = 300 K and the junction barrier height V_B = 0.8 eV.

We have also measured the variation of varistor capacitance with applied dc bias. Figure 19 gives C(V)/C(0) for a varistor with $V_{BR} \sim 230$ V. The measured capacitance drops by about 30% until the varistor breakdown region is reached, whereupon a sharp increase in C is observed. Values of C at higher voltages could not be obtained due to the presence of large dc currents in the device. It is interesting to note that the measured behavior is quite similar to the predicted behavior plotted in Figure 20.

We may summarize the features of the conduction model presented above as follows. The model predicts:

1. Coefficients of nonlinearity α as high as 50 or even 100
2. A breakdown voltage insensitive to temperature and electron concentration in the ZnO grains
3. Varistor leakage currents which increase with temperature and electron concentration in the ZnO grains
4. A grain-to-grain breakdown voltage at 1 mA/cm^2 of 3–3.5 V per grain which is insensitive to the metal oxide additives and the firing and processing procedures
5. A capacitance that initially decreases with increasing applied dc bias and then sharply rises as the breakdown voltage is approached

The theory has no adjustable parameters and uses only known material parameters such as the ZnO band gap, measurable extrinsic ZnO

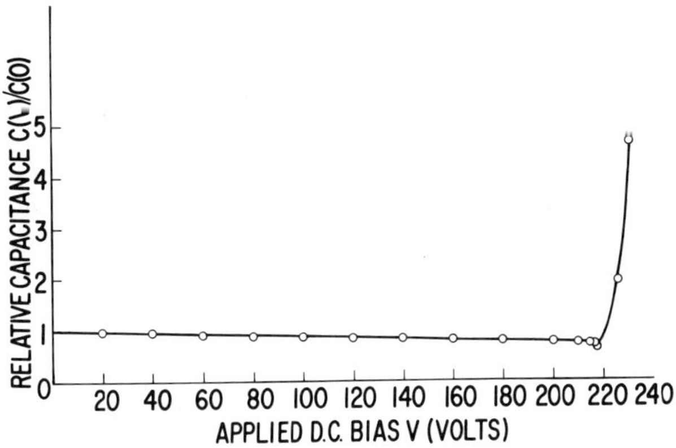

Figure 19 Experimentally measured variation of the varistor capacitance with applied dc bias voltage. Note the steep rise for voltages in the varistor breakdown region. GE MOV® varistor room temperature measurement frequency = 10^4 Hz.

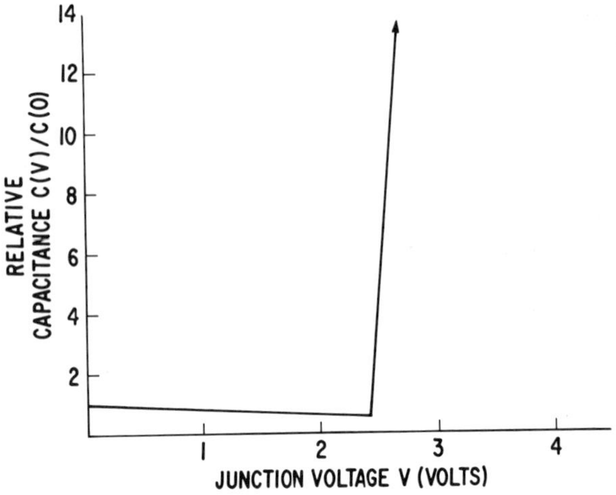

Figure 20 Calculated variation of the varistor junction capacitance with applied dc bias voltage V. A steep rise in the capacitance is predicted to occur concurrent with the initiation of hole creation. (From Ref. 16.)

grain donor concentration, and varistor low-voltage leakage current activation energy. The highly nonlinear varistor conduction behavior results from electron tunneling "triggered" by hole creation in the ZnO when the conduction band in the grain interior drops below the top of the valance band at the grain interface.

The good agreement between experimental results and calculated varistor behavior notwithstanding, it would be wise to add the following note of caution: In ceramic technology in general and in the case of ZnO varistors in particular, understanding and theory have often followed experiment and practice. We can therefore expect the description above to be modified, improved, and altered as more experimental data are incorporated into the model.

VI. CONCLUSION

ZnO varistors are finding increasing acceptance as a simple cost-effective way to protect electrical systems from transient voltage surges. The devices are rugged and reliable. By virtue of their ceramic nature they can be fabricated into a variety of sizes and shapes, and this feature facilitates a high degree of user flexibility.

Varistors are inherently multijunction devices and any transient surge energy to be absorbed when the device acts in a protective mode is distributed between the many ZnO-intergranular barrier heterojunctions. The multijunction feature of varistor behavior lies at the heart of the high current and energy absorption capability of the material. Indeed, ZnO varistors have recently been incorporated into lightning arrestors to protect electric utility power transmission lines [24] and in this application the material is called on to withstand voltages in the many-100-kV range and pass currents in excess of 10 kA. At present it appears quite probable that the uses and applications of ZnO varistors will continue to increase as the technology matures.

REFERENCES

1. D. E. Castleberry, *IEEE Trans. Electron Devices, ED-26,* 1123 (1978); D. E. Castleberry and L. M. Levinson, *1980 SID Int. Symp.*, Paper 18.5.
2. A. Chiang and D. G. Fairbairn, *1980 SID Int. Symp.*, Paper 11.5.
3. F. D. Martzloff and G. D. Hahn, *IEEE Trans. Power Appar. Syst., PAS-89* (6), 1049–1056 (July–Aug. 1970).
4. R. W. Fox, *Electron. Des. 22,* 52 (May 24, 1974).
5. M. Matsuoka, *Jpn. J. Appl. Phys., 10,* 736 (1971).
6. General Electric Semiconductor Products Dept., *Transient Voltage Suppression Manual*, 2nd ed., J. C. Hey and W. P. Kram, eds., 1978.
7. D. R. Clarke, *J. Appl. Phys., 48,* 4372 (1977).
8. W. G. Morris and J. W. Cahn, in *Grain Boundaries in Engineering Materials*, J. L. Walter et al., eds., Claitors, Baton Rough, La., 1975.
9. J. Wong, *J. Appl. Phys., 46,* 1653–1659 (1975).
10. L. M. Levinson and H. R. Philipp, *J. Appl. Phys., 47,* 1117 (1976).
11. L. M. Levinson and H. R. Philipp, *J. Appl. Phys., 46,* 1332 (1975).
12. W. G. Morris, *J. Vac. Sci. Technol., 13,* 926 (1976).
13. J. D. Levine, *Crit. Rev. Solid State Sci., 5,* 597 (1975).
14. J. Bernasconi, H. P. Klein, B. Knecht, and S. Strässler, *J. Electron. Mater., 5,* 473 (1976).
15. L. M. Levinson and H. R. Philipp, *J. Appl. Phys., 47,* 3116 (1976).
16. G. D. Mahan, L. M. Levinson, and H. R. Philipp, *J. Appl. Phys., 50,* 2799 (1980).

17. H. R. Philipp and L. M. Levinson, *J. Appl. Phys.*, *48*, 1621 (1977).
18. L. M. Levinson and H. R. Philipp, unpublished.
19. H. R. Philipp and L. M. Levinson, *J. Appl. Phys.*, *47*, 3177 (1976).
20. J. T. C. van Kemenade and R. K. Eijnthoven, *J. Appl. Phys.*, *50*, 938 (1979).
21. K. Mukae, K. Tsuda, and I. Nagasawa, *Jpn. J. Appl. Phys.*, *16*, 1361 (1977).
22. L. M. Levinson and H. R. Philipp, *IEEE Trans. Parts. Mater., Packag.*, *PHP-13*, 338 (1977).
23. L. M. Levinson and H. R. Philipp, *Appl. Phys. Lett.*, *24*, 75 (1974).
24. E. C. Sakshaug, J. S. Kresge, and S. A. Miske, *IEEE Trans. Power Appar. Syst.*, *PAS-96*, 647–656 (1977).

7

Multilayer Ceramic Technology

WAYNE S. YOUNG
IBM Corporation, San Jose, California

I. INTRODUCTION

The packaging of integrated circuits (ICs) has usually made use of a chip carrier or module to which the IC chip(s) would be joined. The module would in turn be joined to a PC board, becoming the second step in packaging the IC. Such modules have traditionally been made using ceramic substrates as a base material.

Over the past 20 years, improvements have been made in these modules through the use of finer-grained, smoother substrates, better thick- and thin-film metallizing techniques and most recently through the use of multiple wiring layers within the module itself. Such multilayer-ceramic (MLC) modules have been made using two completely different processes: sequential and laminated. The purpose of this chapter is to discuss the laminated MLC process, but a short review of the sequential process will be included.

II. THE SEQUENTIAL MLC

In the sequential process, a fired substrate is used as a base. Upon it, alternate layers of metallurgy and dielectric are deposited and fired. The layering and firing process is repeated, building a three-dimensional structure. The dielectric used was originally glass [1], but later use of glass-ceramics [1,2] or devitrifiable glasses [1,3] became conventional. The metals used include Ag, Au, Pt, Pd, and even Cu [4]. Such a module will usually see a firing for each layer put on it. Many companies and organizations have been involved in building sequential MLCs, including GE [2,5] RCA [4], Teledyne [6], Honeywell [7], Hewlett–Packard [8], Magnavox [1], Bell Labs [9], and the Air Force [3]. Sequential MLCs as large as 16.2 cm × 2.54 cm have been described by Marshall and Rode [8]. Kurzweil and Loughran describe paste evaluation techniques to give 4-mil-wide lines on 8-mil centers, in sequential MLCs [5].

Nonconventional sequential MLCs have been developed by Pactel Corp. [10] using a polyimide dielectric on stainless steel substrates, by DuPont [11] using porcelainized steel substrates, and by Singer Corp. [12] using porcelainized steel or anodized aluminum substrates. Line widths as small as 1 mil on 2-mil centers may be possible [10].

No further description of the sequential MLC process will be given in this chapter. The reader is referred to Agajanian's bibliography [13] for references on all aspects of electronics packaging for 1970 through 1975.

III. THE LAMINATED MLC

In the laminated process, the module is fired only once, instead of seeing a sequence of firings, one for each layer. The process begins with the casting of sheets of unfired (green) ceramic, which are punched with via holes, screened with metallurgy, laminated into a monolithic three-dimensional structure, and fired in a furnace. The ceramic and the metallurgy both densify simultaneously in the same firing cycle.

A nomenclature problem arises because both sequential and laminated modules have been described in the literature as multilayer ceramic (MLC) modules or substrates. The former have also been called sequential MLCs [9], multilayer substrates [8], and screen printed MLCs [1]. The latter have been called screened multilayer ceramics [33], laminated ceramics [18], and monolithic multilayers [1]. In this chapter we refer to a laminated MLC substrate simply as "MLC."

IV. MLC DEVELOPMENT

Development of laminated MLC substrates has been ongoing since J. L. Park patented his tape-casting technique in 1961 [14], but the term "Father of MLC" really goes to Bernard Schwartz, who headed a group developing ceramic packages for RCA, which holds many of the basic patents. He conceived the idea of laminating green layers together while watching a plastic lamination machine at a chemical trade show. The first MLC packages RCA developed were for multilayer capacitors [65,66]. They built the "micromodule" capacitor, which was 0.3 in. square and contained 11 layers. This was the beginning of the multilayer ceramic capacitor industry. Further research led to the development of substrates for devices, in the form of "micromodule" single-device ceramic modules in the early 1960s. Good reviews of the early technology development have been provided by Schwartz and Wilcox [15] and Theobald et al. [16].

A number of other companies have been involved in developing laminated MLC processes, including IBM [15,17–24], 3M (American Lava) [16,25–27], Western Electric [28], Texas Instruments [29], DuPont [30], Kyocera [31,32], Hitachi [33], Ceramic Systems [34], and Dielectric Systems [35], to name a few that have published results.

During the 20 years of MLC development, the sophistication of MLC technology has increased dramatically. The first commercial

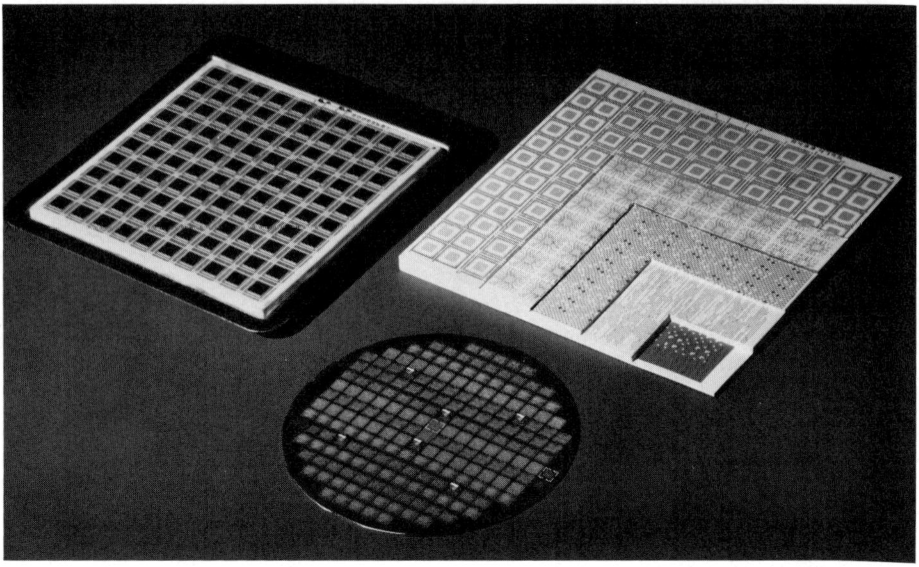

Figure 1 90-mm MLC substrate: fired (left), mounted with 100 chips, and unfired (right), showing internal layers. (Courtesy of IBM, East Fishkill, N.Y.)

MLCs were one step above a planer substrate in density and appeared about 15 years ago. Today, the most complex MLC substrate announced for commercial production appeared in November 1980 for use in IBM 3081 computers and is shown in Figure 1. It is called the Thermal Conduction Module (TCM), and has been reviewed by Seraphim and Feinberg [67], Blodgett [68], and Patrusky [69]. It is 90 mm square, up to 33 layers thick (6.55 mm), has 1800 pins on the bottom, and can accept from 100 to 118 chips reflow-soldered to the top. The TCM substrate has about 350,000 vias connecting various layers, contains 130 m of internal wiring, and typically interconnects 25,000 circuits plus 60,000 array bits [36]. This is the same number of circuits on one module that an entire IBM System/370 model 145 has.

V. MLC DESIGN

The proper design of an MLC substrate is the first of a number of critical steps in successfully fabricating the product. The tolerances achievable and the processing problems encountered are a direct result of the design of the substrate as well as other factors, such as tooling and process parameters. Several authors have published data and ground rules on successful MLC design [16,18,19,23,25,27, 30,33]. Improper design might include not balancing the metallurgy from top to bottom in the substrate, which causes camber during firing, or not balancing the patterns across each layer, which leads to fired dimensional distortion and cracking. Electrical opens in the internal metallurgy can be caused by improper overlap in line/via connections, and so on. Shorts can arise from designing lines and vias with insufficient spacing, and so on, almost adinfinitum.

The successful design of an MLC substrate also depends on complete understanding of the green and fired properties of the ceramic and the metallurgy(s) used and the careful matching of shrinkage curves between the metal and ceramic during firing. The size and thickness of the substrates which can be successfully built rests largely on those parameters, as well as the mechanical properties of the finished metal/ceramic composite.

One thing that makes design so critical is that it is very difficult to separate design faults from process problems during the development of a given MLC substrate configuration.

VI. THE LAMINATED-MLC PROCESS

The basic process for building MLC substrates is illustrated in Table 1, although each different manufacturer uses his own process variations and peculiarities [15,20,29,34]. Depending on the module complexity, some steps may be deleted or others added, especially after sintering.

A. Artwork Generation

One of the most critical and important process steps is the generation of the artwork, which is a direct result of the design phase of the module itself. As Figure 2 shows, an MLC substrate is composed of a number of distinct layers which normally include: top surface metallurgy, as shown in Figure 3 (for chip attach and wire bonding); redistribution or fan-out layers, as shown in Figure 4; signal distribution layers (x and y); power distribution layers (ground planes); and bottom surface pads (for pin attachment) [29,37]. Each of these layers requires a unique punching tape for via punching and possibly an inspection mask for checking the via pattern. Each layer also requires a set of screening masks for filling the vias and personalizing the pattern required on that layer (see Table 1). The quality and integrity of the tapes, masks, screens, and so on, governs the tolerance

Table 1 The Laminated MLC Process

1.	Artwork generation
2.	Raw material procurement
3.	Raw material qualification
4.	Ceramic slurry preparation
5.	Metal paste preparation
6.	Ceramic tape casting
7.	Tape blanking and inspection
8.	Registration hole punching
9.	Via punching
10.	Hole inspection
11.	Metallurgy screening
11a.	Paste drying
12.	Screen layer inspection
12a.	Screening rework
13.	Stacking
14.	Lamination
15.	Sizing
16.	Inspection
17.	Sintering
18.	Refiring and/or flattening
19.	Plating
20.	Shorts/opens testing
21.	Pinning
22.	Top surface wiring
23.	Chip attachment
24.	Electrical testing
25.	Encapsulation

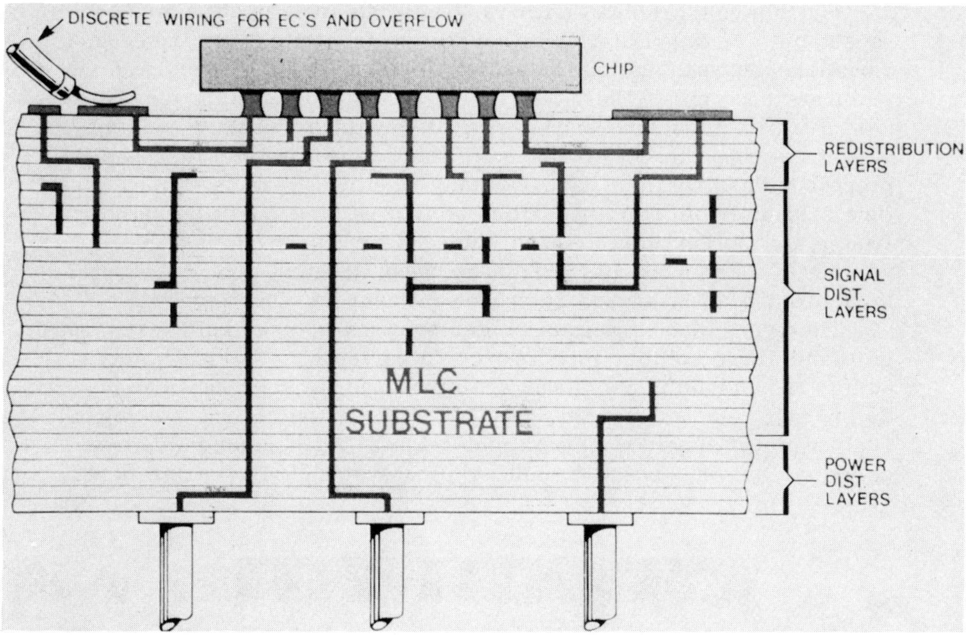

Figure 2 Schematic view of an MLC substrate. (Courtesy of IBM, East Fishkill, N.Y.)

and quality seen in the substrate. Screening masks are fabricated using photoresist processes and artwork generated by computers.

B. Raw Material Procurement, Qualification, and Preparation

The raw materials used in MLC substrates include the ceramic and metal powders and any organic additives, solvents, or binder components used to make the slurry for tape casting and the various metal pastes. The most common MLC ceramic is a 90–96% alumina, although others, such as alumina-mullite, have been reported [23]. Most metals have been tried, with tungsten or molybdenum being the most popular for cofired MLC substrates [18,19,21]. Recent advances include work on an alumina-lead borosilicate glass-ceramic with gold or silver-palladium metallurgy [70].

The biggest problems with raw materials include obtaining a pure, repeatable starting material with uniform properties, and minimizing variations between various lots of the material [36]. Many manufacturers accomplish the latter by purchasing or making very large lots of a given material and storing it for future use. Changing lots of ceramic can completely alter the process control of an MLC process unless the lots have virtually identical particle shapes, particle size distributions, impurities/additives, and surface chemistry (water absorption, etc.). The subject is complex enough that entire volumes [38] and courses [39] have been written about it. Part 2 in Reference 38 is a good reference for alumina characterization.

There are also numerous binder systems in use, both water soluble and those soluble only in organic solvents. Various binders, additives, and binder systems are discussed in the literature [20, 28,38–41]. As an example, the IBM-MLC ceramic is made using a 90:10 ratio of Bayer α-alumina and a glass frit, blended with a polyvinyl butyral–based organic binder system [24]. After firing,

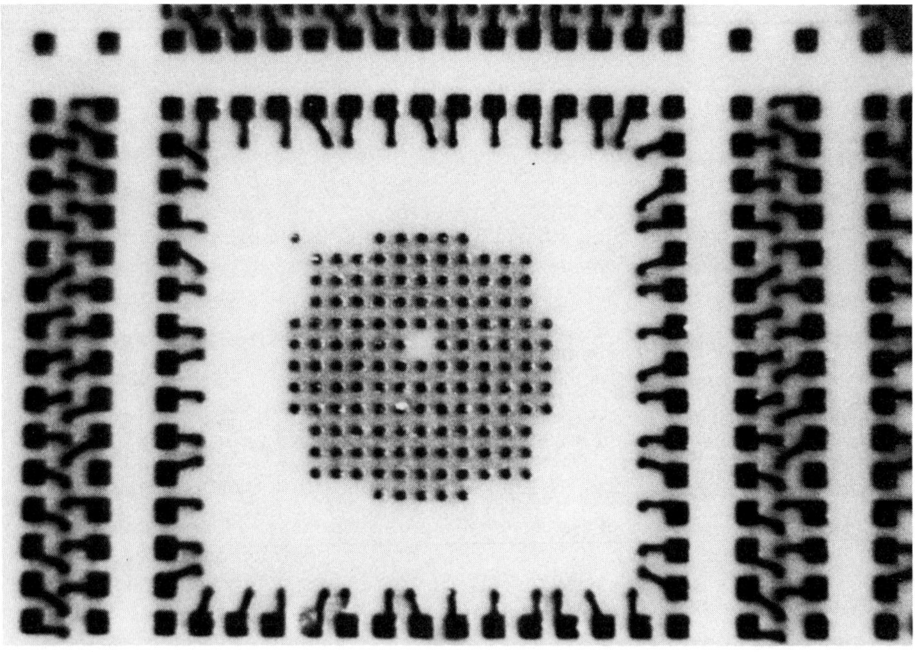

Figure 3 Surface structure of one chip site on an MLC, showing vias for chip attachment (center pattern) and pad/line/via groups for testing and possible wire bonding. (Courtesy of IBM, East Fishkill, N.Y.)

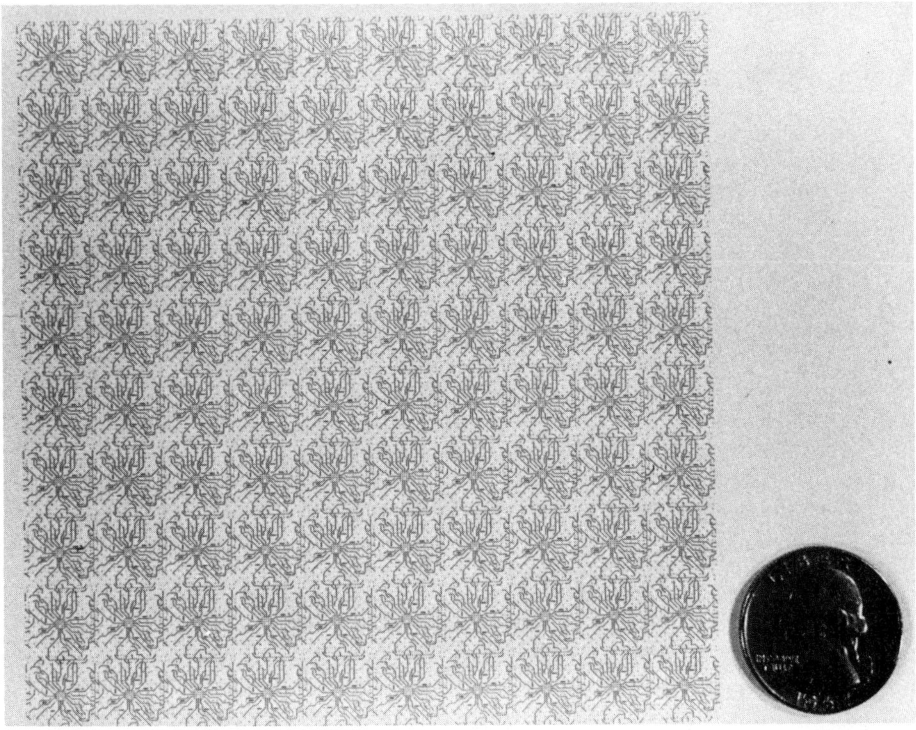

Figure 4 Typical fan-out layer for a 100-chip-site MLC substrate. (Courtesy of IBM, East Fishkill, N.Y.)

the ceramic is equivalent to a 94% alumina. The binder system contains a plasticizer and two solvents in addition to the PVB [22].

Preparation of raw materials includes milling the powders to uniform particle size distribution and sometimes, the classification of the powders to remove fines or agglomerates. Ball milling [28], vibratory milling [41], and jet milling [22] are all successful techniques in use. Often, two-stage milling is used, with the binder system being added to the second stage, which becomes essentially a blending step. The product of these steps is a slurry ready for tape casting.

The processing of paste is similar to the ceramic, although the viscosity is much greater and the quantities smaller. The process can include milling and/or classifying the powders, and blending the binder system with them. The ultimate paste used on each layer will likely be unique to that layer (ground plane versus signal layer, etc). Paste must be carefully optimized throughout the substrate before the best shrinkage tolerances can be realized [17,19].

C. Tape Casting

The science of tape casting (step 6 in Table 1) has been discussed extensively in the literature [22,28,39,41–43]. Basically, the ceramic slurry is dispensed onto a moving plastic carrier sheet, usually by passing under a doctor blade(s). The solvents are driven off and the ceramic cures as it is carried through a drying area, after which the ceramic "green" sheet is stripped from the plastic carrier, inspected for defects and thickness, and rolled onto spools. Smaller laboratory batch casters are also used, but are poor on repeatability. This tape must be cast under the most stringent cleanliness and process controls to achieve acceptable dimensional tolerances after firing. These conditions must be maintained through lamination (step 14). The optimum green sheet properties [22,28,62] include:

High tensile strength and yield strength
Flexibility enough for handling and via punching
Dimensionally stable and free from defects and inclusions
Uniform density across the sheet
High compressibility (for good lamination)
A stable, easily volatilized binder system (in reducing atm)
A solvent system that dries rapidly without pinholes
Sintering shrinkage that is uniform and reproducible

The thickness of a cast sheet is controlled by such factors as the doctor blade gap, the temperature, the carrier speed, the slurry viscosity, the depth of the slurry behind the doctor blade, and the shrinkage of the tape as it dries [22,42]. Widths of cast tape range from 100 to 300 mm or greater. IBM's continuous caster deposits 215-mm-wide sheets on a Mylar carrier, which is carried through five drying ovens before being inspected, stripped, and spooled. Other features on many casters include an enclosed slurry pool, multiple doctor blades, and curved beds [28,42].

Green sheet tests could include density, compressability, yield strength, bond strength, tensile strength, shrinkage, binder distribution, screening stability, and paste absorptivity [22].

After spooling, the green sheet is blanked into squares and any blanks not meeting cleanliness or thickness specs are removed (step 7 in Table 1). One important factor in achieving optimum sintered dimensional tolerances is the control of green sheet thickness (i.e., ±0.01 mm on a 0.2-mm-thick sheet). The green sheets are then punched with four registration holes (step 8), which will align the sheet for punching and screening. Because green sheet properties vary between the casting direction and the transverse direction [44], the layers are rotated during stacking to eliminate the effect of thickness variation on the fired dimensions of the substrate.

D. Via Punching

Green sheet blanks begin personalization at via punching (step 9 in Table 1) and at this point, layer-by-layer control of the blanks must be exercised through lamination. Virtually every layer in a substrate is unique from all other layers and has a unique hole pattern. There is some reference to punching holes in green sheets in the literature [28]. IBM uses a 0.13-mm nominal via diameter and punches up to 25,000 holes in any given sheet using precision cluster punches which can punch up to 100 holes simultaneously [24,36]. Registration of the green blank must be held to less than ±0.1 mm tolerance through punching, screening, and lamination [22].

After punching, each punched sheet is inspected for missing or plugged via holes. This is done by inspection masks and light tables.

E. Screening and Drying

Each punched layer is screened (step 11) with the appropriate paste. Pastes used for ground planes will differ from those used for fine lines, and so on. In the IBM process, molybdenum paste is extruded through a direct-contact mask process onto the green blank. Via holes are filled simultaneously with the line pattern being put down. The paste is then dried at about 75°C for up to 90 min [45].

The literature on thick-film screening is extensive [46–53], and some of the best sources are paste suppliers [49,50]. Several sources deal with screening techniques and problems [51–53]. This bibliography is by no means complete and the reader is referred to [13] for more information. Despite the availability of references, a good screening process is an "art," with the "tricks of the trade" jealously guarded by the developers and carefully optimized for the paste/ceramic system used.

Some of the parameters that have to be optimized to achieve the best possible screening include the number of passes (back and forth), the speed, paste rheology, paste solvent/green sheet interaction, which side of the green sheet is screened, tooling tolerances, and the frequency with which a mask is cleaned [45]. With good tooling and process control, line widths of 0.1 mm on 0.2-mm spacing (4 mils on 8-mil spacing) can be achieved with high yields.

After screening, each layer must again be 100% inspected (step 12). This can be done using simple pattern comparators or automated using computer-controlled laser scanners. For complex patterns and large green blanks, the latter is the only reliable method. Defects in a screened pattern can be hand repaired and the sheet reinspected. In this manner, only perfect screened green blanks will be used in a substrate and the open/short defects in the fired substrate can be minimized. This is one big advantage of the laminated MLC over other types.

F. Lamination and Sizing

The layers are now carefully stacked onto a lamination fixture (step 13) and laminated together (step 14 in Table 1) in a large, precision flat-plate press. During stacking, care must be taken to get every layer in its proper position and orientation in the stack, or all previous care is wasted. A typical lamination cycle used at IBM is 75°C and 25 MPa pressure [24]. The layers become an unfired or "green" laminated substrate.

The individual "green" substrates can either be sized (step 15) automatically during lamination, using lamination dies which shear the sheets as the press closes, or they can be cut on special precision saws after lamination.

One aspect of the laminated MLC process which is unique is the ability to form more than one substrate per laminate, as dimensions permit. This is called "multi-up" processing. For example, using 185-mm square green blanks, 16–25 mm square (fired), 9–35 mm square, or 4–50 mm square substrates can be obtained from each laminate [24,54]. Only 1–90 mm IBM substrate described earlier is made per laminate [36].

After lamination, the edges of the green laminates are lightly sanded to eliminate sharp corners and burrs, and they are inspected for contamination, voids, or other surface damage in preparation for firing (or sintering).

G. Sintering

During the sintering cycle, first the binders are broken down and volatilized and then the metallurgy and ceramic sinter into a fully dense substrate. The furnace used may be either a continuous type or a periodic or "batch" type. Large substrates usually achieve better dimensional tolerances in the latter, although the continuous kilns have more throughput. Figure 5 shows a stack of sintering tiles containing green laminates ready for sintering. The tiles are molybdenum, which allow for more uniform temperature control because of their good thermal conductivity. The atmosphere in the furnace is carefully controlled to have oxygen enough to burn the binders out cleanly and prevent loss of SiO_2 from the ceramic [55, 61]. At the same time, it is reducing enough to prevent oxidation of the metallurgy [19–21,24]. When molybedenum or tungsten is the metallurgy used, a hydrogen or hydrogen/nitrogen atmosphere is

Figure 5 Load of "green" substrates on setter tiles ready for insertion into a sintering furnace. (Courtesy of IBM, East Fishkill, N.Y.)

required [17,56]. The dew point of the hydrogen is controlled by passing it through a temperature-controlled bubbler [21]. If metals more noble than molybdenum are used, nonreducing atmospheres could be employed.

The IBM green laminate shrinks about 17% during the sintering cycle. During the initial phase, the residual solvents evaporate and the organics begin to decompose [37,54]. The hydrogen is initially dry as the cycle is begun, but wet hydrogen is introduced during heat up to accelerate the formation of CO and CO_2. Sintering of the metal occurs in the range 1200–1560°C. Matching the shrinkage of the ceramic to the metallurgy during this phase of the cycle is very important to give a good bond between the molybdenum and the ceramic [17,56,57] and prevent predensification cracks in the substrate. The shrinkage of the molybdenum is very sensitive to the dew point of the hydrogen [21], as well as to the particle size, particle shape, and green density of the paste and any nonvolatile additives in the paste (Mn, glass, etc).

When the sintering cycle reaches 1560°C, it is held there for a soak period to assure maximum densification and metal/ceramic bonding [37]. Cooling is done carefully, to avoid thermal shock cracks in the substrates, and the atmosphere changed back to dry hydrogen. The most common crack problem in substrates is caused not only by the cooling cycle, but also by the thermal expansion mixmatch between the ceramic and the metal. This appears first between vias coming to the surface [19]. The problem is controlled by careful selection of materials used in MLCs, by the proper selection of via patterns, and by careful furnace controls.

A second problem in sintering MLC substrates, as well as any technical ceramic, is in obtaining good dimensional tolerances. This is achieved by proper substrate design, good shrinkage matching between the paste and the green sheet, compatible thermal expansions between metal and ceramic (on cooling), and very low thermal gradients in the furnace. This dimensional control problem is discussed in the literature and a few examples are cited here [19,41,58, 59]. Complex substrates require shrinkage tolerances as low as ±0.1%. In addition to distortion in the horizontal plane of the substrates, other dimensional problems, such as camber and waviness, occur. It is possible to reduce these problems by balancing the metallurgy throughout the substrate, and as a last resort, by refiring the substrate at a slightly lower temperature with a weight on it. This is sometimes called "flattening" (step 18 in Table 1).

H. Postsintering Operations

After the substrate is fired, it is necessary to prepare it to receive the integrated circuit chips and connect it to the next layer of packaging, usually a printed circuit board [63]. In the IBM process this includes nickel and gold plating (step 19), electrically testing the substrate for opens and shorts (step 20), adding pins to the bottom of the substrate, and any necessary wire bonding on the top (step 22). Nickel is used as a corrosion protector and helps gold adhesion. It is plated on and diffusion bonded to the molybdenum [37]. The gold is used to enhance wettability to solders and pin brazes.

The testing of a complex MLC requires sophisticated equipment and is speeded up with computer control and multiple-probe sets [54]. On the 90-mm substrate, there are 1800-pin pads on the bottom and 100 or 118 chip sites on the top, each chip site having vias for the chip, which are surrounded by via/line/pad groups used for probing and rewiring. This makes over 250,000 probe contacts that need to be made and verified to adequately test this substrate [36]. A typical chip site is shown in Figure 3. The 1800 pins required on the I/O side are Kovar and are brazed to the pads using a low-temperature noble metal braze [24].

MLC substrate design flexibility can also allow for some rewiring of internal layers to occur on the top surface, after firing. This would be used to change circuits inside a substrate, or perhaps to correct some internal defects found during testing (step 20). This rewiring is step 22, and can be done using the via/line/pad groups found around the chip sites in Figure 3. By cutting the line with a laser, the pad is spearated from the via. Using wire bonding, the pad can be joined to any other pad. This technique is shown in Figure 6, which illustrates a smaller 50-mm substrate with six chips and a number of wires on it. Figure 7 shows a cross section of part of an MLC substrate showing various wiring layers within it.

I. Chip Joining and Encapsulation

The final steps of joining the chips and doing the final testing and encapsulation change an MLC substrate into an MLC module. Depending on the manufacturer, chips can be joined to the substrate by wire bonds, beam or ribbon leads, or flip-chip solder balls. The latter process allows all the chips to be simultaneously attached to the substrate in a single reflow cycle [24,36]. The module is then tested and encapsulated (steps 24 and 25). The encapsulation is usually accomplished by embedding the chips in a polymer, or by enclosing the chips under a cap. If protection of the pin side is needed, there are glasses available which can be applied before the pins are attached [64]. If enhanced cooling is needed to prevent circuit damage, liquid or gas cooling jackets can be added to the module [37]. For the IBM modules, the latter process is used for

Table 2 Laminated MLC Development

Year reported	Ceramic	Metallurgy	Min. via diameter/min. spacing (mils)	Min. line width/min. spacing (mils)
1961	Al_2O_3	Mo, Mo-Mn	5/10	5/10
1967	Al_2O_3	Mo	–	8/–
1968	–	–	8/25	15/–
1969	Al_2O_3	Pd	5/10	4/8
	Al_2O_3	Noble	10/20	5/10
1971	Al_2O_3	Mo	10/25	8/–
	Al_2O_3	W/Mo/Pd/Pt	5/10	4/8
1972	Al_2O_3	Mo-Mn	5/10	8/16
1973	Al_2O_3	W	8/16	5/10
1975	Al_2O_3	W/Mo/Mo-Mn	5/20	4/9
1977	Al_2O_3 BeO	W/Mo	7/20	4/8
1978	Al_2O_3 (A-16)	W/Mo/Pd/Pt	10/–	–
1980	Al_2O_3	Mo	5/10	–
	Al_2O_3	Mo	4/8	4/10

Shrinkage tolerance	Layers	Size	Other	Ref.
±0.2%	3–24	0.31 in.^{2}	Micromodule	65,66
—	6	0.25 in.^{2}		15
—		1 in. × 2 in.	300 vias per in.^{2}	35
—	15	0.5 in.^{2}		16
±0.5%	6	1 in. × 2 in.		30
—	5–10	0.5 in.^{2}		18
±0.5%	10–20	3 in.^{2}		25
±0.6%	11	1/2 in.^{2}		20
—	10	5 in.^{2}	Includes capacitors	33
±0.5%	11	4.5 in. × 4 in.	Includes capacitors	34
±0.5%	15–20	2.75 in. × 3.75 in.		26,27
±0.1%	—	—		28
—	23	2 in.^{2} (50 mm^2)	9 chips	24
±0.1%	33	3.5 in.^{2} (90 mm^2)	100 chips (350,000 vias)	37,45

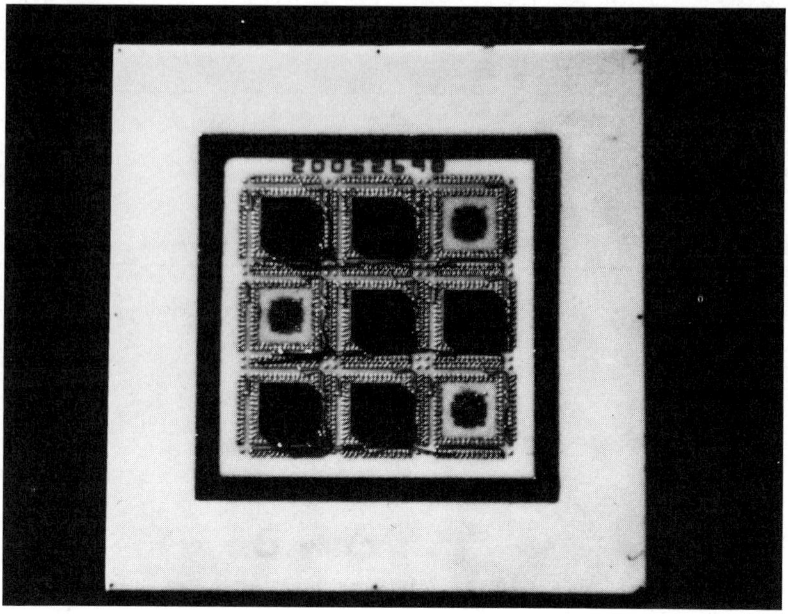

Figure 6 50-mm MLC with six chips and top surface wires. (Courtesy of IBM, East Fishkill, N.Y.)

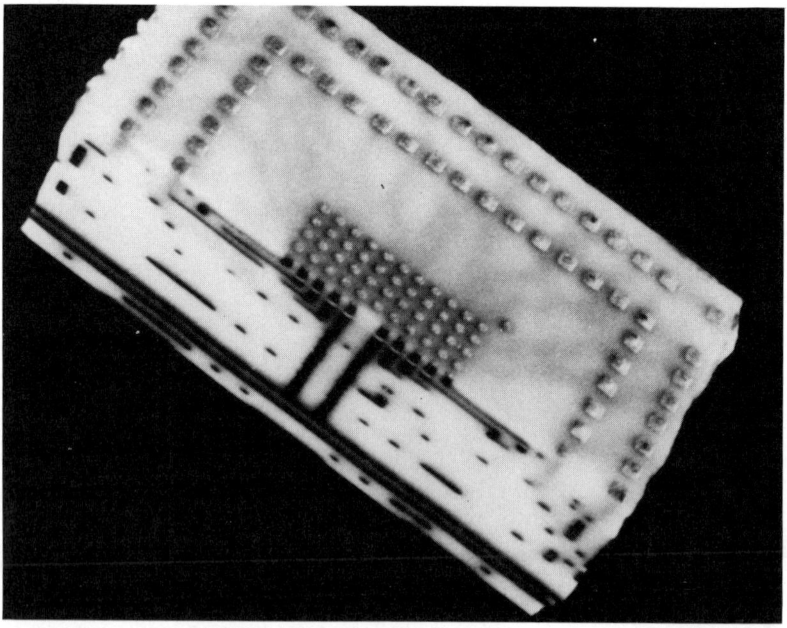

Figure 7 Cross section of a typical MLC chip site. (Courtesy of IBM, East Fishkill, N.Y.)

cooling the 90-mm substrates and Kovar caps are used on the 50-mm and smaller ones [24].

J. Process Commentary

There are a lot of different variations of this process used by various manufacturers of MLC substrates. The substrate family built at IBM is used here as one example among the many listed in the references that typifies the complex process steps that this technology incorporates and the sophisticated product this technology is capable of producing. Every manufacturer will have its own variations on this basic process and areas where its expertise is highly developed. For example, the screened multilayer process [28,33] uses only one green sheet, upon which successive layers of conductor and alumina paste are screened, before the entire substrate is cofired. The bibliography in this chapter does not contain all the published work relating to the various process steps used in MLCs but will hopefully give the reader a broad overview of the technology and also an in-depth look at the process itself.

Table 2 shows the progress made in the technology since 1968, as reported in the literature. Vias have decreased in size by 1/2, with via spacing decreasing by even more. Line widths of screened paste have also been reduced, with comparable reductions in spacing. Shrinkage tolerances now approach ±0.1%, giving good dimensional control in these complicated ceramic/metal composites. The size of reported substrates has gone up to 5 in. square (125 mm) and the number of layers reported has increased to 33. Via density has increased from 300/in.^{2} (50/cm^2) to more than 10 times that amount. Up to 118 chips can be put on one MLC substrate. The technology has made tremendous gains, with more promised in the future.

VII. SUMMARY

The laminated MLC substrate technology has made significant progress in the last 20 years. From simple laminations of green tape to build planar substrates, the technology has advanced to the point that chips containing over 25,000 logic circuits can typically be wired on one single MLC substrate. Sintered inside such a substrate, made up of 30 or more discrete layers laminated together, would be over 100 m of screened lines and 350,000 separate vias. The technology lends itself to computer control of the various process steps as well as the design and the artwork, thus becoming one of the most sophisticated ceramic technologies yet developed. Results of work from almost 20 different companies have been reported, making MLC technology one of the future growth areas in the ceramic industry.

ACKNOWLEDGMENTS

The author would like to thank Rhoda Stoltenberg for preparing the manuscript, Bill Swiss and Craig Hawkins for their advice and encouragement, Jack Butler for the photos, and Don Barbour and Bernie Schwartz for editing the text.

REFERENCES

1. L. K. Keyes et al., *Int. Hybrid Microelectron. Symp.*, Beverly Hills, Calif., Nov. 1970, p. 6.4.1.
2. J. C. Gioia, *Solid State Technol.*, May 1971, p. 43.
3. J. E. McCormick and D. W. Calabrese, *NEPCON Proc.*, 1969, p. 500.
4. R. F. Kolc, *Int. Microelectron. Conf. Proc.*, Anaheim, Calif., 1978, p. 86.
5. K. Kruzweil and J. Loughran, *IEEE Electron. Components Conf.*, 1973, p. 212.
6. R. F. Redemske, *14th IEEE Comput. Soc. Int. Conf.*, San Francisco, Calif., Feb. 1977, p. 253.
7. C. W. Umbaugh, *14th IEEE Comput. Soc. Int. Conf.*, San Francisco, Calif., Feb. 1977, p. 263.
8. J. Marshall and F. Rode, *Int. Microelectron. Symp. Proc.*, Minneapolis, Minn., 1978, p. 26.
9. A. D. Close, *Solid State Technol.*, May 1971, p. 38.
10. S. Lebow, *30th IEEE Electron. Components Conf.*, 1980, p. 307.
11. *Mach. Des.*, Nov. 23, 1978, p. 10.
12. W. Hatfield and D. Wicher, *28th IEEE Electron. Components Conf.*, Anaheim, Calif., 1978; *Circuits Manuf.*, July 1978, p. 7.
13. A. H. Agajanian, *Solid State Technol.*, Oct. 1975, p. 56; Sept. 1976, p. 87.
14. J. L. Park, Jr., "Manufacture of Ceramics," U.S. Patent 2,966,719, 1961.
15. B. Schwartz and D. L. Wilcox, *Proc. Electron. Components Conf.*, Washington, D.C., 1967, p. 17.
16. P. R. Theobald, M. P. Davis, and J. T. Bailey, *Hybrid Microelectron. Symp.*, 1969, p. 447.
17. D. A. Chance, *Metall. Trans.*, *1*, 685 (1970).
18. D. L. Wilcox, *Solid State Technol.*, Jan. 1971, p. 40; Feb. 1971, p. 55.

19. D. A. Chance and D. L. Wilcox, *IEEE Proc.*, *59*, 1455 (1971).
20. H. D. Kaiser, F. J. Pakulski, and A. F. Schmeckenbecher, *Solid State Technol.*, May 1972, p. 35.
21. W. S. Young, *J. Less-Common Met.*, *32*, 321 (1973).
22. W. R. Swiss, *Circuits Manuf.*, Aug. 1979, p. 43.
23. A. H. Kumar and J. B. Niklewski, *Am. Ceram. Soc. Bull.*, *58*, 1179 (1979).
24. A. J. Blodgett, Jr., *Proc. Electron. Components Conf.*, San Francisco, Calif., 1980, p. 283.
25. B. M. Hargis, *Solid State Technol.*, May 1971, p. 47.
26. M. L. Burch and B. M. Hargis, *Ceramic Chip Carrier*, 3M Corp., 1977.
27. *Guidelines for Designing Multilayer Substrates*, 3M Corp., 1977.
28. R. E. Mistler, D. J. Shanefield, R. B. Runk, in *Ceramic Processing Before Firing*, G. Onoda, Jr., and L. Hench, eds., Wiley, New York, 1978.
29. C. E. Williams, *Am. Ceram. Soc. 81st Annu. Meet.*, Cincinnati, Ohio, #4-E-79, 1979. [See *Am. Ceram. Soc. Bull.*, *58*, 363 (1979).]
30. W. R. Keller et al., *Proc. Electron. Components Conf.*, 1969, p. 52.
31. E. L. Rogers, *Am. Ceram. Soc. Pacific Coast Reg. Meet.*, San Francisco, Calif., 20-E-80P, 1980. [See *Am. Ceram. Soc. Bull.*, *59*, 830 (1980).]
32. *Fine Ceramics* (bulletin), Kyocera International, San Diego, Calif.
33. T. Ihochi, *IEEE Electron. Component Conf.*, Washington, D.C., 1973, p. 204.
34. J. D. Welterlen and S. D. Anderson, *Int. Microelectron. Symp.*, Orlando, Fla., 1975, p. 358.
35. E. G. Fagersten, *Hybrid Microelectron. Symp.*, 1968, p. 401.
36. *IBM News*, Special Edition, Nov. 1980.
37. A. J. Blodgett, Jr., and D. R. Barbour, *IBM J. Res. Dev.*, *26*, 30 (1982).
38. G. Onoda, Jr. and L. Hench, eds., *Ceramic Processing Before Firing*, Wiley, New York, 1978.
39. G. Onoda, Jr. and D. J. Shanefield, "Organic Additives and Ceramic Processing Principles," a course for the Center for Professional Advancement, Chicago, Il., Apr. 30, 1980.
40. A. G. Pincus and L. E. Shipley, *Ceram. Ind. Mag.*, Apr. 1969, p. 106.
41. D. J. Shanefield and R. E. Mistler, *Am. Ceram. Soc. Bull.*, *53*, 416, 564 (1974).
42. R. B. Runk and M. J. Andrejco, *Am. Ceram. Soc. Bull.*, *54*, 199 (1975).
43. J. C. Williams, in *Treatise on Materials Science and Technology*, F. Y. Wang, ed., Academic Press, New York, 1976.

44. F. V. DiMarcello, P. L. Key, and J. C. Williams, *J. Am. Ceram. Soc.*, *55*, 509 (1972).
45. K. S. Desai, private communication.
46. E. N. Urfer, *Am. Ceram. Soc. Bull.*, *52*, 713 (1973).
47. K. Kruzweil and J. Loughran, *IEEE Trans. Parts, Hybrids, Packag.*, *PHP-9*(4)/216 (1973).
48. L. F. Miller, *Solid State Technol.*, *17*, 54 (1974).
49. *Thick Film Products Catalogue*, Engelhard Ind., East Newark, N.J.
50. *Thick Film Materials*, E. I. DuPont, Wilmington, Del.
51. J. Bradigan, *Insul./Circuits*, Jan. 1980, p. 33.
52. *Technical Notes 2.1.2* (9/73), E. I. DuPont, Wilmington, Del.
53. R. M. Rosenberg, this volume.
54. *The Circuit*, IBM, East Fishkill, N.Y., Feb. 1979, p. 5.
55. R. A. Gardner, *J. Solid State Chem.*, *9*, 336 (1974).
56. L. W. Bean, *Br. Ceram. Soc. Trans.*, 1971, p. 121.
57. A. G. Pincus, *Ceram. Age*, Mar. 1954, p. 16.
58. J. E. Fenstermacher, *Am. Ceram. Soc. Bull.*, *48*, 775 (1969).
59. J. C. Williams, *Am. Ceram. Soc. Bull.*, *56*, 580 (1977).
60. *Thick Film Technology 2.4.1* (3/72), E. I. DuPont, Wilmington, Del.
61. W. R. Swiss, private communication.
62. R. A. Gardner and R. W. Nufer, *Solid State Technol.*, *17*, 38 (1974).
63. G. G. Werbizky, P. Winkler, and F. W. Haining, *Electronics*, Aug. 2, 1979.
64. A. H. Kumar and R. R. Tummula, *Am. Ceram. Soc. Bull.*, *57*, 738 (1978).
65. H. W. Stetson, B. Schwartz, and W. H. Liderback, two papers presented at *Electron. Div. Meet.*, *Am. Ceram. Soc.*, San Francisco, Calif., Oct. 26, 1961. [See *Am. Ceram. Soc. Bull.*, *40*, 584 (1961).]
66. "Micromodule Crystal Units," U.S. Signal Corps Contract DA-36-039-SC-85046, Mar. 31, 1961.
67. D. P. Seraphim and I. Feinberg, *IBM J. Res. Dev.*, *25*, 617 (1981).
68. A. J. Blodgett, Jr., *Sci. Am.*, *249*, 86 (July 1983).
69. B. Patrusky, *Think*, Jan./Feb. 1983, p. 23.
70. Y. Shimada, K. Utsumi, M. Suzuki, H. Takamizawa, M. Nitta, and S. Yano, *IEEE Electron. Comput. Conf.*, Florida, May 1983, p. 314.

8

Highly Conductive Ceramics

HARRY L. TULLER
Massachusetts Institute of Technology, Cambridge, Massachusetts

I. INTRODUCTION

For the better part of this century, ceramics have been viewed primarily from the perspective of their insulating or dielectric properties. Applications such as high-voltage standoffs, capacitors, spark plugs, and the like followed naturally. Within the last several decades, however, this picture of ceramics has changed rapidly. It has now become evident that ceramics exhibit the full spectrum of electrical properties spanning the gap from metallic to insulating electronic conductors and from fast ionic to insulating ionic conductors. This rapid expansion in the types of phenomena exhibited by ceramics has both challenged theorists to develop more sophisticated models to characterize these properties and technologists to develop novel devices based on them. A cursory glance at both the recent scientific and patent literature attests to the fact that these challenges have been taken up in all seriousness.

Electrical conductors can be subdivided into various categories: by the nature and sign of the charge carriers, by the magnitude of electrical conductivity, by the carrier mobility mechanism, by their temperature dependence, and so on. The categories one chooses will depend on those aspects which are of most immediate concern (e.g., for an electronics engineer, the magnitude of resistance in a circuit element).

The full range of electrical conductivities of general interest covers roughly 25 orders of magnitude, as illustrated in Figure 1. The bounds for the three categories of electronic conductors—metals, semiconductors, and insulators—are somewhat vague, as are the distinctions between the ionic conductors—fast ion conductors, solid electrolytes, and insulators. Later in the course of this chapter more precise definitions will be presented. Nevertheless, from the standpoint of applications these bounds are generally satisfactory.

Many of the applications listed in Figure 1 rely heavily on the enhanced chemical and thermal stability exhibited by ceramics over that of metals or polymers. Examples include MHD electrodes operating in plasmas at temperatures above 1200–1500°C and ionically conductive membranes separating molten sodium metal and sulfer in high-energy density batteries operating at 300°C. Other applications, such as sensors, on the other hand, rely on the high sensitivity exhibited by some ceramics to variations in their environments (e.g., atmosphere, temperature, and pressure).

The aim of this chapter is to (1) describe the key features necessary for reasonably high ionic and electronic conduction in ceramics; (2) describe how the crystal, electronic, and defect structures affect transport properties; (3) review experimental results and interpretations for a number of representative systems; and (4)

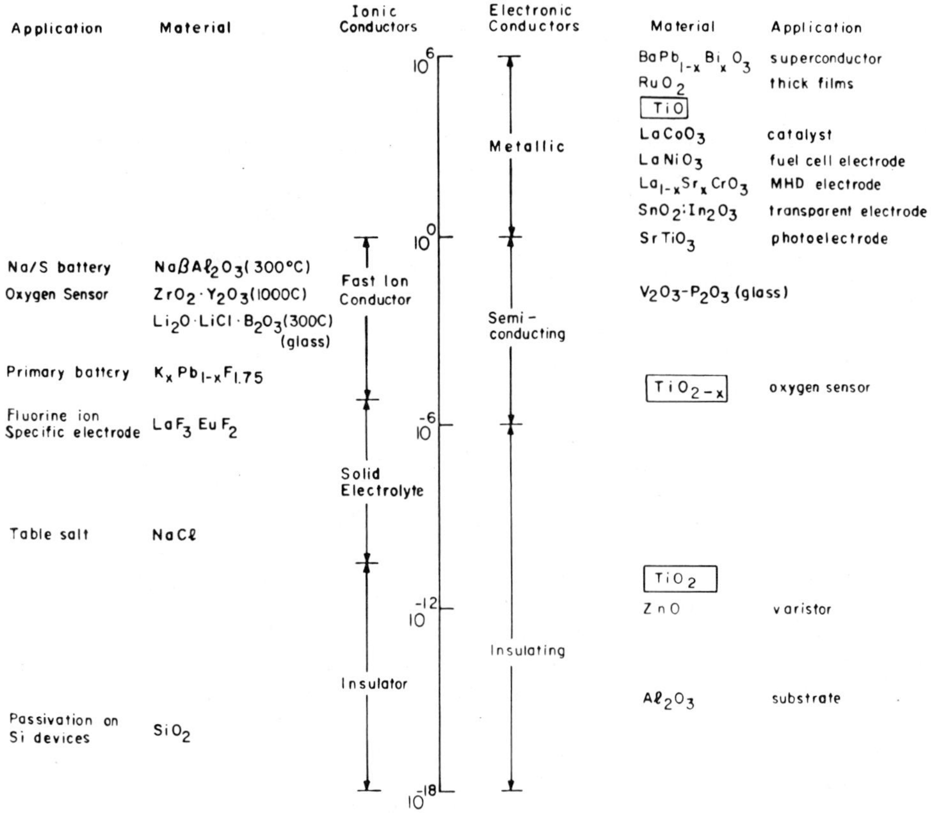

Figure 1 Logarithmic scale of the electrical conductivities of materials categorized both by magnitude and carrier type (i.e., ionic versus electronic). Examples of the various categories are given together with some typical applications.

discuss briefly some of the considerations important in choosing and optimizing ceramics for specific applications (e.g., energy conversion, sensor technology, etc.).

Although in attempting to treat both highly conducting electronic and ionic ceramics in one chapter, one is obliged to skim or miss entirely some important aspects of each, I believe that in the synthesis some of the commonality becomes more evident, particularly with regard to the distinction between the different categories of conductors and in relation to the analysis of defect structures. The subject of mixed ionic-electronic conductors has been treated explicitly by the author in a previous publication [1].

II. ELECTRONIC CONDUCTION IN OXIDES

The band theory of Block and Wilson [2] provides us with simple rules for predicting the conductive nature of solids. Partially filled bands correspond to metallic conduction, while completely filled or empty bands signify either semiconducting or insulating character. In Figure 2 we illustrate the three possibilities while emphasizing that the distinction between an intrinsic semiconductor and insulator is only a qualitative one (i.e., the magnitude of the energy gap separating the occupied valence band from the empty conduction band). This scheme, which enables us to predict the density of mobile carriers contributing to conduction in a given material, has been highly successful in analyzing the electronic properties of the majority of conventional semiconductors and metals. As we will later discover, serious discrepancies often exist between predicted and observed properties when this theory is applied to conduction in metal oxides.*

The electrical conductivity σ, which is the proportionality constant between current density j and the electric field E, is given by

$$\frac{j}{E} = \sigma = ne\mu \tag{1}$$

*Although ceramics include large numbers of refractory compounds (e.g., nitrides, carbides, borides, etc.), in this chapter the primary emphasis will remain on oxides. Most of the principles discussed, however, may be applied to these other systems as well.

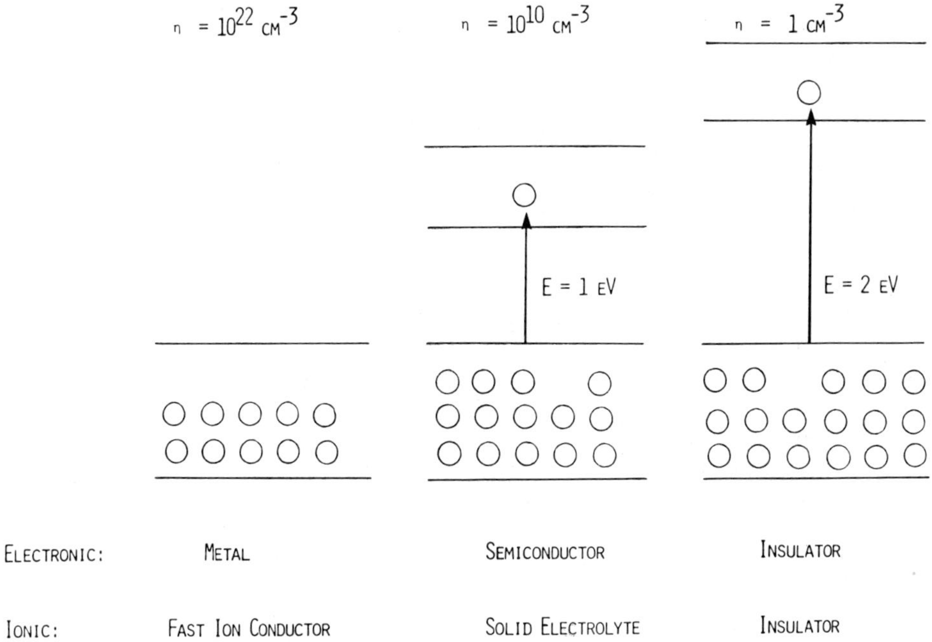

Figure 2 The three categories of electronic and ionic conductors are distinguished by the density of free carriers available at a temperature T. For the electronic conductors a comparison is made of the relative occupancy of valence and conduction bands for the various categories. For the ionic conductors, the bands correspond to the relative occupancy of ionic sublattices instead (e.g., the lower band to the cationic sublattice and the upper band, cation interstitial sites). The numbers at the top of each column specify the carrier density expected at $T \simeq 500$ K.

in which n is the carrier density (number/cm^3), μ the mobility (cm^2/V-sec), and q the electronic charge. The many orders of magnitude difference in σ which distinguish metals, semiconductors, and insulators generally resides in differences in n rather than μ. Thus in the one half-filled band of a metal such as sodium, a high density of electrons, on the order of 10^{22} cm^{-3}* coexist with an equal

---

*This is not strictly true since only those electrons in the vicinity of the Fermi energy ($\sim$kT) have easy access to unoccupied energy levels.

number of unoccupied states, while in a semiconductor or insulator the ratio of occupied to unoccupied levels is far from 1 [e.g., n(300 K) in Si $\sim 10^{10}$ cm^{-3}].

To predict the numbers of electrons that become excited across the band gap from valence to conduction band at a given temperature, one must know the density-of-states functions, which describe the number of states per unit energy interval in the vicinity of the band edges and the probability of their occupancy, which is given by the Fermi function,

$$f(E) = \left[1 + \exp\left(\frac{E - E_f}{kT}\right)\right]^{-1} \tag{2}$$

in which E is the energy of interest, E_f the Fermi energy, and T the temperature in kelvin. Utilizing a density-of-states function derived for the case of a spherical Fermi surface and convoluting it with the Fermi function, the following important relations for the density of electrons in the conduction band and holes in the valence band are derived [3]:

$$n = N_c \exp -\left(\frac{E_g - E_f}{kT}\right) \tag{3}$$

$$p = N_v \exp -\left(\frac{E_f}{kT}\right) \tag{4}$$

in which E_g is the band gap energy, and the effective density of states in the conduction band, N_c, and the valence band, N_v, are defined by

$$N_{c,v} = 2\left(\frac{2\pi m^*_{e,h} kT}{h^2}\right)^{3/2} \tag{5}$$

in which m_e^* and m_h^* are the electron and hole effective masses and k and h are the Boltzmann and Planck constants, respectively.

Taking the product of Eqs. (3) and (4), we obtain

$$np = N_c N_v \exp\left(\frac{-E_g}{kT}\right) \tag{6}$$

which is independent of the Fermi energy and may be viewed as the equilibrium constant corresponding to the reaction

$$O \rightarrow e' + h^{\bullet} \tag{7}$$

which describes the thermal generation of an electron-hole pair by excitation of an electron across the band gap.

For an intrinsic semiconductor

$$n = p = n_i = (N_c N_v)^{\frac{1}{2}} \exp \frac{-E_g}{2kT} \tag{8}$$

and it is thus quite clear how a factor of 2 in band gap could easily distinguish a semiconductor from an insulator in terms of Figure 1 at a given temperature T.

A certain minimum concentration of atomic defects are predicted on the basis of statistical thermodynamics in all solids. Generally, large concentrations of additional defects are added either intentionally or nonintentionally during the processing of materials. Due to the distortions in potential they create in their vicinity, discrete energy levels form within the energy gap, often in vicinity of the band edges. Impurities with an excess or deficiency of electrons relative to the host lattice may thus donate electrons to the conduction band or accept electrons from the valence band, thereby contributing to the electron-hole density far in excess of that expected by the intrinsic process given in Eq. (8). Impurity ionization energies typical in elemental semiconductors are of the order of 0.01 eV, consistent with a modified hydrogenic model in which the ionization energy (13.6 eV) is greatly reduced as a consequence of the dielectric constant, which enters the equation as ε^{-2}. A schematic depicting the energy levels in extrinsic semiconductors is shown in Figure 3.

Other defect levels often exist far from the band edges. These deep levels are formed when the wave functions of the defects do not mix effectively with those of the host. Such levels contribute little directly to the excess carrier density but can serve to compensate the influence of shallow levels by trapping their excess electrons or holes.

With a single type of donor or acceptor, two extrinsic conduction regimes are obtained as a function of temperature. At low temperatures (i.e., $kT < E_i$; E_i = ionization energy of impurity) the temperature dependence of carriers is given by

$$n = (N_i N_c)^{\frac{1}{2}} \exp \frac{-E_i}{2kT} \tag{9}$$

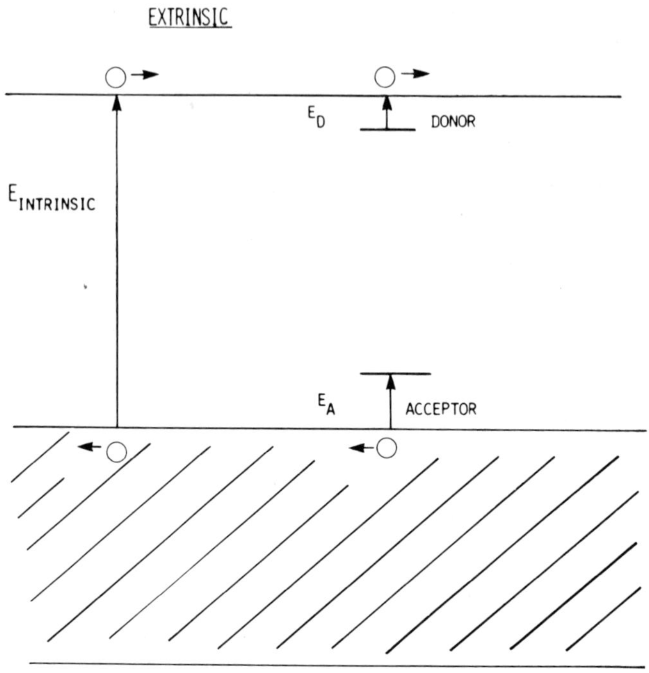

Figure 3 The distinctions between extrinsic and intrinsic conductors are emphasized here (i.e., the much reduced energies required to generate free carriers and the generation of only one type of carrier in the extrinsic case).

in which N_i is the donor impurity density. A similar expression holds for holes, with N_i now being the acceptor density. For $kT \gtrsim E_i$, all impurity levels are effectively ionized and the carrier density is temperature independent and equal to the impurity density. Finally, at sufficiently high temperatures the intrinsic carrier density will surpass that of the impurity-generated carriers and a temperature dependence of the form given in Eq. (8) is obtained.

Predictions based on the models above often hold true for metal oxides. Thus TiO_2 with an empty 3d conduction band approximately 3 eV above the valence band is highly insulating in its undoped form $\sigma \sim 10^{-12}$ $(\Omega\text{-cm})^{-1}$, while when doped with pentavalent niobium, with an ionization energy of only $\sim$0.05 eV, or rendered oxygen deficient, it becomes an n-type semiconductor with $\sigma \sim 1$ $(\Omega\text{-cm})^{-1}$ at room temperature (see Figure 1). Very wide band gap oxides such as MgO (Eg $\sim$ 6.7 eV), which apparently accommodate only deep traps, remain highly insulating to temperatures above 1200°C. On the other hand, oxides such as ReO_3, which possess a partially filled 5 d conduction band, exhibit metallic conduction as expected from the Block-Wilson band picture. Furthermore, the change in oxidation state of Ti from, for example, 4+ to 2+ in TiO, is seen to convert an insulator to a metal by partially populating Ti 3d levels. Changes in the electrical conductivities of the titanium oxides due to valence changes in the Ti ion are emphasized in Figure 1.

A. Narrow-Band Conduction

In contrast to those examples in which the electronic properties of oxides are consistent with the predictions above, there are many more that conflict with normal expectations. MnO, for example, has five d electrons, resulting in a partially filled band. Contrary to expectations, rather than being metallic, this material is highly insulating. Similar arguments hold for both CoO and NiO in this group, as well as many other oxides among the perovskites (e.g., $LaMnO_3$, $LaFeO_3$), corundum structure oxides (e.g., Cr_2O_3, Fe_2O_3), and so on [4].

To begin to understand the sources of these discrepancies, a number of features characteristic of the metal oxides must be taken into account. First, as pointed out by Mott [5], when overlap between nearest-neighbor electronic wave functions is small, electron-electron correlations become sufficiently important so as to result in the localization of the "conduction" electrons around their ion cores, resulting in the formation of a "Mott insulator." Since transition and rare earth metals with weakly extended d and f orbitals often contribute to the formation of conduction or valence bands in oxides, it is therefore not surprising that such materials may not satisfy predictions of classical band theory in which electron-electron correlations may be neglected.

The self-trapping of carriers in partially filled but narrow bands is visualized in terms of a lattice distortion induced by the carrier polarizing the charged ions in its vicinity. When the distortion is highly localized, the carrier and the accompanying distortion is known as a small polaron. Motion of this quasi-particle proceeds by an activated hopping process and takes the form [6]

$$\mu = \left[\frac{(1 - c)\, ea^2 \nu_0}{kT}\right] \exp\left(\frac{-E_H}{kT}\right) \qquad (10)$$

in which E_H is the hopping energy, $(1 - c)$ the fraction of sites unoccupied, a the jump distance, and ν_0 the attempt frequency. As might be expected, small polaron mobilities are considerably reduced and are generally reported to be on the order of 10^{-4} and 10^{-2} cm^2/V-sec at elevated temperatures—hundreds to thousands of times smaller than in normal band conduction.

Where the distortion is inadequate to self-trap the carrier, but still serves to slow the carrier, a designation of "large polaron" is applied. The large polaron mobility at temperatures above the Debye temperature is given by [7]

$$\mu = \mu_0' T^{-\frac{1}{2}} \tag{11}$$

and is expected to be $\sim$ 1–100 cm^2/V-sec at elevated temperatures. Under these circumstances the bandwidths are sufficiently wide to support itenerant electronic transport and thus agree with expectations based on the band model. Examples of both cases (i.e., small versus large polaron) will be presented shortly.

Second, because the structures of oxides vary considerably, symmetry considerations must be examined to establish the nature and degree of splitting of bands, related to certain hybridized atomic orbitals, into subbands. A tight binding approach as discussed by Honig [8,9] is very effective in highlighting the implications of these features on transport in oxides.

Last, the influence of nonstoichiometry on the electronic carrier density needs to be examined. Changes in stoichiometry (i.e., the metal-to-oxygen ratio), induced by changes in the oxygen or metal vapor partial pressure in the gas phase, can result in the creation of large numbers of donor- or acceptorlike states in the band gap and thus render a wide-band-gap oxide, semiconducting. Furthermore, the influence of donor or acceptor impurities may be markedly affected by relatively small changes in the stoichiometric ratio. Examples of the influence of stoichiometric imbalance on electronic properties will also be provided.

B. Band Structure Determinations for Narrow-Band Oxides

Aside from the Block-Wilson approach, which examines the implications of applying periodic constraints on a quasi-free electron, a tight-binding model that begins with electrons localized near the ion cores has also been successfully applied in predicting the major features of the quantum mechanical band picture in solids. Because of the localized nature of the d and f levels in many of the oxides of interest here, it is natural to examine the implications of the tight-bindng model on band formation and transport in such oxides. The analysis, which follows closely that of Goodenough [10] as interpreted by Honig et al. [8,9], is presented next.

In the tight-binding approach one assumes that the crystal wavefunction may be written as a linear combination of the unperturbed atomic functions and that the presence of the neighboring atoms is reflected only as a weak perturbation in the potential seen by the atom of interest. In the solution of Schrödinger's equation, one thus obtains for the energy term the initial energy plus a small perturbation given by [11]

$$E = E_0 - \alpha - \beta \sum_m \exp(ik \cdot r_m) \tag{12}$$

in which E_0 is the unperturbed energy; α the so-called "self" term, which reflects the shift in energy of the level due to the change in potential seen by the central atom; and β the overlap integral, which is a measure of how strongly the wavefunctions of the neighboring atoms overlap with the central atom. Given the weak interactions assumed in the tight-binding model, one generally assumes little error in truncating the summation beyond the nearest-neighbor atoms.

For a simple cubic lattice (in which $r_m = a$) one may readily show that the last term in Eq. (12) can take on values between $\pm 6\beta$, thus broadening the initially discrete level into a band with width 12β, as illustrated in Figure 4. Clearly, as the overlap of wavefunctions increases, so does the width of the band. The effective mass may be calculated by examining the curvature of the E versus k relationship (in the vicinity of k = 0), that is,

$$m^* = \hbar^2 \left(\frac{d^2E}{dk^2} \right)^{-1} = h^2 (2\beta a^2)^{-1} \tag{13}$$

and one readily observes the inverse relationship between m* and the bandwidth.

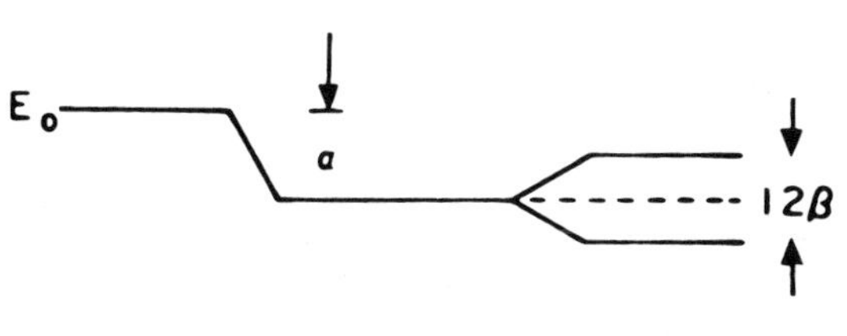

Figure 4 The degenerate electronic states of isolated atoms are observed to shift lower and broaden into a band of energies with bandwidth 12β as atoms are brought together in compound formation.

Certain rules need be obeyed in deriving the crystalline energy bands from the interaction of cationic and anionic atomic wavefunctions [8]. First, energy bands split under the influence of crystal fields (e.g., d levels split into a lower threefold degenerate t_{2g} level and a higher twofold degenerate e_g level under octahedral symmetry). Second, meaningful overlap resulting in bond formation requires both compatible symmetry and proximity in energy. Strong interactions lead to splitting of resulting hybrid energies into bonding and antibonding levels, with the degree of splitting proportional to the degree of interaction. Last, as indicated in Eq. (12), the widths of the bands also increase with increasing overlap. Thus bandwidths increase as one proceeds from δ-type to π-type to σ-type orbital overlaps and for a given type of bond as one proceeds from lower- to higher-lying energies. With these guidelines in hand we can proceed to examine the energy diagrams for a number of oxide systems.

C. Band Structures and Electrical Conduction—Perovskites

For simplicity we begin with ReO_3, which crystallizes in the perovskite structure ABO_3 illustrated in Figure 5 but with the A site

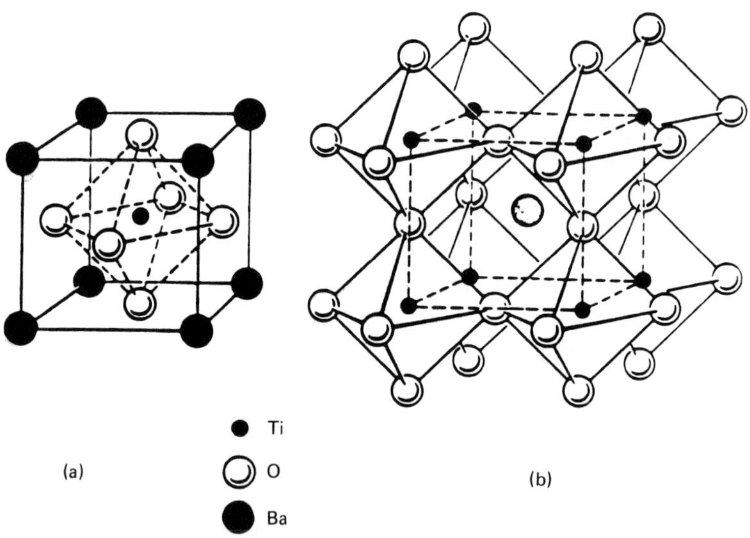

Figure 5 Ion positions in ideal perovskite structure with $BaTiO_3$ used as an example. (From Ref. 10.)

empty. Due to the octahedral surroundings of the Re ion, its 5d levels split into the t_{2g} and e_g sublevels. The O_{2p} orbitals also split ($2p_\pi$, $2p_\sigma$), with the $2p_\sigma$ orbital lying lower in energy as its lobes point directly toward the electropositive neighbors. The orientations of the various orbitals along the edges of the unit cell, which correspond to the directions of greatest overlap, are illustrated in Figure 6.

Due to their σ-type overlap (e.g., Figure 6a) a series of bonding and antibonding σ-like bands are formed from admixtures of 6p, 6s, $5e_g$, $2p_\sigma$, and 2 s states. Several of the σ bands with antibonding character are illustrated in Figure 7, with σ* (e_g) and σ* corresponding to the bands with predominant $5e_g$ and 6s character,

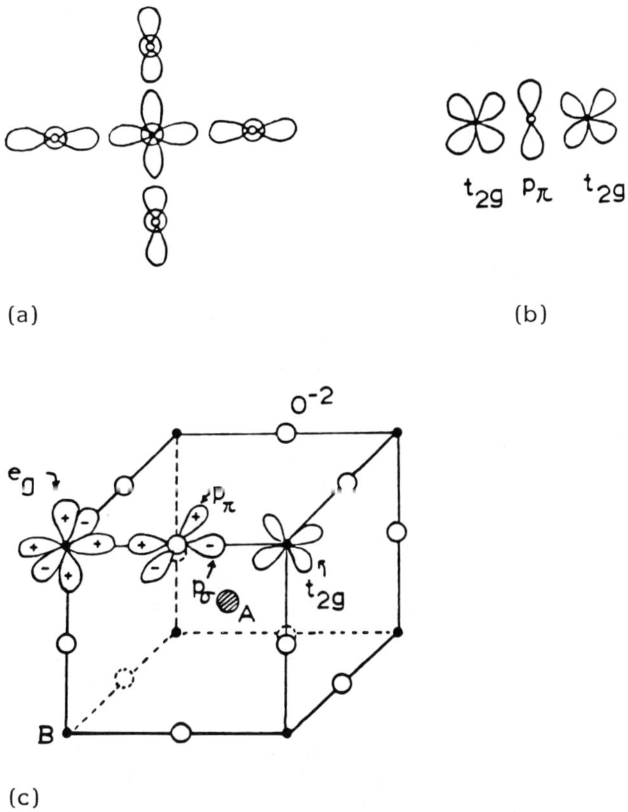

Figure 6 Orbital overlap: (a) σ bonding; (b) π bonding. (c) Relative orientations of the cation 3d orbitals and oxygen 2p orbitals, leading to both σ and π types of bonding in the perovskite structure. (From Ref. 38.)

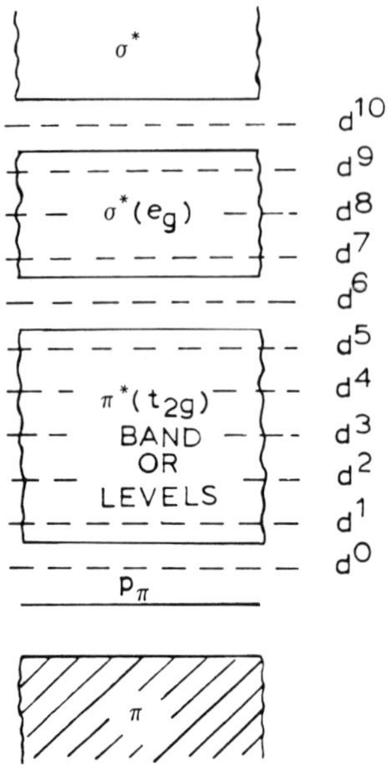

Figure 7 Predicted Fermi levels for the B cation electronic band configurations in the perovskite structure. (After Ref. 9.)

respectively. Figure 6b illustrates the weaker π-type bonding that occurs between $5t_{2g}$ and $2p_{\pi}$ orbitals, resulting in the formation of the π-like bands also illustrated in Figure 7. In contrast to the schematic band diagrams of Figure 7, the higher-lying σ-like bands can be expected to be wider than the π-like bands below them, in agreement with Eq. (12). This has important implications for the nature of transport in the respective bands, as we shall soon discover.

Since the number of states available for occupancy are conserved, one need only fill them in sequence to establish the position of the Fermi energy. The total number of valence electrons in the formula unit ReO_3 equals 25, of which 24 fill the lower bands of anionic character. This leaves one electron in the $5t_{2g}$ band, resulting in a partially filled band. ReO_3 is indeed metallic with a conductivity within a factor of 2 of Cu at room temperature [9]. On the other hand, WO_3, with a closely related structure to ReO_3 but with a d^0 electronic configuration, is found to be a semiconductor whose conductivity is, however, very sensitive to its degree of nonstoichiometry [12].

Extension of the model above to the case of the perovskites in which the body center position is filled with the A atom (ABO_3) is straightforward since the energy-band diagram described above for ReO_3 is little perturbed, given the negligible overlap along the A-B and A-O bond directions (see Figure 6). The bands associated with the A stoms (e.g., Sr) lie well above the other bands so far considered. Consequently, the valence electrons associated with the A atoms will fall to the lower empty d-like energy states and must be included when establishing the position of the Fermi energy.

Based on the energy scheme presented in Figure 7 for perovskites (similar considerations result in band distributions for the other crystal structures), all compounds with electronic configurations corresponding to $d^1 - d^5$ and $d^7 - d^9$, which correspond to either partially filled $\pi^*(t_{2g})$ or $\sigma^*(e_g)$ bands, should be metallic, while d^6 or d^{10} configurations should be semiconducting or insulating. This of course ignores the influence of the bandwidth on transport. Below a critical value b_c, the carriers must localize and render the expected metal semiconducting or insulating.

Goodenough [10] compiled a table classifying the energy bands found in perovskite oxides in terms of their localized or collective natures. An update of this table is presented in Figure 8, in which the compounds are grouped in columns according to their electronic configurations and in rows according to the valence states of the two sets of cations. In general, oxides with small net spin and high B cation valence state enjoy the greatest degree of overlap.

In order for a band to support metallic conduction, its bandwidth b must be greater than b_c. For some compounds $b^\sigma > b^\pi > b_c$, so a partially filled t_{2g} or e_g band will lead to metallic conduction; this holds true [e.g., $SrVO_3$ ($\pi^*1\sigma^*0$) and $LaNiO_3$ ($t_{2g}^6\sigma^*1$)]. The latter is an example where only the upper σ^* band of the solid is sufficiently wide to remain delocalized. For high spin conditions and low B ion valence, both b^σ and b^π drop below b_c, causing both bands to localize and rendering oxides such as $LaFeO_3$ (t_{2g}^3, e_g^2) and $LaCrO_3$ (t_{2g}^3, e_g^0) semiconducting.

The perovskite $LaCoO_3$ is interesting since it exhibits a transition from localized to collective electronic transport with either temperature (T > 937°C) [14] or composition [15]. Introduction of Sr, for example to give $La_{1-x}Sr_xCoO_3$, results in the formation of Co^{4+} to maintain charge neutrality. Formation of Co^{4+} with intermediate spin configuration of $t_{2g}{}^4 e1_g$, leads to band broadening and with it metallic conduction for $0.1 < x < 0.5$ [13,14]. $La_{0.5}Sr_{0.5}CoO_3$, for example, exhibits a conductivity $\sigma > 10^4$ $(\Omega\text{-cm})^{-1}$ [16]. Van Buren and de Wit [17], however, interpret their thermopower measurements to indicate some carrier dependence on temperature up to T = 500°C and suggest, therefore, that these materials fall somewhere between semiconductors and metals.

A familiar example of metallic conductivity induced by d-band filling is the group of alkali tungsten bronzes (e.g., Na_xWO_3) in

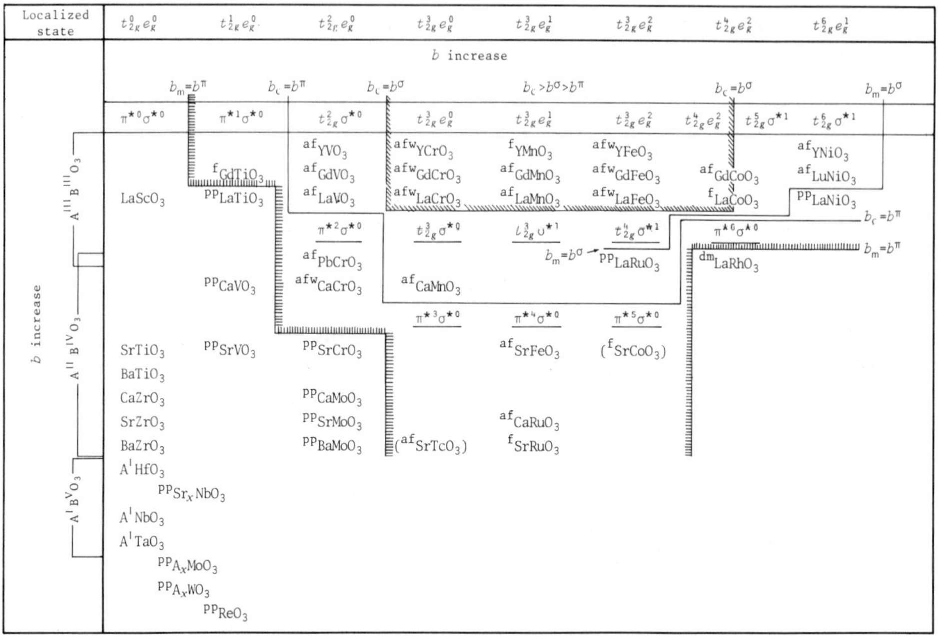

Figure 8 Energy-band classification of perovskite oxides and analogs, indicating degree of localization. af, Antiferromagnetic; afw, antiferromagnetic with parasitic ferromagnetism; f, ferromagnetic; dm, diamagnetic; pp, Pauli paramagnetism. (From Ref. 13.)

which the alkali ions donate their electrons to the initially empty d-band of WO_3. Of particular note is the ability, at times, to reversibly insert and remove such ions from the material by electrochemical means at relatively low temperatures. This has led to the application of similar such materials as "solid solution" electrodes, which serve simultaneously as a source and sink of reactants and as electronically conductive electrodes in solid-state batteries [18]. Similar considerations apply to electrochromic displays based on the insertion of, for example, protons, into initially insulating and transparent WO_3 to form conducting and highly colored H_xWO_3 [19]. Note that the examples above require both high electronic and ionic conduction for their operation.

Superconductivity with a critical temperature as high as 13 K has been reported by Sleight et al. [20] in the system $BaPb_{1-x}Bi_xO_3$. For $BaPb_{0.8}Bi_{0.2}O_3$, for example, the conductivity increases abruptly from a 3×10^3 to above 10^6 $(\Omega\text{-cm})^{-1}$ at 11 K.

The end member $BaPbO_3$, reported to be a metal [22], first appears to be another example of the failure of simple energy-band

Table 1 Listing of a Large Number of Oxides According to Their d-Band Filling and Electronic Conductivity

Electronic configuration, d^n	Monoxides		Sesquioxides		Dioxides	
	Metal	Semiconductor	Metal	Semiconductor	Metal	Semiconductor
d^1-d^5	TiO, VO, NbO, TaO	MnO	Ti_2O_3,[a] V_2O_3[a]	Cr_2O_3, Fe_2O_3	VO_2,[a] CrO_2, MoO_2, WO_2, RuO_2, ReO_2, OsO_2, RhO_2, IrO_2	NbO_2,[a] MnO_2
d^6		FeO		Rh_2O_3		PtO_2
d^7-d^9		CoO, NiO, PdO				
d^{10}		CdO[b]		In_2O_3,[b] Th_2O_3[b] Ga_2O_3	$PbO_2\Delta$[c]	GeO_2, SnO_2[b]

[a] Metal-to-semiconductor transitions.

[b] Degenerate semiconductors.

[c] $\sigma^* - 6s$ overlap.

models to predict electrical behavior given the filled d^{10} configuration of the Pb^{4+} ion. Sleight et al. [20] explain $BaPbO_3$ metallic conduction on the basis of Pb 6s–O 2p overlap due to the high degree of covalency, while Vest and Honig [9] suggest, instead, $\sigma^*(e_g)$ and σ^* overlap (see Figure 7).

A spectrum of electrical conductivities ranging from insulating to metallic is thus observed for the perovskite oxides (see Figure 16, Reference 9), which are now for the most part understood on the basis of Goodenough's criteria, as represented in Figure 8. These criteria relate only to intrinsic electronic properties, however. Many intrinsically insulating oxides (e.g., $SrTiO_3$) exhibit interesting semiconducting properties when doped or driven from stoichiometry by atmosphere treatment. These characteristics are treated in some detail in a later section.

D. Other Oxides

The electrical properties of a large number of binary monoxides, sesquioxides, and dioxides have also been reviewed by Honig [8], Vest and Honig [9], and Rao and Subba Rao [4], the results of which are summarized in Table 1. These oxides are subdivided according to their d-level filling and type of conductivity. The majority of the monoxides crystallize in the rocksalt structure (PdO, tetragonal), the sesquioxides, the corundum structure (Mn_2O_3, In_2O_3, Tl_2O_3, C-type cubic) and the dioxides, the rutile structure, or distortions thereof. The lower symmetry of the rutile and corundum structures results in further splitting of the d levels, making more difficult interpretation of band structure than in the perovskites. Nevertheless, on the whole, these results also appear to satisfy Goodenough's criteria. Generally, the oxides with elements Cr, Mn, Fe, Co, and Ni tend to remain semiconducting even with partially filled bands due to inadequate overlap and band broadening.

III. SEMICONDUCTOR-METAL TRANSITIONS

A number of the oxide systems listed in Table 1 have been found to exhibit abrupt semiconductor-metal transitions (e.g., $V_2O_3 \sim 8$ orders of magnitude jump in conductivity at 150 K) in the vicinity of some characteristic temperature T. A large number of theories to explain this phenomenon have been proposed and are summarized by Vest and Honig [9] and Rao and Subba Rao [4]. Some, like the Mott transition model [23], may be understood on the basis of our previous discussions. Simply, one assumes that the spacing between atoms decreases as the temperature increases, thereby resulting in increased electron-electron overlap. At some critical interatomic distance a switchover from localized to collective behavior would then be expected. Most authorities, however, discount this model on the basis that the likelihood that many materials would lie so close to the critical atomic spacing is very small indeed.

Other models include transitions induced by crystallographic distortions or magnetic ordering. In either case this would correspond to bands originally split due to lower crystal symmetries or magnetic ordering reuniting as the distortions in symmetry disappear or magnetic order is eliminated above the Néel temperature. A number of additional models have been proposed but are beyond the scope of this chapter. The reader is directed to References 4, 8, 9, and 23 to 25 for further information.

A. Extrinsic Semiconductors—A Defect Chemical Approach

We have already encountered semiconduction due to the localization of carriers in partially filled bands as a result of small polaron formation. In these materials, because large concentrations of carriers, albeit trapped, already find themselves in the "conduction" band, variations in impurity concentrations or stoichiometry can be expected to have only secondary effects on conduction. In this section we concern ourselves instead with solids that satisfy conditions characteristic of intrinsic insulators (i.e., large gap separating filled and empty bands) but may be induced to become highly conductive by control of doping and stoichiometry. Interest in this group of semiconducting ceramics is growing due to their importance in increasing numbers of electronically active devices (e.g., varistors and PTC thermistors).

B. Controlled Valence Semiconductors

Let us begin by considering the case of CeO_2, which reportedly has a moderately large band gap of ~ 5.5 eV separating the O_{2p} valence band from the empty Ce4f conduction band [26]. Using Eq. (8), we calculate an intrinsic carrier density of $\sim 5 \times 10^8$ cm^{-3} at 1273 K. This may be compared with the 2×10^{20} cm^{-3} excess electrons obtained by doping with approximately 1% niobium or uranium [27,28], a value roughly 12 orders of magnitude greater than the intrinsic level. Conductivities as high as ~ 1 $(\Omega\text{-cm})^{-1}$ are obtained in this way. Given the strong temperature dependence of Eq. (8), the discrepancy between intrinsic and extrinsic grows rapidly larger as the temperature is decreased.

Assuming full ionization of the Nb or U donors, the weakly activated temperature dependence of the conductivity is ascribed to an activated, small polaron mobility with hopping energy of approximately 0.3 eV, in agreement with results obtained for nonstoichiometric CeO_2 [6]. Because the electrons in the Ce 4f band are localized, it becomes convenient to think of them as forming Ce^{3+} ions in a sea of Ce^{4+} ions with transport occurring by an activated exchange of an electron from an "occupied" Ce^{3+} site to an "empty" Ce^{4+} site. Furthermore, following this line of reasoning, a maximum in the conductivity should occur when the concentrations of Ce^{3+} and Ce^{4+} ions are equal, followed by a drop back toward zero as all the Ce ions become trivalent. Under these circumstances

$$\sigma \propto c(1 - c) \tag{14}$$

in which c is the fraction occupied (e.g., Ce^{3+}) and (1 − c) the fraction unoccupied (e.g., Ce^{4+}). Clearly, conditions of c = 0 (empty band) or c = 1 (full band) should result in no conduction.

The above illustrates the importance of inducing mixed valence so as to provide for conduction in insulating oxides. In CeO_2 this mixed valence was a result of doping with a higher valent dopant such as Nb^{5+}. To maintain charge neutrality, this ionized donor was compensated by a negative charge, the self-trapped electron. In terms of presently accepted defect notation [29], this condition is given by

$$[Nb^{\bullet}_{Ce}] = [Ce'_{Ce}] = [e'] \tag{15}$$

in which subscripts refer to lattice sites and dots and primes refer to net charge relative to the lattice. The square brackets denote defect density.

Next consider FeO*, which according to Table 1 possess a filled $\pi(t_{2g})$ band by virtue of its $3d^6$ electronic configuration. Since all the Fe atoms are Fe^{2+}, motion of an electron from one Fe ion to its neighbor would require the formation of a Fe^{1+}, Fe^{3+} pair, a process requiring the elevation of an electron across the $\pi(t_{2g}) - \sigma(e_g)$ band gap. If, on the other hand, a monovalent acceptor such as Li were added to the crystal, it would induce the formation of Fe^{3+}

*For the purposes of this argument, we ignore the fact that strictly stoichiometric FeO does not exist.

ions or effectively holes in the $\pi(t_{2g})$ band. This neutrality relation is given by

$$[Li'_{Fe}] = [Fe^{\bullet}_{Fe}] = [h^{\bullet}] \tag{16}$$

and corresponds to the case in which the term c in Eq. (14) is close to 1. This method of creating mixed valence for conduction by altervalent doping is known as the "controlled valency" technique [30] and has been applied extensively to many ceramic systems, including NiO:Li [30], CoO:Li [30], TiO_2:Nb [31], $SrTiO_3:Nb_{Ti}$ [32], $SrTiO_3:La_{Sr}$ [33], $UO_2:Y_U$ [34], $Y_3Fe_5O_{12}:Ca_{Fe}$ [35], and $LaCrO_3:Sr_{La}$ [36], as well as many others. Figure 9 illustrates the early work of Verwey and coworkers on the NiO-Li_2O system, in which they examined the influence of Li by controlling its level over nearly two orders of magnitude.

We conclude this section by noting that although NiO and CoO possess $3d^8$ and $3d^7$ configurations, respectively, they exhibit little conduction as long as all the Ni or Co ions are divalent. Apparently,

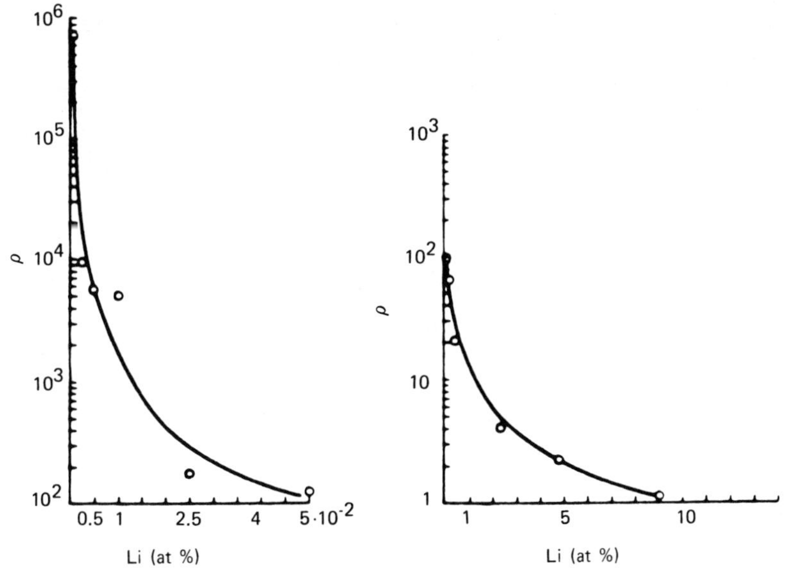

Figure 9 Logarithmic dependence of resistivity in NiO on percentage Li acceptor additions. Note two orders of magnitude variation in Li concentration. (After Ref. 30.)

only by creating some fraction trivalent ions by doping with Li do we enable electron transfer between ions without engaging strong electron-electron correlation effects.

C. Nonstoichiometric Semiconductors

Three mechanisms for defect formation in oxides need to be considered simultaneously. These are (1) thermally induced intrinsic electronic and intrinsic ionic disorder (e.g., Schottky and Frenkel defects), (2) redox, and (3) impurity-induced defects. The first two categories of defects are predicted from statistical thermodynamics [37] and the latter form to satisfy electroneutrality. Examples of typical defect reactions in the three categories are given in Table 2.

Intrinsic electronic disorder [Eq. (17)] has already been discussed earlier. Two types of intrinsic ionic disorder are represented in Eqs. (18) and (19) with the first corresponding to a pair of cations and anions, leaving the interior of the crystal—Schottky disorder—and the second to a cation leaving a normal site—Frenkel

Table 2 Typical Defect Reactions

Defect reactions	Mass action relation	
$O \rightarrow e' + h^{\bullet}$	$n \cdot p = K_e(T)$	(17)
$O \rightarrow V_M'' + V_O^{\bullet\bullet}$	$[V_M''][V_O^{\bullet\bullet}] = K_s(T)$	(18)
$O \rightarrow V_O^{\bullet\bullet} + O_i''$	$[V_O^{\bullet\bullet}][O_i''] = K_F(T)$	(19)
$\frac{1}{2} O_2 \rightarrow O_i'' + 2h^{\bullet}$	$[O_i''] p^2 = K_{ox}(T) P_{O_2}^{\frac{1}{2}}$	(20)
$O_O \rightarrow V_O^{\bullet\bullet} + 2e' + \frac{1}{2} O_2$	$[V_O^{\bullet\bullet}] n^2 = K_R(T) P_{O_2}^{-\frac{1}{2}}$	(21)
$N_2O_3 \xrightarrow{MO_2} 2N_M' + 3O_O + V_O^{\bullet\bullet}$	$\dfrac{[N_M']^2 [V_O^{\bullet\bullet}]}{a_{N_2O_3}} = K_N$	(22)

disorder. Generally, only one of these two mechanisms need be considered for a given material, the choice being determined by the crystal structure (i.e., open structures encourage Frenkel disorder, while close-packed structures encourage Schottky disorder). Both of these mechanisms leave the stoichiometric balance intact.

The oxidation-reduction behavior, as represented by Eqs. (20) and (21), does result in an imbalance in the ideal cation-to-anion ratio and thus leads to "nonstoichiometry." In both cases equilibration with the gas phase by exchange of oxygen between the crystal lattice and the gas is observed to result in the generation of electronic carriers.

Consider the two-dimensional representation of an oxide with cubic rocksalt structure illustrated in Figure 10a. On the left, oxygen from the gas phase enters the lattice interstitially, becomes negatively charged, and thereby creates two holes to maintain charge neutrality. On the right, on the other hand, oxygen leaves the lattice, thereby freeing its two electrons while simultaneously creating a vacant oxygen site, $V_O^{\cdot\cdot}$. The redox reactions, (20) and (21) are written with the assumption that the defect states are fully ionized. As with conventional acceptors and donors, the electronic carriers tend to recombine with the atomic defects as the temperature is lowered. For the doubly charged species considered here, two energy levels per defect are formed in the forbidden gap between conduction and valence band. These are shown sketched in Figure 10b.

Deviations from stoichiometry in the direction of oxygen excess (MO_{1+x}) or deficiency (MO_{1-x}) thus form defect states that act identically in every way to impurity-related acceptor or donor states, respectively. Some of the special features associated with such reactions are: (1) the nature of the semiconductor can be changed from p type to n type under isothermal conditions by control of the oxygen partial pressure (or equivalently, the metal vapor pressure); (2) the atomic defects that make up the acceptor and donor states can be quite mobile under some operating conditions, in contrast to the situation in conventional semiconductors. The latter aspect has special relevance for our upcoming treatment of ionic conduction in oxides.

Given that most oxides of interest here have relatively large band gaps (e.g., TiO_2, $E_g \sim 3$ eV) and correspondingly large Schottky and Frenkel energies, the likelihood of intrinsic disorder controlling electrical behavior over an extensive range of conditions is practically nil. Therefore, either redox (atmosphere) or impurities will control the defect equilibria, depending on the stoichiometry of the oxide. Furthermore, we will discover that the controlled valency concept discussed earlier can also be markedly influenced by stoichiometry; that is, electronic compensation induced by altervalent

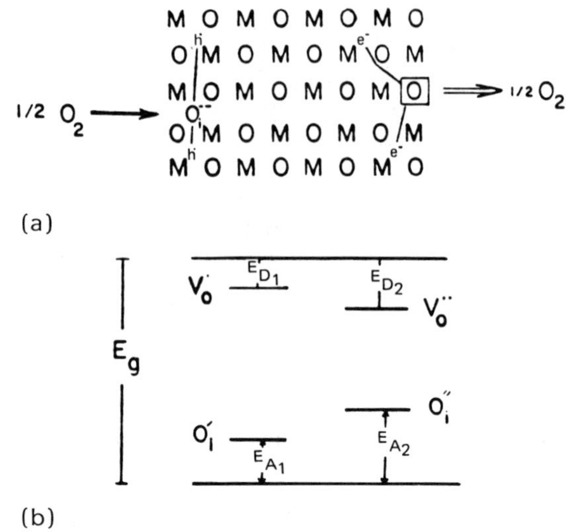

Figure 10 (a) Schematic of a two-dimensional square lattice. Entry of excess oxygen into lattice on left results in formation of an oxygen interstitial and two holes, while exit of an oxygen atom into the gas phase at the right results in formation of an oxygen vacancy and two electrons. (b) Representation of the energy gap E_g separating valence and conduction bands. The two upper discrete energy levels in the gap correspond to oxygen vacancy donor states, while the two lower levels correspond to oxygen interstitial acceptor states.

impurities may be converted to ionic compensation [see Eq. (22)], thereby rendering a material initially semiconducting, insulating. This principle is instrumental in the formation of grain boundary–controlled active devices.

In general, the electrical behavior of solids depends on defects formed in response to both impurities and deviations from stoichiometry. At or near stoichiometry, impurities predominate, while under strongly reducing or oxidizing conditions, defects associated with deviations from stoichiometry control. To characterize the electrical response of a metal oxide to temperature and atmosphere excursions, a series of simultaneous reactions of the form represented by Eqs. (17) to (22) must therefore be considered. Furthermore, assuming the addition of an acceptor impurity N'_M, a representative electroneutrality equation for the case considered in Figure 10 would be

$$2[O_i''] + [N_M'] + n = 2[V_O^{\cdot\cdot}] + p \tag{23}$$

Note that intrinsic Frenkel disorder is assumed to predominate, so that Eq. (18) may be ignored in subsequent discussions. Furthermore, it may be recognized that the remaining equations are not independent of each other (i.e., $K_{OX}K_R = K_F K_E^2$). One may therefore remove one of the equations [e.g., Eq. (20)] in solving for the unknown defect densities.

A piecewise solution to such problems is commonly attempted by sequentially choosing conditions for which only one term on each side of Eq. (23) need be considered. For example, under heavily reducing conditions, Eq. (23) may be simplified to read

$$n = 2[V_O^{\cdot\cdot}] \tag{24}$$

Combining this with Eq. (21), one obtains

$$n = 2[V_O^{\cdot\cdot}] = \left[\frac{K_R(T)}{2}\right]^{1/3} P_{O_2}^{-1/6} \tag{25}$$

and from Eqs. (17) and (19),

$$p = K_E(T) \left[\frac{K_R(T)}{2}\right]^{-1/3} P_{O_2}^{1/6} \tag{26}$$

$$[O_i''] = K_F(T) \left[\frac{K_R(T)}{2}\right]^{-1/3} P_{O_2}^{1/6} \tag{27}$$

The three other defect regimes most likely to occur with successively increasing P_{O_2} include $2[V_O^{\cdot\cdot}] = [N_M']$, and $p = 2[O_i'']$. A diagram depicting the atmosphere dependencies of the defects over the four defect regimes is presented in Figure 11. Here the carrier densities of the individual defects have been multiplied by their mobilities to give the partial conductivities as a function of P_{O_2} and in turn of carrier density.

We emphasize the predominant partial conductivity in each defect regime in Figure 11 with a solid curve. The two solid curves represent two different ways in which the total conductivity may vary with

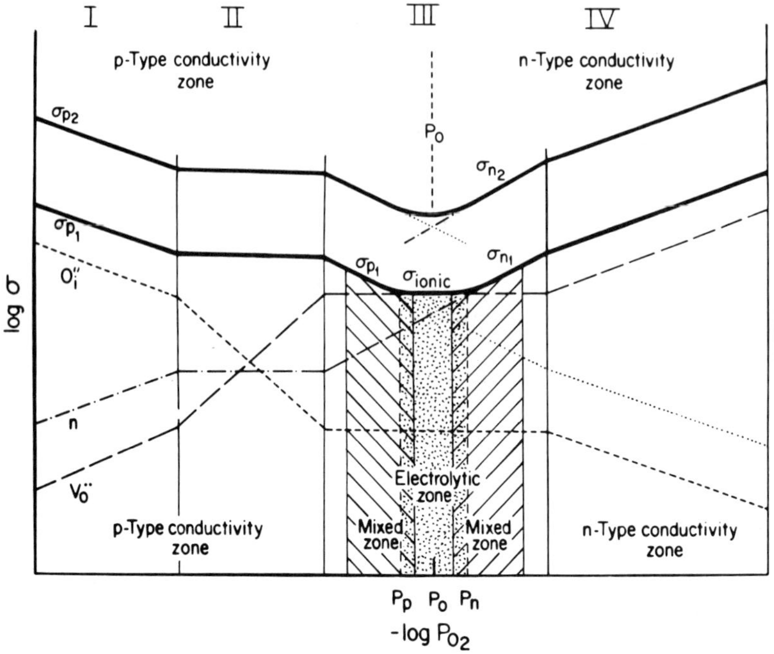

Figure 11 Conductivity isotherms. Solid curves correspond to the following cases: (1) μ_n, $\mu_p > \mu\,(V_O^{\cdot\cdot})$; (2) μ_n, $\mu_p \gg \mu(V_O^{\cdot\cdot})$. In case 1 five conductivity zones are obtained with the central zone (electrolytic) corresponding to predominantly ionic conduction. P_p and P_n correspond to the P_{O_2} values at which $\sigma_p = \sigma_{ion}$ and $\sigma_n = \sigma_{ion}$, respectively, while P_0 corresponds to the P_{O_2} at which $\sigma_n = \sigma_p$. Overlap of electrolytic and mixed zones is used to illustrate imprecise nature of these zones. (From Ref. 1.)

atmosphere. The upper curve corresponds to the case where μ_n, μ_p are sufficiently greater than $\mu(V_O^{\cdot\cdot})$ that even in the defect regime where $V_O^{\cdot\cdot}$ is the predominant defect (i.e., $2[V_O^{\cdot\cdot}] = [N_M']$, the total conductivity remains electronic. In the lower curve, the carrier mobility inequality is not nearly so pronounced, so that at the P_{O_2} (= P_0) at which electronic defects are at a minimum (i.e., $n = p$, conduction is predominantly ionic. Under these circumstances the oxide acts as a solid electrolyte, the topic of a later section. Aside from the electrolytic zone, the neighboring zones on either side are designated as mixed zones within which both ionic and electronic conduction are of comparable magnitude.

Focusing on the upper curve of Figure 11 for the moment we make the following useful observations:

1. Where they are P_{O_2} dependent, electrons and holes always exhibit opposite P_{O_2} dependencies, as might be expected from Eq. (17). Electrons follow a $P_{O_2}^{-1/x}$, holes a $P_{O_2}^{+1/x}$ dependence.
2. In each defect regime, the majority carrier exhibits a unique P_{O_2} dependence.
3. The "controlled valency" feature holds in only one of the defect regimes, namely where $p = [N_M']$.

The first observation enables us to identify the type of electronic carrier (n or p), while the second gives us the defect regime. By establishing the bounds of the controlled valency regime, it allows us to predict the conditions of temperature and P_{O_2} for which the material will remain highly conducting. In general, it should be noted, only one or two of these defect regimes are experimentally accessible at a given temperature. By following the guidelines above, it becomes possible to identify the defect regimes one is operating under and with the aid of the defect relations derive the controlling equilibrium constants, K(T).

To illustrate the predictions above, I show recent results obtained in our laboratory [38] on a single crystal of $Ba_{0.03}Sr_{0.97}TiO_3$ in Figure 12. Here log σ is plotted versus log P_{O_2} for a series of isotherms between 800 and 1100°C at intervals of 50°C (note that the log P_{O_2} axis is the reverse of that in Figure 11). An obvious feature of the data is the minimum in σ at higher P_{O_2} values, indicating a transition from p-type to n-type conductivity. This therefore places the data in the defect regime described by the simplified neutrality relation $[N_M'] = 2[V_O^{\cdot\cdot}]$ with $P_0 \sim 10^{-7}$ atm at 800°C. The solutions for n and p in this region are given by

$$n = \left(\frac{K_R(T)}{[N_M']}\right)^{\frac{1}{2}} P_{O_2}^{-\frac{1}{4}} \tag{28}$$

$$p = K_e(T)\left(\frac{K_R(T)}{[N_M']}\right)^{-\frac{1}{2}} P_{O_2}^{+\frac{1}{4}} \tag{29}$$

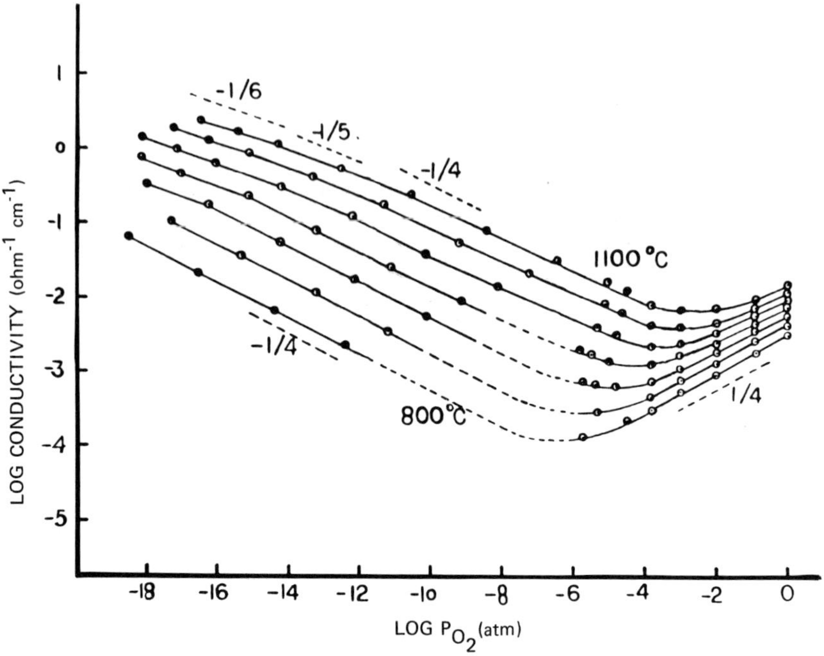

Figure 12 Log σ is plotted as a function of log P_{O_2} for a series of isotherms for the compound $Ba_{0.03}Sr_{0.97}TiO_3$. (From Ref. 38.)

The observed +1/4 slope at high P_{O_2} and $-1/4$ slope at low P_{O_2} are in agreement with these predictions.

Thermoelectric power measurements, performed concurrently with the electrical conductivity measurements, are illustrated in Figure 13. They show a sharp transition from n to p type at the same P_{O_2} values at which the conductivity goes through its minima, again strengthening the applicability of the defect model.

In the upper left-hand corner of Figure 12, corresponding to reducing conditions at the highest temperatures, the slope is observed to decrease from $-1/4$ toward $-1/6$. This we interpret as signifying the transition from one defect regime to the bordering defect regime characterized by the neutrality relation, $n = 2[V_O^{\cdot\cdot}]$ [see Eq. (25)]. A similar transition in slope is observed in the thermoelectric power.

Given a well-defined defect model, it is possible to derive a number of the equilibrium constants and carrier mobilities from these data. From a plot of the conductivity minimum versus 1/T [see Eq. (6)] a value for the thermal band gap of 3.35 ± 0.4 eV (at O K) is obtained. From a combined analysis of the electrical conductivity and thermoelectric power following the approach of Jonker [39], we obtained [38] mobility values for the electrons and holes (μ_e = 0.23 cm^2/V-sec and μ_h = 0.10 cm^2/V-sec at 1000°C). Finally, from a plot of log σ_n versus 1/T at constant P_{O_2} in the $P_{O_2}^{-\frac{1}{4}}$ regime, an expression for K_R [see Eq. (28)] given by

$$K_R(T) = 1.03 \times 10^{69} \exp(-5.18 \pm 0.03 \text{ eV}/kT) \text{ cm}^{-9} \text{ atm} \qquad (30)$$

is obtained.

The influence of atmosphere on σ in this material and in other titanates is notable. For example, σ (1100°C) in Figure 12 increases nearly three orders of magnitude, to ∿3 (Ω-cm)$^{-1}$ in going from the conductivity minimum to $P_{O_2} \simeq 10^{-17}$ atm. Even higher conductivities

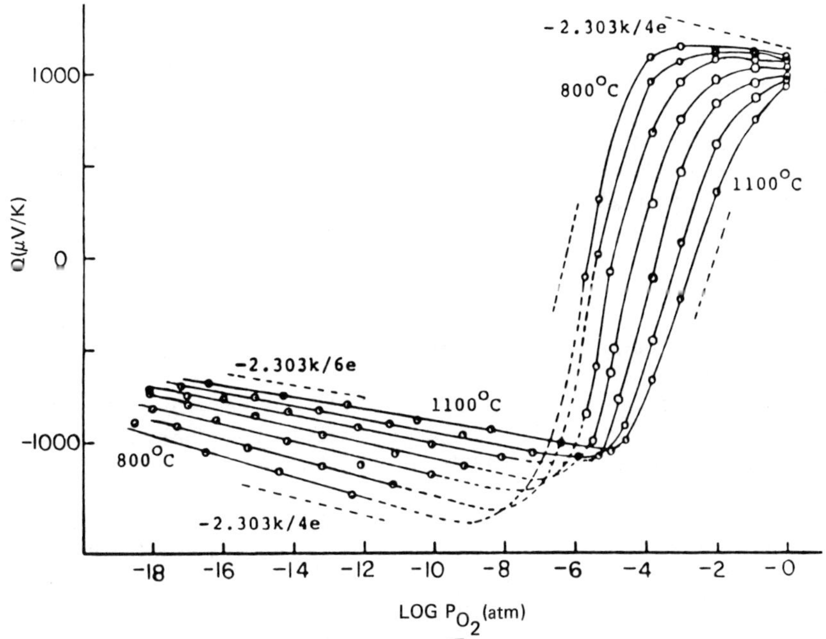

Figure 13 Thermoelectric power (Q) of $Ba_{0.03}Sr_{0.97}TiO_3$ is plotted as a function of log P_{O_2} under conditions identical to those utilized in Figure 13 for conductivity measurements. (From Ref. 38.)

may be obtained by annealing in H_2. In practice, these high conductivities in the titanates may be maintained by quenching to room temperature [40,41]. For this reason and because of their relative stability in basic solutions, they have been of great interest as photoanodes in photoelectrochemical cells [42–44].

In many oxides, quenching does not retain the high-temperature conductivities. Let us consider what requirements must be satisfied for high conduction. First, the excess donors or acceptors induced in the material as a result of nonstoichiometry must not be allowed to diffuse out of the solid during cooling. Thus the more rapid the diffusivity of the $V_O^{\cdot\cdot}$ in $SrTiO_3$, for example, the more rapid need be the quench. Second, the ionization energy of the donor or acceptor states will deionize resulting in a greatly depressed conductivity. The high dielectric constants characteristic of the titanates ensure low ionization energies. The same does not hold true for many other oxides.

An interesting consequence of the first requirement above being only partially fulfilled is the existence of a number of technically important grain boundary–controlled electronic devices. Consider that diffusion generally occurs much more rapidly along grain boundaries than through the bulk. Thus during a cool-down, while the donor or acceptor defects may be maintained in the core of the grain, partial equilibration may occur in the vicinity of the grain boundaries, resulting in a depletion of carriers. Such depletion layers result in the formation of barriers that block the flow of carriers between conductive grains. Examples of devices that exhibit such features include varistors (ZnO) [45], PTC thermistors ($BaTiO_3$) [46], and boundary layer capacitors ($SrTiO_3$). Detailed discussions concerning a number of these devices may be found in other chapters of this book.

D. Extrinsic Broad-Band Semiconductors

Although large numbers of well-known oxides possess broad s- and p-derived conduction bands with large associated carrier mobilities (e.g., MgO [47] and Al_2O_3 [48]) only a handful can be made highly conducting. This is related to the difficulty in forming shallow donor or acceptor levels in many oxides [47].

Large gap oxides that can be prepared as highly conducting semiconductors include In_2O_3 [49], SnO_2 [50], ZnO [51], Tl_2O_3 [52], and CdO [53]. Conductivities as high as 10^4 $(\Omega\text{-cm})^{-1}$ have been reported for a number of these systems, while mobilities of 1200 cm^2/V-sec were reportedly measured on CVD-prepared antimony-doped SnO_2 single crystals [54].

Carrier densities up to levels of 10^{20} cm^{-3} may be induced in either one of the two ways described above. Well-known examples of controlled valency include Sn-doped In_2O_3 [55] and Sb-doped SnO_2 [56,57]. High densities of carriers may also be induced by deviations from stoichiometry. In the case of ZnO this is believed to be by the creation of excess Zn in the form of zinc interstitials [45], while in SnO_2 oxygen vacancies are believed to be the source of shallow donors [54].

The combination of large band gaps and high carrier mobilities provide for relative transparency in the visible even for conductive materials. The high carrier mobility ensures that the free carrier absorption remains low even at electrical cnnductivity levels of $\sim 10^2 - 10^3$ $(\Omega\text{-cm})^{-1}$. ITO (indium tin oxides) films are especially noted as examples of transparent electrodes.

IV. IONIC CONDUCTION IN OXIDES

Optimized ionic conduction is a well-known characteristic of molten salts and aqueous electrolytes wherein all ions move with little hindrance within their surroundings. This leads to ionic conductivities as high as $10^{-1} - 10^1$ $(\Omega\text{-cm})^{-1}$ in molten salts at temperatures of 400–900°C [58]. Typical ionic solids, in contrast, possess limited numbers of mobile ions which are hindered in their motion by virtue of being trapped in relatively stable potential wells. Ionic conduction in such solids easily falls below 10^{-10} $(\Omega\text{-cm})^{-1}$ for temperatures between room temperature and 200°C. In the following sections we examine the circumstances under which ionic conduction in solids approaches that found in liquid electrolytes. Solids that fall in this category are most popularly called "fast" ion conductors (FICs), although other terms, such as "superionic" and "optimized" ionic conductors, are also sometimes used. Although the term "solid electrolyte" also in principle describes FICs, one commonly reserves the term for ionic conductors with intermediate magnitudes of ionic conduction, as I have done in Figures 1 and 2.

Motion of ions is described by an activated jump process for which the diffusion coefficient is given by [59]

$$D = D_0 \exp\left(\frac{-\Delta G}{kT}\right)$$

$$= \gamma\ (1 - c)\ Za^2\ \nu_0 \exp\left(\frac{\Delta S}{k}\right) \exp\left(\frac{-E_m}{kT}\right) \tag{31}$$

where a is the jump distance, ν_0 the attempt frequency, and E_m the migration energy. The factor $(1 - c)Z$ defines the number of neighboring unoccupied sites, while γ includes geometric and correlation factors. Note the similarity to Eq. (10). Since the ion mobility is defined by $\mu = qD/kT$ and the density of carriers of charge q is N_0, where N is the density of ion sites in the sublattice of interest, the ionic conductivity becomes

$$\sigma_{ion} = \gamma\left(\frac{N_q}{kT}\right)^2 c\ (1 - c)\ Za^2\ \gamma_0 \exp\left(\frac{\Delta S}{k}\right) \exp\left(\frac{-E_m}{kT}\right)$$

$$= \frac{\sigma_0}{T} \exp\left(\frac{-E}{kT}\right) \tag{32}$$

This expression shows that σ_{ion} is nonzero only when the product $c(1 - c)$ is nonzero (i.e., for c not equal to 1 or zero) [see also Eq. (14)]. Since all normal sites are fully occupied (i.e., $c = 1$) and all interstitial sites are empty (i.e., $c = 0$) in a perfect classical crystal, this is expected to lead to highly insulating characteristics.

The classical theory of ionic conduction in solids is thus described in terms of the creation and motion of atomic defects, notably vacancies and interstitials. Figure 2, already used earlier to describe intrinsic electronic disorder, may be equally well applied to describe intrinsic ionic disorder with the notable difference being that the lower band now corresponds to, for example, normally occupied sites and the upper band normally unoccupied interstitial sites rather than energy states. Similarly, the designated energies now correspond to the formation of a Frenkel pair rather than an electron-hole pair. It is clear from the figure that the concentration of ionic carriers is a sensitive function of the Frenkel energy [see Eq. (19)] and might therefore distinguish between solids with measurable ionic conductivity (i.e., a solid electrolyte from those which are highly insulating even at relatively elevated temperatures). Similar comments are applicable to solids with predominantly Schottky disorder [i.e., Eq. (18)].

As with electronic conductors, addition of altervalent impurities may result in a marked enhancement of carriers over that of intrinsic levels. Figure 3, for instance, may now be viewed as corresponding to the considerably reduced energies required to dissociate an impurity-defect pair as compared to intrinsic defect generation. For example, E_A might correspond to the energy required to dissociate an acceptor-anion vacancy pair or E_D the energy to dissociate a donor-anion interstitial pair. Such dissociative effects have been extensively reported in both the halide and oxide literature [60].

In general, therefore, two energies contribute to ionic conduction: a defect energy, E_D, which may either be related to the Frenkel or

Schottky formation energy, or to a dissociation energy and a migration energy E_M. The value of E in Eq. (32) therefore takes on different values in three characteristic temperature regimes. These include:

1. $E = E_M + E_A/2$: extrinsic associated regime at low T
2. $E = E_M$: extrinsic fully dissociated regime at intermediate T
3. $E = E_M + E_F/2$: intrinsic defect regime at elevated T (e.g., for Frenkel equilibrium)

For optimized ionic conduction to exist, two criteria must be satisfied simultaneously. First the term c in Eq. (32) must approach 1/2. This corresponds to nearly all of the ions on a given sublattice being mobile. Second, the crystal structure must be so arranged as to enable easy motion of ions from one equivalent site to the next. This is reflected in exceptionally low values for the migration energy E_M. In the next section we discuss the conditions under which these criteria are satisfied.

V. FAST ION CONDUCTION

A number of routes leading to exceptionally high ion carrier densities in solids have been identified over the last several decades. These are subdivided into two major categories below (i.e., structurally disordered solids and highly defective solids).

A. Structurally Disordered Solids

In contrast to the idealized picture of crystal structures presented above, many solids exist in which a sublattice of sites is only partially occupied. Strock [61] already came to such a conclusion in the 1930s in relation to the Ag sublattice in the high-temperature form (α phase) of AgI. More recent neutron diffraction studies [62] differ with regards to the number of Ag sites which are in fact equivalent. The special feature of partial occupancy of sites is nevertheless retained.

Other notable systems characterized by sublattice disorder include Nasicon ($Na_3Zr_2PSi_2O_{12}$), sodium beta alumina ($1.2\ Na_2O-0.11\ Al_2O_3$) and $LiAlSiO_4$, which exhibit fast ion transport in three, two, and one dimensions, respectively. Hundreds of other structurally disordered conductors may be found listed in recent review articles on the subject [18,63]. Figure 14 illustrates the log $\sigma - 1/T$ relations for representative FICs.

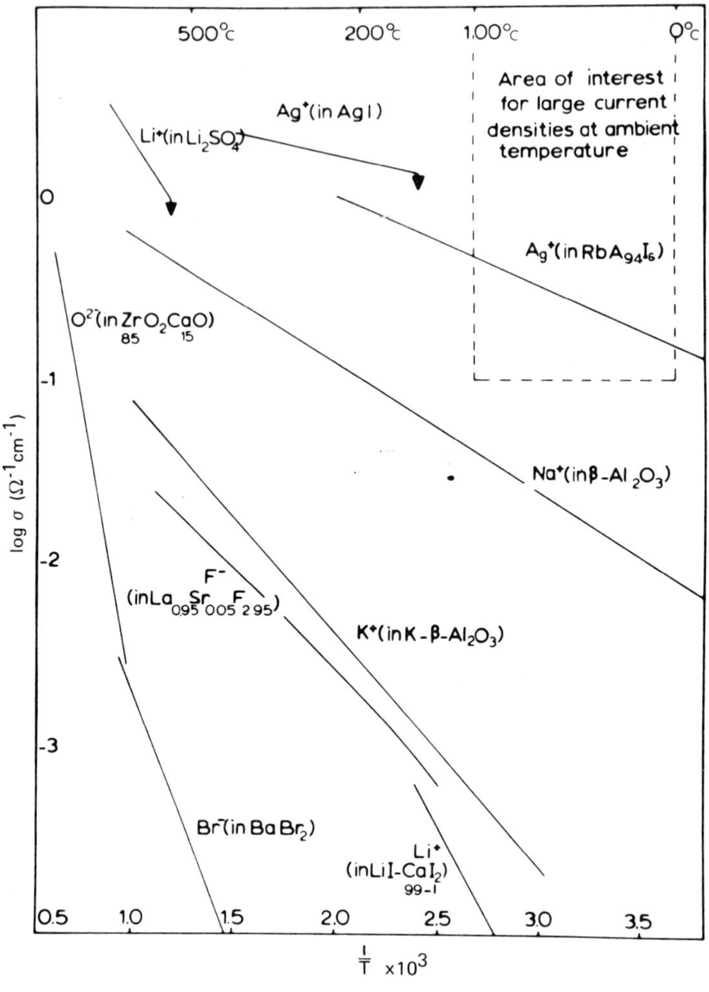

Figure 14 Temperature dependence of ionic conductivity of a number of representative FICs, illustrating the fact that some of these materials achieve ionic conductivities approaching those of liquids only at elevated temperatures. [From W. Van Gool, *Ann. Rev. Mater. Sci.*, *4*, 311–335 (1974).]

By virtue of the coexistence of occupied and unoccupied sites on a given sublattice, FIC conductors resemble metals in that the order of 10^{21} to 10^{22}-cm^{-3} carriers already exist at low temperatures. This correspondence is illustrated in Figure 2.

Similarities between FICs and liquid electrolytes are also often noted, the most important of which is that the disordered sublattice in the solid resembles the disordered nature of ions in a liquid. For this reason one often hears about sublattice melting occurring during phase transitions in solids, which convert solids from conventional ionic conductors to FICs (e.g., $\beta \rightarrow \alpha$ transition in AgI at approximately 150°C). Nevertheless, most investigators now believe that transport in FICs occurs via correlated jumps between well-defined sites rather than the liquidlike motion characteristic of aqueous or molten salt electrolytes.

The major structural features of FICs may be subdivided into the following two elements:

Framework: a highly ordered, immobile sublattice which provides continuous open channels for ion transport

Mobile carrier sublattice: a random distribution of carriers over an excess number of equipotential sites

Although perhaps not explicitly stated, the framework must provide tunnels of dimensions comparable to those of the mobile ions so that major network ion displacements and thus significant strain contributions to the migration energy do not occur. Aside from strain effects, Flygare and Huggins [64] have demonstrated theoretically the importance of coulombic and polarization interactions between the mobile and network ions in establishing the magnitude of the migration energy and in turn the ion mobility. They showed that ions of intermediate radius would be expected to be most mobile given that small ions tend to "stick" to walls of the tunnels due to high polarization energies, while large ions are limited in their motion by repulsive or equivalently large strain contributions. This is illustrated in Figure 15. Thus for each structure an optimum mobile ion size exists, as has since been experimentally verified in, for example, sodium beta alumina [65].

The latter feature of optimized ion size matched to an immobile ion network with well-defined geometry and polarizability is perhaps not as well emphasized in some of the FIC literature as it should be. It is, for example, possible to find structurally disordered solids with their associated high intrinsic carrier densities, which nevertheless exhibit unexceptional magnitudes of ionic conductivities [e.g., $\sigma(25C) \sim 10^{-2}$ $(\Omega\text{-cm})^{-1}$ for Na^+ in β-alumina, while $\sigma(25C) \sim 10^{-11}$ $(\Omega\text{-cm})^{-1}$ for H_3O^+ in β-alumina [58]. It is for this reason that certain authors prefer the term "optimized ion conductors" to FICs

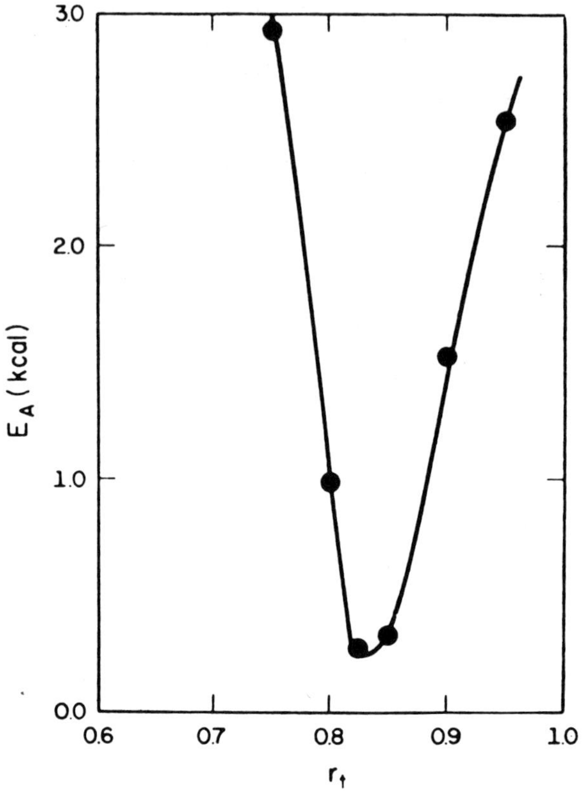

Figure 15 Calculated dependence of migration energy on cation radius in α AgI. (After Ref. 64.)

to emphasize the importance of optimizing both carrier density and ion mobility for obtaining exceptional levels of ionic conductivity at temperatures well below the melting point of the solid. This brings us to a description of a number of transport parameters which are characteristics of FICs.

Although no precise criterion now exists for categorizing FICs, the following conduction parameters are useful in emphasizing the major features:

1. Unusually large magnitude of ionic conductivity [$\sigma > 10^{-2}$ $(\Omega\text{-cm})^{-1}$] well below the melting point
2. Low activation energies ($E \simeq 0.05-0.5$ eV)
3. Low preexponentials [$\sigma_o \simeq 10^2-10^4$ $(\Omega\text{-cm})^{-1}$ K]

The first two features have already been discussed above, in that the combination of low migration energies and high c(1 −c) terms in Eq. (32) lead to anomalously high ionic conductivities. Based on the large c(1 −c) term it is therefore all the more surprising to find the preexponential σ_0 to be orders of magnitude lower in FICs than in conventional conductors. Since the terms γ, N, Z, and a in Eq. (32) can be expected to differ little in various solids, this leaves only γ_0 and ΔS as the possible sources of this discrepancy. Given the disordered nature of these solids and the shallow potential wells for the mobile ions, low values for these parameters are not at all surprising.

An important feature, so far ignored, has been correlation effects between carriers which must arise as the concentration of mobile ions increases to the levels observed in FICs. Calculations by Wang et al. [66], for example, have demonstrated that cooperative motions of ions can lead to significantly lower calculated migration energies than those based on consideration of isolated jumps alone. Other studies that examine alternative models for correlated motion may be found in the literature [65].

In the next section I discuss more recent studies on structurally disordered FICs, which in some ways resemble even more closely analogous liquid electrolytes.

B. Fast-Ion-Conducting Glasses

One of the often mentioned criteria for FIC in solids (see above) is the existence of a highly ordered framework which provides channels for the ready motion of ions in the complementary, disordered sublattice. Recent reports of FIC in glasses [67] raise serious doubts concerning the relevance of this feature.

The amorphous state is viewed as being liquidlike. Glasses are known to lack long-range order, with short-range order typically extending to, at most, a few atom spacings. Because the covalently bonded framework in oxide glasses is characterized by variable bond angles and lengths, it is often referred to as a random network. Although amorphous structures are generally open and rigid, they do not provide a coherent network with well-defined and periodic interconnected polyhedra. As we shall demonstrate, although highly oriented channels may be helpful in FIC, they are certainly not essential.

In a recent study, our research group [68—70] investigated the ionic conductivity of a number of lithium borate crystalline compounds and their corresponding glasses. The ionic conductivities of lithium diborate ($Li_2B_2O_7$), lithium metaborate ($LiBO_2$), and boracite ($Li_4B_7O_{12}Cl$) crystals and glasses are shown plotted versus reciprocal temperature in Figure 16. Two situations characteristic of

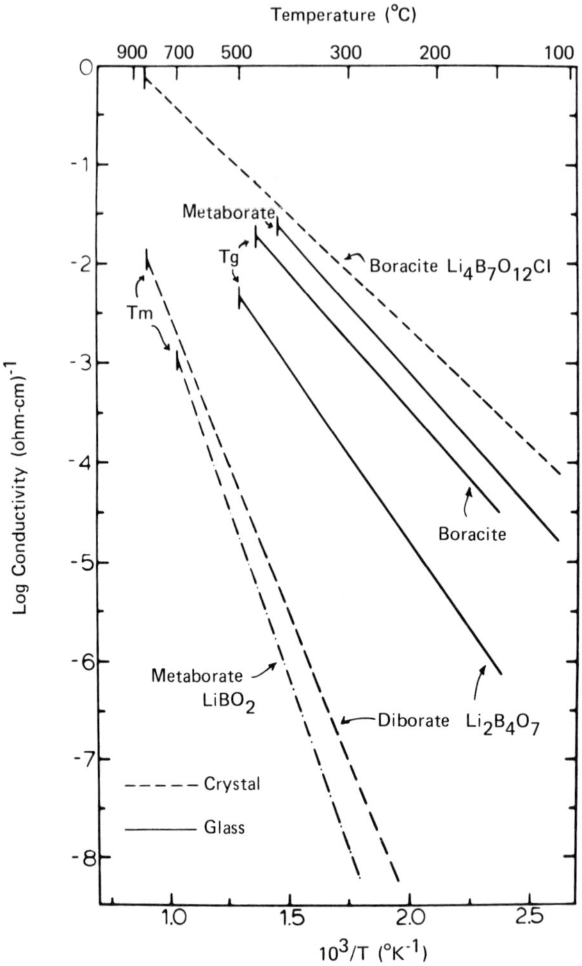

Figure 16 Log σ versus $10^3/T$ for a series of lithium borate crystalline compounds and their glassy counterparts. Note that while in the case of boracite, the difference in electrical properties is small, for the others the glassy state exhibits greatly enhanced conductivities. (From Ref. 70.)

many similar comparisons [67] are readily distinguished. The first comparison between the FIC boracite crystal and the corresponding glass shows only a small relative difference in both the magnitude of conductivity [$\sim 8 \times 10^{-3}$ versus 2×10^{-3} $(\Omega\text{-cm})^{-1}$ at 300°C] and the activation energy (0.50 versus 0.59 eV). The second comparison, which involves the di- and metaborate compositions, is dramatically different. Here the glasses are observed to possess Li ion conductivities many orders of magnitude greater than the crystals, for example, a factor of 10^6 at 500 K for the case of the metaborates and markedly lower activation energies ($\sim$0.5 versus 1.1 eV).

Because disorder in at least a subgroup of Li ions already exists in crystals of the first type, it is reasonable to assume that long-range disorder produced by fusion might have only a minor influence on Li transport. The fact that this is borne out by experiment supports the view that other features, such as highly ordered frameworks, can play only secondary roles in FIC.

On the other hand, the crystals in the second group represent examples of more conventional ionic conductors in which the Li ions fully occupy the sublattice of equivalent alkali sites and provide cohesion to the overall framework. These crystals therefore depend on defect processes for carrier generation. Under these circumstances the creation of structural disorder during fusion of the glasses may be expected to result in the formation of a large excess of nearly equivalent sites. These presumably provide for the greatly enhanced ionic conductivities observed in the glasses above.

Many FIC glasses, including silver and alkali borates, phosphates, and molybdates, exhibit a strong dependence of ionic conductivity on halide additions. Some authors speculate [71] that the large salt anions enhance ionic transport in glasses by lowering electrostatic and/or polarization barriers. They further suggest that the observed activation energy for conduction is made up, in large part, of an association energy that is reduced on addition of the halide anions.

In a recently completed set of studies [68,69,72] we have demonstrated that at least for the case of alkali borates with chlorine additions, the predominant influence of the chlorine addition is on the strain component of the migration energy. Figure 17 illustrates the influence of chlorine additions on a series of glasses based on the diborate composition, in which the total lithium content is maintained constant. The nearly-linear increase of log σ with chlorine addition is simply a reflection of the corresponding nearly-linear decrease in activation energy. Complementary measurements of glass transition temperature and density demonstrated that the chlorine additions were markedly influencing the structure of these glasses by both weakening and dilating the network, a conclusion in conflict with earlier spectroscopic studies [73]. In a subsequent study

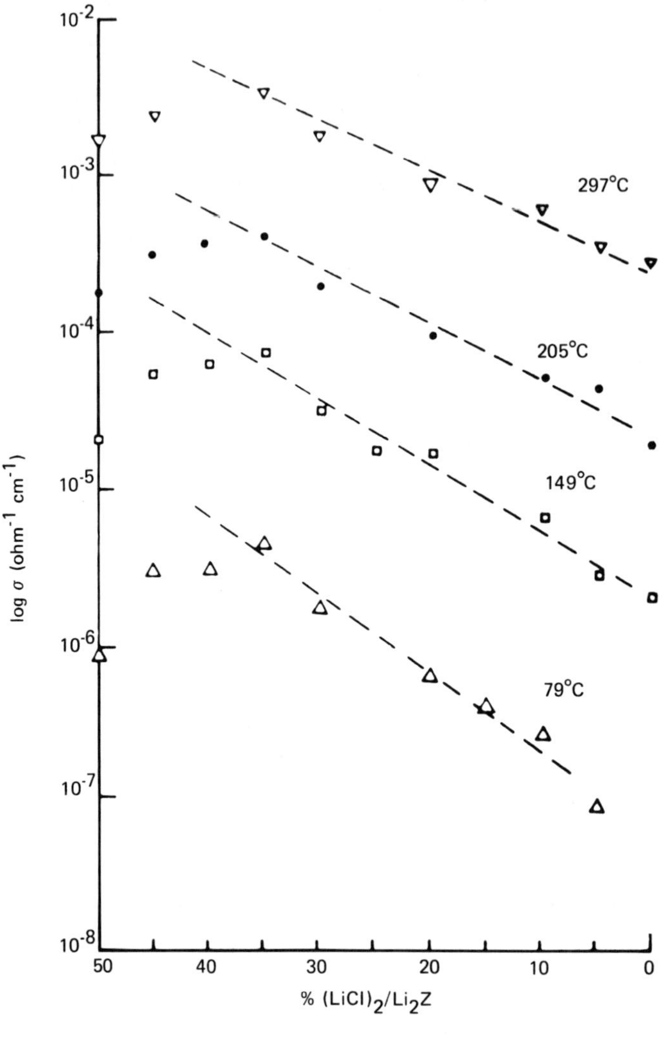

Figure 17 Log conductivity as derived from complex impedance measurements, is plotted as a function of nominal chlorine content for the glass composition 36.4 mol % Li_2Z–63.6 mol % B_2O_3 at four temperatures. Note that the liquid-liquid immiscibility boundary occurs at roughly 35% $(LiCl)_2/Li_2Z$. (From Ref. 68.)

[72] in which measurements were extended to the much larger Na ion in a similar glass matrix, the influence of glass structure on transport was demonstrated even more clearly. Comparison of Na^+ and He^0 atom migration in these glasses illustrated most directly the predominant influence of strain to the overall activation energy. Chlorine additions thereby lower these energies by dilating the bottlenecks through which the ions must pass.

To conclude this section we examine the general trends in σ_0 and in E for FIC glasses. High ionic conductivities are accompanied by low activation energies: ∿0.2–0.3 eV for Ag ion conductors and 0.4–0.7 eV for alkali conductors. Typical values for σ_0 are ∿ 5 × 10^4 $(\Omega\text{-cm})^{-1}$ K for Ag and ∿1 – 10 × 10^5 $(\Omega\text{-cm})^{-1}$ K for alkali conductors. These values for σ_0 cover the range between the lowest values obtained for the best crystalline FICs [10^2–10^3 $(\Omega\text{-cm})^{-1}$ K] and the much higher values obtained for classical ionic conductors [e.g., NaCl (10^6 $\Omega\text{-cm})^{-1}$ K]. This suggests that the distinction between FIC and classical conductors occurs only in degree rather than form.

Furthermore, in contrast to predictions based on Eq. (32), σ_0 in many FIC glasses are virtually insensitive to composition variations in which the number of mobile interstitial cations are presumably increased and for which the ionic conductivity increases orders of magnitude. A dramatic case in point was reported by Kawamoto et al. [74], in which σ_0 was found to be independent of Ag content in As_2S_3-, $GeS{-}GeS_2$-, and P_2S_5-based glasses under conditions for which σ (25°C) increased by approximately six orders of magnitude. Needless to say, the classical picture that predicts σ_0 to be proportional to carrier concentration appears to be far from the truth in this case. Because composition may be readily varied over wide limits in glasses, such tests of predictions of diffusion theory may often be more easily tested in glasses than in FIC crystalline materials. Figure 18 compares a sampling of the large variety of FIC oxide glasses which have already been identified against a number of well-known crystalline FICs. This figure reemphasizes the wide spectrum of ionic conductivities which lie between so-called ionic insulators and the most idealized FICs.

C. Highly Defective Solids

Anomalously high concentrations of ionic carriers may even be induced in solids which are intrinsically insulating. This may be accomplished, in some systems, by inducing marked deviations from the ideal stoichiometric cation-to-anion ratio. In the following we briefly discuss two approaches for generating such "highly defective solids."

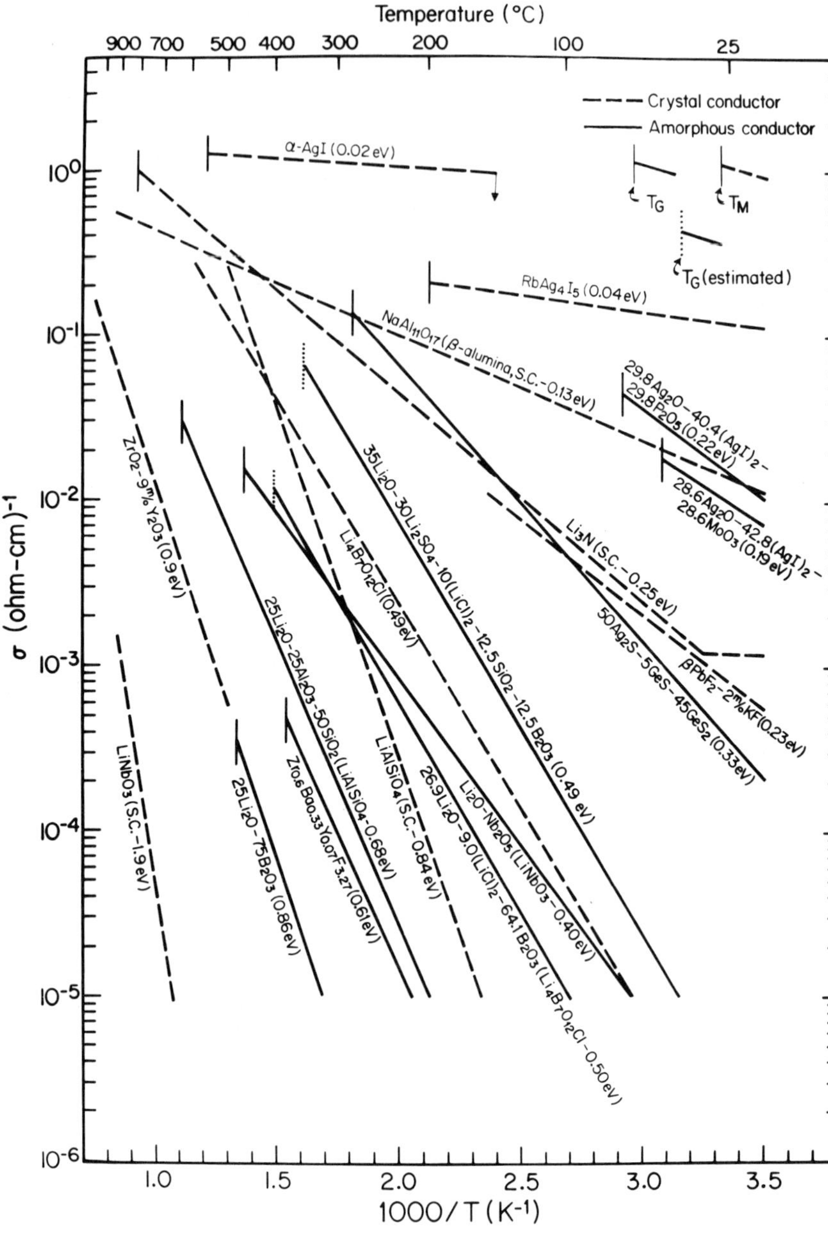

Figure 18 The ionic conductivities as a function of reciprocal temperature of a number of fast Ag, Li, and F conducting glasses are compared with some of the better known crystalline FICs. Glass transition and melting temperatures are indicated where appropriate. (From Ref. 67.)

We already know that ionic defect densities may be greatly enhanced above intrinsic levels by doping with altervalent impurities. However, the solubility limit of such impurities often is limited to only tens or hundreds of ppm in many compounds. This corresponds to roughly $10^{17}-10^{18}$ defects/cm^3, a value 10^3-10^4 times smaller than in typical FICs. Many compounds do exist, however, in which the solubility limit is extensive, reaching the 10–20% level even at reduced temperatures. Perhaps the most familiar example of such a system is stabilized zirconia, which due to its wide solid solubility with lower-valent cations such as Ca^{2+} and Y^{3+}, exhibits exceptionally high oxygen ion conductivity [$\sigma \sim 10^{-1}$ $(\Omega\text{-cm})^{-1}$] at temperatures approaching 1000°C.

As discussed above, high carrier densities must be coupled with high ion mobilities in order to attain high magnitudes of ionic conduction. The cubic fluoritelike structure exhibited by stabilized zirconia (ZrO_2), ceria (CeO_2), and thoria (ThO_2), for example, supports high oxygen ion mobility due to the low four-fold coordination of cations around the oxygens, coupled with the interconnected

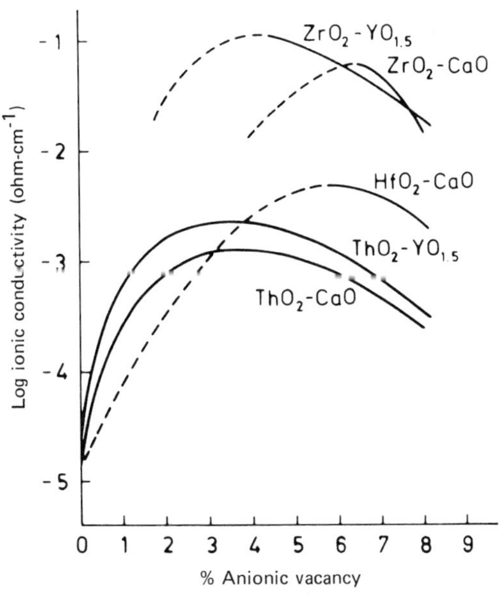

Figure 19 Log conductivity is plotted for a number of oxygen ion conductors as a function of lower-valent cation substituent. All systems exhibit a maximum in the ionic conductivity at some characteristic composition. [From B. C. H. Steele and C. B. Alcock, *Trans. Metall. Soc., AIME, 233*, 1359 (1965).]

nature of the face-shared polyhedra which surround the oxygen sites. Migration energies as low as ∿0.6 eV are reported for oxygen vacancy motion in ceria-based solid solutions [75]. High fluorine ion mobility is also observed in fluorite CaF_2 and related crystal systems.

Ionic conductivities do not generally increase linearly with foreign atom additions. At the levels of defects being discussed here, defect-defect interactions become important, generally leading to defect ordering. This results, as illustrated in Figure 19, in a maximum in ionic conductivity at some level of doping which depends on the particular system being investigated. Nowick et al. [75,76] have demonstrated, in a series of studies, that the deviations from ideality are caused initially by composition-dependent activation energies rather than preexponentials, a feature already observed earlier for a number of FIC glasses.

The formation of ionic defects that accompany excursions in composition away from stoichiometry due to redox reactions may also be large. CeO_2, for example, may be readily reduced to $CeO_{1.8}$ at 1000°C [77], resulting in oxygen vacancy concentrations of 5×10^{21} cm^{-3} (i.e., c = 0.9). Although such excursions from stoichiometry are generally accompanied by electronic defects of the same order of magnitude [see Eq. (21)], this need not be the case, as was recently demonstrated by the author for the ThO_2-CeO_2 system [78].

VI. APPLICATIONS

Highly conducting ceramics are rarely of interest in applications for their highly conducting nature alone. Conventional metals, which are easily fabricated into complex geometries, readily satisfy this criterion. Where ceramics are highly desired are in applications where the conductive material must stand up against a highly corrosive environment often at elevated temperatures. Examples include MHD electrodes operating at 1500°C in a plasma [79], fuel cell electrodes operating at 1000°C in oxidizing environments [80], and electrodes operating in molten Na-S batteries [81].

Semiconducting titanate electrodes are being investigated as active elements in semiconductor—liquid junction photoelectrochemical cells. A major feature of these materials is their stability under conditions where conventional semiconductors corrode [42]. For obvious reasons, ITO transparent electrode films, already mentioned above, play a critical role in the design and operation of various solar cells.

Metallic oxides are often used in the fabrication of conductive thick films. Oxides are required since conventional metals would oxidize under the high-firing conditions required for substrate adhesion and "curing." Metallic oxides commonly used in these applications include RuO_2 and $Bi_2Ru_2O_7$ [82]. Other pyrochlore-type ruthenates have been found to be excellent redox catalysts [83].

The strong temperature and atmosphere dependence of the electrical conductivity of some oxides has already been noted. This makes semiconducting oxides such as TiO_2, SnO_2, and others attractive as sensing elements. Chapter 5 discusses these applications in some detail.

Fast ionic conducting ceramics and solid electrolytes are beginning to find extensive application in various electrochemical devices. Some of these include fuel cells and electroyzers utilizing stabilized zirconia [80], high-energy-density batteries with Na β alumina as the electrolyte [84], and auto exhaust sensors [85] which again utilize stabilized zirconia as the active element. Many other such applications are now under consideration and are reviewed in books on the subject (see, e.g., Ref. 86).

VII. CONCLUSION

Although ceramics are often more commonly known for their insulating properties, in actuality they exhibit an extensive range of electrical properties competing with metals and liquid electrolytes for the highest electronic and ionic conductivities, respectively. Although progress has been made in the last decades in clarifying the source of the often complex electrical behavior observed in many ceramics, much remains to be done. One can point to few oxides indeed for which the electrical behavior comes even remotely close to the understanding one now has for the conventional semiconductors such as Si or GaAs. A great deal more activity is, however, now being generated in these materials both by those interested in gaining an improved fundamental understanding and by the many groups interested in their application. Based on this attention, one can expect significant progress in this field in coming years.

ACKNOWLEDGMENT

I would like to acknowledge the National Science Foundation (contract 84-18896 DMR) for the continued support they have provided me for research in areas related to the topics discussed in this chapter.

REFERENCES

1. H. L. Tuller, in *Nonstoichiometric Oxides*, O. T. Sorensen, ed., Academic Press, New York, 1981, pp. 271–335.
2. A. H. Wilson, *Theory of Metals*, 2nd ed., Cambridge University Press, Cambridge, 1954.
3. C. K. Kittel, *Introduction to Solid State Physics*, 5th ed., Wiley, New York, 1976.
4. C. N. R. Rao and G. V. Subba Rao, *Phy. Status Solidi, (a), 1*, 597 (1970).
5. N. F. Mott, *Proc. Phys. Soc. (Lond.), A62*, 416 (1949).
6. H. L. Tuller and A. S. Nowick, *J. Phys. Chem. Sol., 38*, 859–867 (1977).
7. J. Appel, *Solid State Phys., 21*, 193 (1968).
8. J. M. Honig, in *Electrodes of Conductive Metallic Oxides*, Part A, S. Trasatti, ed., Elsevier, Amsterdam/Oxford, New York, 1980, pp. 1–96.
9. R. W. Vest and J. M. Honig, in *Electrical Conductivity in Ceramics and Glass*, Part B, Marcel Dekker, New York, 1974, pp. 343–452.
10. J. B. Goodenough, *Prog. Solid State Chem., 5*, 145 (1971).
11. J. P. McKelvey, *Solid State and Semiconductor Physics*, Harper & Row, New York, 1966.
12. J. M. Berak and M. J. Sienko, *J. Solid State Chem., 2*, 109 (1970).
13. H. Tamura, H. Yoneyama, and Y. Matsumoto, in *Electrodes of Conductive Metallic Oxides*, Part A, S. Trasatti, ed., Elsevier, Amsterdam/Oxford, New York, 1980, pp. 261–299.
14. P. M. Raccah and J. B. Goodenough, *Phys. Rev. 155*, 932–943 (1967).
15. P. M. Raccah and J. B. Goodenough, *J. Appl. Phys., 39*, 1209–1210 (1968).
16. G. H. Jonker and J. H. Van Santen, *Physica, 29*, 120 (1953).
17. F. R. van Buren and J. H. W. de Wit, *J. Electrochem. Soc., 126*, 1817–1820 (1979).
18. D. H. Whitmore, *J. Cryst. Growth, 39*, 160–179 (1977).
19. P. G. Dickens, D. J. Murphy, and T. K. Halstead, *J. Solid State Chem., 6*, 370 (1973).
20. A. W. Sleight, J. L. Gillson, and P. E. Bierstedt, *Solid State Commun., 17*, 27–28 (1975).
21. R. D. Shannon and P. E. Bierstedt, *J. Am. Ceram. Soc., 53*, 635 (1970).
22. M. Suzuki, Y. Enomoto, T. Murakami, and T. Inamura, *Jap. J. Appl. Phys., 20-4* (Suppl.), 13–16 (1981).
23. N. F. Mott, *Adv. Phys., 16*, 49 (1967).
24. D. Adler and H. Brooks, *Phys. Rev., 155*, 826 (1967).

25. N. F. Mott, *Rev. Mod. Phys.*, *12*, 328 (1967).
26. W. H. Stehlow and E. L. Cook, *J. Phys. Chem. Ref. Data*, *2*, 163–200 (1973).
27. I. K. Naik and T. Y. Tien, *J. Electrochem Soc.*, *126*, 562–566 (1979).
28. T. Stratton, Doctoral thesis, Department of Materials Science and Engineering, Massachusetts Institute of Technology, Sept. 1983.
29. F. A. Kröger and H. J. Vink, in *Solid State Physics*, Vol. 3, F. Seitz and D. Turnbull, eds., Academic Press, New York, 1956, p. 307.
30. E. J. W. Verwey, P. W. Haaijman, F. C. Romeijn, and G. W. van Oosterhaut, *Philips Res. Rep.*, *5*, 173 (1950).
31. J. Gantron, J. F. Marucco, and P. Lamasson, *Mater. Res. Bull.*, *16*, 579 (1981).
32. J. P. R. Frederkse, W. R. Thurber, and W. R. Hosler, *Phys. Rev.*, *134*, A442 (1964).
33. U. Balachandran and N. G. Eror, *J. Electrochem. Soc.*, *129*, 1021 (1982).
34. N. J. Dudney, R. L. Coble, and H. L. Tuller, *J. Am. Ceram. Soc.*, *64*, 627 (1981).
35. P. K. Larson and R. Metselaar, *Phys. Rev. B.*, *14*, 2520–27 (1976).
36. D. B. Meadowcraft, *Br. J. Appl. Phys. (J. Phys. D) Ser. 2*, *2*, 1225–1233 (1969).
37. F. A. Kröger, *The Chemistry of Imperfect Crystals*, 2nd ed., North-Holland, Amsterdam, 1974.
38. G. M. Choi, M. S. thesis, Department of Materials Science and Engineering, Massachusetts Institute of Technology, June 1983; G. M. Choi and H. L. Tuller, submitted for publication.
39. G. H. Jonker, *Philips Res. Rep.*, *23*, 131 (1968).
40. H. L. Tuller, J. M. Kowalski, and F. D. Gealy, in *Grain Boundaries and Interfaces in Ceramics (Advances in Ceramics*, Vol. 7), M. F. Yan and A. H. Heuer, eds., American Ceramic Society, Columbus, Ohio, 1984, p. 281.
41. B. Odekirk, U. Balachandran, N. G. Eror, and J. S. Blakemore, *Mater. Res. Bull.*, *17*, 199 (1982).
42. J. M. Kowalski and H. L. Tuller, *Ceram. Int.*, *7*, 55 (1981).
43. G. Campet, M. P. Dare–Edwards, A. Hamnet, and J. B. Goodenough, *Nouv. J. Chim.*, *4*, 501 (1980).
44. H. P. Maruska and A. K. Ghosh, *Solar Energy Mater.*, *1*, 237 (1979).
45. M. H. Sukkar and H. L. Tuller, in *Grain Boundaries and Interfaces in Ceramics*, (*Advances in Ceramics*, Vol. 5), M. F. Yan and A. H. Heuer, eds., Americal Ceramic Society, Columbus, Ohio, 1984, p. 71.

46. R. Wernicke, *Phys. Status Solidi (a)*, *47*, 139–144 (1978).
47. D. R. Sempolenski, W. D. Kingery, and H. L. Tuller, *J. Am. Ceram. Soc.*, *63*, 669 (1980).
48. B. V. Dutt, J. P. Hurrell, and F. A. Kröger, *J. Am. Ceram. Soc.*, *58*, 920 (1975).
49. R. Groth, *Phys. Status Solidi*, *11*, 69 (1966).
50. H. J. Van Daal, *Solid State Commun.*, *6*, 5 (1968).
51. P. W. Li and K. I. Hagemark, *J. Solid State Chem.*, *12*, 371 (1975).
52. V. N. Shukla and G. P. Wirtz, *J. Am. Ceram. Soc.*, *60*, 253 (1977).
53. F. P. Koffyberg, *J. Solid State Chem.*, *2*, 176 (1970).
54. C. G. Fonstad, A. Linz, and P. H. Rediker, *J. Electrochem Soc.*, *116*, 1269 (1969).
55. F. T. J. Smith and S. L. Lyn, *J. Electrochem Soc.*, *128*, 2388 (1981).
56. J. A. Marley and R. C. Dockerty, *Phys. Rev.*, *140*, A304 (1965).
57. M. K. Paria, H. S. Maiti, *J. Mater. Sci.*, *17*, 3275 (1982).
58. G. C. Farrington, *Sensors Actuators*, *1*, 329 (1981).
59. J. B. Goodenough, in *Solid Electrolytes*, P. Hagenmuller and W. Van Gool, eds., Academic Press, New York, 1978, p. 393.
60. J. Hladik, ed., *Physics of Electrolytes*, Vol. 1, Academic Press, New York, 1972.
61. L. W. Strock, *Z. Phys. Chem.*, *B25*, 441 (1934); *32*, 132 (1932).
62. R. Cava and B. J. Wuensch, *Solid State Commun.*, *24*, 411–416 (1977).
63. P. Mcgeehin and A. Hooper, *J. Mater. Sci.*, *12*, 1 (1977).
64. W. H. Flygare and R. A. Huggins, *J. Phys. Chem. Sol.*, *34*, 1199 (1973).
65. M. S. Wittingham, *Electrochim. Acta*, *20*, 575 (1975).
66. J. Wang, M. Kaffari, and D. Choi, *J. Chem. Phys.*, *63*, 772 (1975).
67. H. L. Tuller, D. P. Button, and D. R. Uhlmann, *J. Non-Cryst. Solids*, *40*, 93 (1980).
68. D. P. Button, R. P. Tandon, H. L. Tuller, and D. R. Uhlmann, *J. Non-Cryst. Solids*, *42*, 297 (1980).
69. D. P. Button, R. P. Tandon, H. L. Tuller, and D. R. Uhlmann, *Solid State Ionics*, *5*, 655 (1981).
70. D. P. Button, L. S. Mason, H. L. Tuller, and D. R. Uhlmann, *Solid State Ionics*, *9*, *10*, 585 (1983).
71. J. L. Souquet, *Solid State Ionics*, *5*, 77 (1981).
72. D. P. Button, P. K. Moon, H. L. Tuller, and D. R. Uhlmann, *Glastech. Ber.*, 56K, 856 (1983).
73. M. Irion, M. Couzi, A. Levasseur, J. M. d'leu, and J. C. Brethous, *J. Solid State Chem.*, *21*, 285 (1980).

74. Y. Kawamoto, N. Nagura, and S. Tsuchihasi, *J. Am. Ceram. Soc.*, *57*, 489 (1974).
75. D. Y. Wang, D. S. Park, J. Griffiths, and A. S. Nowick, *Solid State Ionics*, *2*, 95 (1981).
76. R. Gerhart–Anderson and A. S. Nowick, *Solid State Ionics*, 5, 547 (1981).
77. H. L. Tuller and A. S. Nowick, *J. Electrochem Soc.*, *126*, 209 (1979).
78. H. H. Fujimoto and H. L. Tuller, in *Fast Ion Transport in Solids*, P. Vashista, J. N. Mundy, and G. K. Shenoy, eds., Elsevier/North-Holland, Amsterdam, 1979, p. 649.
79. H. K. Bowen and B. R. Rossing, in *Critical Problems in Energy Production*, C. Stein, ed., Academic Press, New York, p. 311.
80. H. Obayaski and T. Kudo, in *Solid State Chemistry of Energy Conversion and Storage*, J. B. Goodenough and M. S. Wittingham, eds., American Chemical Society, Washington, D.C., 1977, p. 316.
81. S. A. Weiner, in *Solid State Chemistry of Energy Conversion and Storage*, J. B. Goodenough and M. S. Wittingham, eds., American Chemical Society, Washington, D.C., 1977.
82. G. E. Pike and C. H. Seager, *J. Appl. Phys.*, *48*, 5152 (1977).
83. H. H. Horowitz, H. S. Horowitz, J. M. Longo, in *Proc. Symp. Electrocatal.*, W. E. O'Grady and F. G. Will, eds., Electrochemical Society, London, Vol. 82-2, 1982, p. 285.
84. B. B. Scholtens and W. Van Gool, in *Solid Electrolytes*, P. Hagenmuller and W. Van Gool, eds., Academic Press, New York, 1978, p. 463.
85. W. J. Fleming, *J. Electrochem Soc.*, *124*, 21 (1977).
86. P. Hagenmuller and W. Van Gool, eds., *Solid Electrolytes*, Academic Press, New York, 1978.

Index